# Mixed-Mode Official Surveys

T0132598

# Statistics in the Social and Behavioral Sciences Series

*Series Editors*

*Jeff Gill, Steven Heeringa, Wim J. van der Linden, Tom Snijders*

For more information about this series, please visit: https://www.routledge.com/Chapman--HallCRC-Statistics-in-the-Social-and-Behavioral-Sciences/book-series/CHSTSOBESCI

# Mixed-Mode Official Surveys

## Design and Analysis

Barry Schouten, Jan van den Brakel,
Bart Buelens, Deirdre Giesen, Annemieke Luiten,
and Vivian Meertens

## CRC Press

Taylor & Francis Group

Boca Raton  London  New York

CRC Press is an imprint of the
Taylor & Francis Group, an **informa** business

A CHAPMAN & HALL BOOK

First edition published 2022
by CRC Press
6000 Broken Sound Parkway NW, Suite 300, Boca Raton, FL 33487-2742

and by CRC Press
2 Park Square, Milton Park, Abingdon, Oxon, OX14 4RN

© 2022 CRC Press

CRC Press is an imprint of Taylor & Francis Group, LLC

Reasonable efforts have been made to publish reliable data and information, but the author and publisher cannot assume responsibility for the validity of all materials or the consequences of their use. The authors and publishers have attempted to trace the copyright holders of all material reproduced in this publication and apologize to copyright holders if permission to publish in this form has not been obtained. If any copyright material has not been acknowledged please write and let us know so we may rectify in any future reprint.

Except as permitted under U.S. Copyright Law, no part of this book may be reprinted, reproduced, transmitted, or utilized in any form by any electronic, mechanical, or other means, now known or hereafter invented, including photocopying, microfilming, and recording, or in any information storage or retrieval system, without written permission from the publishers.

For permission to photocopy or use material electronically from this work, access www.copyright.com or contact the Copyright Clearance Center, Inc. (CCC), 222 Rosewood Drive, Danvers, MA 01923, 978-750-8400. For works that are not available on CCC please contact mpkbookspermissions@tandf.co.uk

Trademark notice: Product or corporate names may be trademarks or registered trademarks and are used only for identification and explanation without intent to infringe.

ISBN: 9781138618459 (hbk)
ISBN: 9781032102962 (pbk)
ISBN: 9780429461156 (ebk)

DOI: 10.1201/9780429461156

Typeset in Palatino
by Deanta Global Publishing Services, Chennai, India

# Contents

# *Preface*

The past two decades have seen the rise of online surveys and along with it a strong and renewed interest in mixed-mode surveys. The main reason for this renewed interest is that the web is a cheap and fast survey mode but has lower response rates than the traditional interviewer-assisted and paper survey modes. As a consequence, the mixing of modes has been, and still is, a natural approach to elevate the lower response rates, especially in the larger, official surveys.

The use of mixed-mode designs is not at all new, but what is new is the much stronger focus on the implications of mixing modes for analysis and estimation. The motivation for this book came from the desire to provide a comprehensive overview of the role of mixed-mode surveys in all survey methodology stages, from the early design stage to the analysis and estimation stages. Such a complete account was not available. The authors of this book, motivated by their experiences in the migration of many surveys to mixed-mode designs, strongly believe that it is important to consider all stages of survey research. In this book, all stages and all options are discussed, as well as strategies to report mode effects in publications.

This book is intended for both survey designers and users of surveys. The authors of the book have very diverse backgrounds, from questionnaire development up to sampling design and nonresponse adjustment. Readers of the book may have equally different backgrounds and may profit from the different chapters.

The authors of this book are all methodologists or developers at a northern-European national statistical institute. Surveys conducted at an NSI typically are larger, often repeated, and concern the general population or large subpopulations. This implies a relatively high budget per sample unit, a focus on a full coverage of the population, and increased attention for comparability of survey statistics across time and between relevant subpopulations.

# Acknowledgements

The authors like to thank Statistics Netherlands, in particular, the management of the Methodology Department, for the opportunity to report research and findings on the design and analysis of mixed-mode surveys. Special thanks go to Jeldrik Bakker and Anouk Roberts for creating some of the figures on mobile devices response and break-off and to Stefan Theunissen for designing the front cover of the book. Chapter 6 of the book was written with the help of Madelon Cremers and Rachel Vis-Visschers. We are also grateful to six unknown referees for careful reading and commenting on former drafts of the manuscript, which proved to be very helpful to improve the content of this book. The views expressed in this book are those of the authors and do not reflect the policy of Statistics Netherlands.

# *Authors*

**Bart Buelens** – Bart Buelens has worked in data analytics after graduating in mathematics and obtaining a PhD in computer science. As a statistician and data scientist at Statistics Netherlands, he conducted research on inference in mixed-mode surveys, model-based estimation, and machine learning. In 2018, he moved to VITO, a Belgian research and technology organization, where he contributes to data science research in the area of sustainability with an emphasis on applied artificial intelligence.

**Jan van den Brakel** – After attaining a master's in biometrics, Jan van den Brakel started as a junior methodologist at the Methodology Department of Statistics Netherlands in 1994. Based on his research work at Statistics Netherlands on design-based inference methods for randomized experiments embedded in probability samples, he finalized his PhD in statistics in 2001. In 2005, he became senior methodologist, responsible for coordinating research into model-based inference methods. Since 2010 he is an extended professor of survey methodology at Maastricht University. His research interests are sampling, design and analysis of experiments, small area estimation, time series analysis, and statistical methods for measuring the effects of redesigns of repeated surveys.

**Deirdre Giesen** – Deirdre Giesen holds a master's in health sciences and in sociology. She has worked as a survey methodologist at Statistics Netherlands since 2000. She is a senior methodologist and responsible for the coordination of cognitive lab/userlab tests at Statistics Netherlands. Part of her work is pretesting and evaluating questionnaires, both for businesses and household surveys. Her research interests include questionnaire testing methodology, mixed-mode mixed-device questionnaire development, and the measurement and reduction of response burden.

**Annemieke Luiten** – Annemieke Luiten holds a master's in psychology and a PhD in survey methodology. She is a senior methodologist at Statistics Netherlands, with a specialty in data collection methodology. Her research areas comprise nonresponse reduction, fieldwork, interviewer behavior, mixed-mode surveys, and the role of sensor measurement in surveys.

**Barry Schouten** – After a master's and PhD in mathematics, Barry Schouten started as a junior methodologist at the Methodology Department of Statistics Netherlands in 2002. In 2009, he became a senior methodologist and coordinator for research into primary data collection. His research interests gradually widened from nonresponse reduction and adjustment to multi-mode surveys, measurement error, and adaptive survey design. In 2017, he became a professor at Utrecht University, holding a special chair on mixed-mode survey designs. He is one of the coordinators of a joint data collection innovation network (WIN in Dutch) between Statistics Netherlands and Utrecht University that was established in 2016.

**Vivian Meertens** – Vivian Meertens holds a master's and PhD in sociology at the University Medical Centre Nijmegen and Interuniversity Centre for Social Theory and Methodology (ICS). After working for some years as an associate professor at the Department of Social

Medical Science, she started as a survey methodologist at Statistics Netherlands in 2007. She is a senior methodologist and has worked on pretesting and evaluating mixed-mode and mixed-device questionnaires for household surveys. Her research interest focuses on developments and pretesting of several European model questionnaires measuring social phenomena to produce (inter)national social statistics.

# Part I

# Introduction

# 1

## Foreword to Mixed-Mode Official Surveys: Design and Analysis

### 1.1 Why a Book on Mixed-Mode Survey Design and Analysis?

The past two decades have seen the rise of online surveys and along with it a strong and renewed interest in mixed-mode surveys. The main reason for this renewed interest is that the web is a cheap and fast survey mode but has lower response rates than the traditional interviewer-assisted and paper survey modes. As a consequence, mixing of modes has been, and still is, a natural approach to elevate the lower response rates, especially in the larger official surveys. In a recent survey among European national statistical institutes (Murgia, Lo Conte, and Gravem, 2018), all but one country indicated to use mixed-mode designs and about half of the countries replied that web is one of the modes in these designs.

The use of mixed-mode designs is not at all new, but what is new is the much stronger focus on the implications of mixing modes for analysis and estimation. Mixed-mode designs have been around since the early days of surveys and have become a prominent option with the advent of telephone surveys. The impact of modes on the various survey errors, such as nonresponse and measurement, is known for a long time and has been the subject of research for several decades. Accounts of these can be found in Groves (1989), De Leeuw (1992), Groves et al. (2002), Biemer and Lyberg (2003), the four editions of Dillman, Smythe, and Christian (2014), Cernat and Sakshaug (2020), Olson et al. (2020), and Toepoel, De Leeuw, and Hox (2020). With the web as the fourth-largest survey mode, next to mailed questionnaires, telephone, and face-to-face interviews, the number of design options increased. However, another mode was introduced with quite different features from the two interviewer-assisted modes. These different features have not gone unnoticed and, since the web is a relatively cheap mode, have led to much more explicit cost–benefit analyses in the survey design. For several years, survey methodology conferences, such as AAPOR and ESRA, had many sessions just on mixed-mode survey design and their impact on survey errors, survey costs, and survey estimates. This interest was unprecedented up to the introduction of the web.

The motivation for this book came from the desire to provide a comprehensive overview of the role of mixed-mode surveys in all survey methodology stages, from the early design stage to the analysis and estimation stages. Such a complete account was not available at the time of writing this book, although strong and informative books have been written about design, such as Dillman, Smyth, and Christian (2014). The authors of this book, motivated by their experiences in the migration of many surveys to mixed-mode designs, strongly believe that it is important to consider all stages of survey research. The options to reduce mode impact range from prevention of effects in questionnaire design to avoidance

DOI: 10.1201/9780429461156-1

of effects in mode choice and mode allocation to adjustment of effects in the estimation. However, the options must and cannot be seen as complementary or separate; they are intertwined and must be considered in conjunction with each other. In this book, all stages and all options are discussed, as well as strategies to report mode effects in publications.

A large wave of transitions of surveys to mixed-mode designs has passed, but this book will still be of use to survey institutes considering migration to surveys employing multiple modes and to scholars investigating design and analysis of the impact of modes (Signore, 2019; Olson et al., 2020). Institutes may consider a redesign of their surveys, may be in the middle of such redesigns, or may face questions about the handling of mode impact after redesigns. Furthermore, the rise of all kinds of online devices implies that mixed-mode surveys are not fixed and stable and will have to be reconsidered continuously for the near and longer-term future. It must also be remarked, however, that mode effects do not occur in mixed-mode designs only. It is in those designs that mode impact is most visible, but the impact existed all along with the introduction of a mode. Hence, even for single-mode surveys, it is important to consider the role of a survey mode and this book may provide useful background.

This book is intended for both survey designers and users of surveys. The authors of the book have very diverse backgrounds, from questionnaire development up to sampling design and nonresponse adjustment. Readers of the book may have equally different backgrounds and may profit from the different chapters. Some chapters are more psychological, some are more methodological, and some are more statistical. Some chapters deal with multiple disciplines and may be of interest to all kinds of readers. The authors have been active in large-scale redesign project teams at Statistics Netherlands and often were responsible for different stages of the redesigns. The book may add value to readers who participate in such multidisciplinary teams as well. It may help them understand the terminology and methodological options.

All the authors of the book are methodologists or developers at a northern-European national statistical institute. Surveys conducted at an NSI typically are larger, often repeated, and concern the general population or large subpopulations. This implies a relatively high budget per sample unit, a focus on a full coverage of the population, and increased attention for comparability of survey statistics across time and between relevant subpopulations. In this type of surveys, mode impact can strongly affect time series. Because comparability in time is deemed a key feature, there is a lot of interest to detect and reduce mode impact. General references to literature are included, but the book does tend toward mixed-mode designs for the general population, high-budget surveys. The northern-European context means a relatively rich administrative data context and a high population coverage of online devices. Both impact the way surveys can and are designed and analyzed. Although theory and takeaway messages translate to all surveys, this must be kept in mind. Readers are referred to best practice documents as prepared by AAPOR, see www.aapor.org, and ESOMAR, see www.esomar.org.

In the next two sections, the origin and type of modes and devices are introduced and an outline is provided of the book.

## 1.2 Modes and Devices

Response of a sampled population unit requires at least four steps: making contact, assessing eligibility, obtaining participation, and completing the questionnaire. This book deals with designs in which multiple modes may be used in parallel to complete the questionnaire. So, when multiple modes are used to make contact, to assess eligibility, or to obtain cooperation, but only one mode is used to complete the survey, then these are not viewed as mixed-mode surveys here. Nonetheless, contact modes will naturally come up as part of mixed-mode designs in various chapters.

In this context, a survey mode is a communication channel that can be used to collect survey data from one or more respondents. The four main contemporary modes are mail, web, telephone, and face-to-face. The oldest modes are mail and face-to-face; they were the main communication channels when large-scale surveys advanced. In the 1960s and 1970s, telephone became a dominant communication channel and found its way to surveys. In the 1990s and 2000s, internet emerged and became the last of the four main modes. The older modes, face-to-face and mail, became less prominent but are still used extensively. The same is true for the telephone mode. To date, all four modes may appear in a single survey. An overview of the history of surveys and survey modes can be found in Bethlehem (2009). The four modes have been implemented in various ways. In the 1980s and 1990s, computer-assisted implementations of telephone (Computer-Assisted Telephone Interviewing or CATI) and face-to-face modes (Computer-Assisted Personal Interviewing or CAPI) were developed and introduced, partly replacing the paper questionnaires that interviewers used to fill in by hand. The web mode is the most diverse of modes as many devices, both fixed and mobile, allow access to the internet. Desktops, laptops, tablets, smartphones, smart TVs, and even some wearables can go online. The four modes are briefly introduced here.

The mail mode amounts to a paper questionnaire and is most often combined with a mailed invitation letter. Sometimes it is implemented as an optional mode by allowing sampled population units to apply for a paper questionnaire, in which case the contact mode can be any mode. Sometimes it is implemented as a drop-off questionnaire when interviewers have been involved in previous waves or rounds. Occasionally, telephone reminders are used to stimulate sample units to complete and submit their questionnaires. Out of the four contemporary modes, mail is the only mode that does not produce any digital data collection process data, also termed paradata (Kreuter et al., 2018), unless telephone reminders are conducted. The date of return of the paper questionnaire is usually recorded, and sometimes respondents write textual remarks on the questionnaire. This means that the mail mode is hard to analyze, and it may be impossible to determine which of the data collection steps have been problematic. As a consequence, it may be hard as well to separate ineligibility from noncontact or refusal and to derive the specific impact of the mail mode on survey errors. Nonreturned paper questionnaires may be the result of a sample unit not living (anymore) on the address, a sample unit not opening invitation letters, or a sample unit not willing or able to respond. Without contacting the sample unit, one cannot determine the exact cause of nonresponse.

The online mode has perhaps the largest variety in forms of contact. An invitation letter with a URL, possibly supplemented by a QR code, is the most commonly used option when contact information is limited to addresses. When recent and accurate email addresses are available, then emails with URLs are commonly used, especially when the email provider is the same as the survey sponsor, e.g., employers, universities, or government. When mobile phone numbers are available, then also text messages may be used to invite sample units. Finally, in self-selection samples, URLs may simply be placed as banners at frequently visited websites. Like the mail mode, the online mode is also poor in paradata for those sample units that never entered the survey landing website. Contrary to the mail mode, however, it can be determined what happened after a sample unit entered the website, but decided to stop prematurely. Still, like the mail mode, the confounding of selection and measurement is hard to analyze.

The telephone mode is often, but not always, implemented as a one-way contact; i.e., the survey institute calls the sample unit to conduct the interview. In line with tailoring and maintaining interaction strategies (Groves and Couper, 1998), interviewers will avoid a refusal and attempt to find a suitable time for the interview. Usually, an advance or prenotification letter is sent in which the interview call and period are announced. In most implementations, it is possible for sample units to call a helpdesk to ask for information or to object to the survey. The telephone mode is relatively rich in paradata as each call can be documented and saved.

The face-to-face survey mode, like the telephone mode, is often a one-way contact at first as interviewers are not available on standby. However, since travel costs are demanding in both time and resources, there is a stronger incentive to schedule visits either by leaving interviewer contact information or by calling sample units (if possible) to make appointments. In practice, the face-to-face mode may have been preceded by almost all possible forms of contact. It is also the richest mode in paradata as interviewers can make observations apart from documenting the outcomes of visits. As such, it provides the best option to analyze the impact of the mode on selection and measurement.

Other modes and new implementations of current modes are likely to emerge in the future. Interactive Voice Response (IVR) and chatbots are signs of such developments. Such developments urge survey designers to be alert when it comes to respondent preferences and expectations about communication. As a consequence, this book will demand updates.

## 1.3 Outline of the Book

This book focuses on a mix of modes for the administration of a survey. It considers, first, the impact of the survey mode on survey errors and, second, the methodology for the reduction of this impact in the design and analysis of surveys. The book is divided into five parts:

I. Introduction

II. Mode Effects

III. Design

IV. Analysis

V. The Future of Mixed-Mode Surveys

In Part I, terminology, types of mixed-mode survey designs, and methodology options are introduced. Part II addresses survey errors and the impact of modes on selection and measurement. Parts III and IV then move to the reduction of mode effects. Part III focuses on the design stage, and Part IV on the estimation and analysis stage. Part V looks ahead and describes multi-device surveys and adaptive mixed-mode designs.

Chapter 2, Designing Mixed-Mode Surveys, in Part I, forms the backbone of the book as a whole. It introduces the terminology of mode effects, distinguishes different types of mixed-mode designs, presents overviews of survey errors and the survey data collection and estimation plan-do-check-act cycle, and introduces the case studies that are used as examples throughout the book. The survey errors and survey process schemes presented in this chapter return in each of the subsequent chapters.

Part II has two chapters: one on mode-specific measurement effects, and one on mode-specific selection effects. Traditionally, survey errors are divided into those affecting representation and those affecting measurement (Groves et al., 2010). The two chapters address the two main survey error types.

Chapter 3, Mode-Specific Measurement Effects, identifies the measurement features of the contemporary modes, describes the relation of these features to answer behavior of respondents and forms a prelude to Chapter 6, which discusses methodology to reduce measurement differences attributable to modes.

Chapter 4, Mode-Specific Selection Effects, focuses on all survey errors that affect representation, i.e., the selection into the survey; explains the relation of the modes to selection; presents results from an international study into the selection effects; and is the starting point for Chapter 5 on the reduction of selection biases.

Part III deals with the methodology to reduce the impact of modes on survey errors. It has two chapters: one on selection, and one on measurement.

Chapter 5, Mixed-Mode Data Collection Design, introduces mode strategies, elaborates the choice and order of modes, discusses the impact on response rates and representativeness, and makes recommendations as to how to optimize response rates across relevant population subgroups.

Chapter 6, Mixed-Mode Questionnaire Design, describes the various cognitive steps in answering a survey and how modes affect these steps, explains the role of modes in questionnaire design, and provides guidelines as to how to structure questionnaire design for multiple modes simultaneously. Questionnaire development goes hand in hand with cognitive testing and usability testing. The chapter explains approaches to perform questionnaire testing that account for the natural context in which respondents fill in questionnaires for the different modes, and gives recommendations as to how to implement such testing procedures.

Whereas Part III is about the design stage of a survey, Part IV is about the estimation and analysis stage. It has three chapters. Obviously, estimation and analysis feedback to design from a plan-do-check-act point of view. One chapter focuses on such feedback through experiments. The other two chapters discuss detection and adjustment of mode effect components.

Chapter 7, Field Tests and Implementation of Mixed-Mode Surveys, discusses the various options to conduct experiments that may assist data collection design and questionnaire design in making further improvements or decisions. The design of such experiments, as well as analysis, is explained, including power analyses and optimal sampling design.

Chapter 8, Re-interview Designs to Disentangle and Adjust for Mode Effects, describes what tactics and techniques may be used to decompose overall method effects that result from the choice of mode(s) into its two main components: mode-specific selection bias and mode-specific measurement bias. It also provides a cost–benefit analysis of experimental designs focused on such decompositions.

Chapter 9, Mixed-Mode Data Analysis, discusses analysis techniques and weighting methods that account for mode differences in respondent answers. In this chapter, it is presumed that no extra experimental data are available and that analyses need to be based on only linked auxiliary data and survey data. It is explained what assumptions are made in order to justify such presumptions.

Part V deals with relatively new developments in mixed-mode survey design and analysis. It considers various devices for accessing online surveys and adaptive designs for mode allocation to population subgroups. The part ends with a general look into the near future.

Chapter 10, Multi-Device Surveys, shifts attention to the various online devices and their impact on selection and measurement. The chapter evaluates device use and device-specific break-off; it discusses decisions to block, accept, or encourage mobile devices, it lists consequences for questionnaire design; and describes mobile device sensor data to replace or supplement survey data.

Chapter 11, Adaptive Mixed-Mode Survey Designs, first describes the components of adaptive survey designs in general, and then restricts attention to the survey mode as a design feature of interest. Since survey modes affect both selection and measurement, it is discussed how mode-specific measurement effects may be incorporated into adaptive survey design.

Chapter 12, Future Challenges and Open Issues, is the closing chapter of the book. It discusses anticipated developments for the near future. Furthermore, it identifies a list of open research questions and research areas.

# 2

## Designing Mixed-Mode Surveys

## 2.1 Introduction

This chapter forms a stepping-stone to the other chapters in this book. It explains the business case of mixed-mode survey designs, the types of mixed-mode designs, what role survey modes play in the various survey errors, how survey modes are key features in survey design, how they may lead to method effects, and what methodological options exist to reduce these so-called 'mode' effects.

It is important to note that this book focuses on survey designs that use two or more modes for *data collection*. Designs using multiple modes for *contacting* respondents, but that use only a single mode to collect the data, are not considered as mixed-mode surveys in this book. Within the online survey mode, different devices are distinguished, and they are discussed in Chapter 10, Multi-Device Surveys. It is often true, however, that multiple modes of data collection also imply multiple modes of contact, if only because interviewers cannot do a survey without making contact. Multiple modes of contact will still come up naturally, therefore. In this book, longitudinal surveys, i.e., surveys collecting data on multiple occasions from the same sample, are not explicitly considered. In such surveys, different modes may be used at different time points. See, for example, Lynn (2009).

The remainder of this chapter reads as follows: Section 2.2 explains the business case of mixed-mode survey designs. Section 2.3 presents the various types of mixed-mode designs. Section 2.4 explains the influences on total survey error components, called mode effects. Section 2.5 introduces the methodological options to reduce such effects and enhance comparability, which are elaborated in later chapters. Finally, Section 2.6 presents the case studies that are used throughout the book to illustrate the methodology.

## 2.2 Why Mixed-Mode Survey Designs?

What are the main motivations for combining modes in surveys? In designing surveys, different quality and costs objectives play a role, and virtually all of these objectives are affected by the choice of modes. To understand the motives for mixed-mode surveys, a look at quality and costs is, therefore, imperative. Essentially, modes may be combined because their mix improves quality at the same cost, or because their mix reduces cost while not affecting quality. In practice, since the quality of a survey is hard to pinpoint, neither setting may hold, and a mix of modes may be chosen because the quality–cost ratio is deemed optimal within the objectives and purpose of the survey.

DOI: 10.1201/9780429461156-2

Let us start from the cost point of view. Can mixed-mode surveys be cheaper than single mode surveys? The answer depends on the reference mode. Compared to a single mode web survey, the answer will often be no. However, compared to a face-to-face interview, the answer will often be yes. In general, the introduction of the web survey mode in a survey design, formerly without web, in the majority of cases leads to a cost reduction. With the emergence of web as a survey mode, this has, indeed, been a main driver for survey redesigns.

Survey costs are, generally, hard to measure exactly, because a survey is seldom run in isolation from other surveys and since part of the costs may be seen as investments that stretch out over a longer time period. Costs are made up of variable costs, i.e., the data collection cost per respondent, and fixed costs following from overhead, such as the development and maintenance of survey instruments and tools for monitoring of data collection. Table 2.1 provides an estimate of the variable costs relative to face-to-face interviewing, for a range of mode strategies in the 2012 first wave of the Dutch Labour Force Survey (LFS), some of which mix modes. The costs are estimated for nine relevant population strata based on age, registered employment, ethnicity, and size of the household (see Calinescu and Schouten, 2015a). It is clear that variable costs differ greatly. The estimated costs for a single mode web design, for example, vary between 3% and 4% of the costs of a single mode face-to-face design.

A single look at variable costs is not sufficient, however; also fixed costs vary, especially between interviewer-assisted and self-administered modes. Interviewer modes tend to have more overhead costs due to interviewer recruitment, training, and planning. These costs are, generally, harder to link to single surveys. For a detailed discussion, see Groves (1989) and Olson, Wagner, and Anderson (2020).

Combining multiple modes also by itself increases fixed costs, since multiple questionnaires (frontend tools) and case management systems (backed tools) need to be developed, coordinated, and maintained. Multi-mode case management systems are much more demanding in terms of logistics and staffing. Nonetheless, the combined costs of

**TABLE 2.1**

Estimated Data Collection Costs per Sample Unit for Five Mode Strategies and Nine Population Strata in the 2012 First Wave of the Dutch Labour Force Survey

| Mode Design | Strata Based on Age, Employment, Ethnicity, and Household Size[a] | | | | | | | | |
|---|---|---|---|---|---|---|---|---|---|
| | 1 | 2 | 3 | 4 | 5 | 6 | 7 | 8 | 9 |
| Web | 0.03 | 0.04 | 0.04 | 0.04 | 0.04 | 0.03 | 0.03 | 0.03 | 0.03 |
| Phone | 0.13 | 0.17 | 0.11 | 0.10 | 0.15 | 0.14 | 0.11 | 0.16 | 0.19 |
| F2F | 1.00 | 1.00 | 1.00 | 1.00 | 1.00 | 1.00 | 1.00 | 1.00 | 1.00 |
| Web → phone | 0.09 | 0.12 | 0.10 | 0.10 | 0.10 | 0.09 | 0.09 | 0.08 | 0.07 |
| Web → F2F | 0.72 | 0.71 | 0.80 | 0.84 | 0.73 | 0.68 | 0.81 | 0.62 | 0.71 |

[a]  Strata: 1 = at least one person is registered unemployed; 2 = at least one person is 65+ and no registered employment; 3 = at least one young person in household and no registered employments; 4 = non-Western household without registered employment; 5 = Western household without registered employment; 6 = at least one young household member and at least one registered employment; 7 = non-Western household and at least one registered employment; 8 = Western household and at least one registered employment; 9 = large households. See Calinescu and Schouten (2015a) for details.

Costs are estimated relative to F2F, which is set as a benchmark. The nine population strata are based on age, registered employment, ethnicity, and size of the household and are homogeneous with respect to the LFS main target variables (Calinescu and Schouten 201).

mixed-mode designs including web may be lower than designs without web. The follow-up question then is whether quality can be maintained. So let us turn to the quality point of view.

From a quality point of view, there are also great differences between modes. Can quality be maintained, or even improved, when mixing modes? The answer to this question is much harder to give as quality has many facets. Modes differ in the extent to which persons have access to them, in the extent to which persons are familiar with them, in the extent to which persons prefer them to provide answers, in the extent to which they motivate respondents to make the required cognitive effort to answer questions correctly, and in the extent to which they can assist respondents to avoid measurement errors. The question of whether mixing modes is beneficial amounts to a judgment about the net effect of all these differences. However, there is more to quality than accuracy of statistics alone. Quality dimensions, such as comparability in time and comparability between the statistics of relevant population groups, are often crucial as well.

As a first example of the impact on quality, consider Table 2.2. Here, the same nine strata are distinguished as in Table 2.1, but now the estimated response rates per design are given. The response rate is an example of a survey quality metric; it sets an upper bound to the nonresponse bias that may be present in survey estimates. The response rates show great variety between the designs, but also between strata. The implications are that different designs lead to different missing data patterns and to different shares of modes between strata. In other words, biases of statistics may be different for the designs, but also comparability of strata may vary for the designs, when modes affect answers that respondents give. In addition, different response rates imply a different precision of statistics when the sample size would be fixed. In practice, the sample size is usually made dependent on the response rate by fixing the number of respondents, potentially per stratum and must, thus, be made larger for designs with lower response rates.

It may be clear from this example that the question of whether the quality of mixed-mode designs is similar, worse, or better is complex and not easy to answer. This book devotes much effort to explain how modes impact quality, how such impact can be measured, and if and how it can be reduced during data collection and analysis. However, from Table 2.2, it is safe to say that mixing modes can be advantageous in terms of quality; response rates for the design that mix web and telephone are higher than for the single web and telephone designs, i.e., they imply lower limits to the maximal nonresponse bias.

The great variety in both quality and costs imply that survey modes are a crucial design choice, perhaps the most influential of all design choices. The strong differences between

**TABLE 2.2**

Response Rates for Five Mode Strategies and Nine Population Strata in the 2012 First Wave of the Dutch LFS

| Strategy | Strata Based on Age, Employment, Ethnicity, and Household Size | | | | | | | | |
|---|---|---|---|---|---|---|---|---|---|
| | 1 (%) | 2 (%) | 3 (%) | 4 (%) | 5 (%) | 6 (%) | 7 (%) | 8 (%) | 9 (%) |
| Web | 23.2 | 23.6 | 15.5 | 10.8 | 27.9 | 27.7 | 17.5 | 36.7 | 22.4 |
| Phone | 20.8 | 41.3 | 15.2 | 8.6 | 31.1 | 23.8 | 14.3 | 33.3 | 37.5 |
| F2F | 52.4 | 58.3 | 51.0 | 41.2 | 51.2 | 54.9 | 46.0 | 56.8 | 61.4 |
| Web → phone | 32.8 | 48.4 | 23.8 | 17.5 | 42.1 | 41.1 | 25.8 | 52.1 | 24.4 |
| Web → F2F | 49.8 | 58.3 | 43.4 | 36.6 | 52.6 | 54.7 | 44.3 | 62.0 | 54.2 |

The nine population strata are the same as in Table 2.2

survey modes in quality and costs, therefore, make them a usual suspect in total survey error evaluations and total quality control assessments (Biemer and Lyberg, 2003; Bethlehem, 2009). Modes contribute to all errors in such evaluations and assessments and, thus, are identified as separate data collection phases. In data collection monitoring dashboards, typically, the various modes are evaluated both separately and cumulatively, e.g., Jans, Sirkis, and Morgan (2013). Weaknesses of modes can be compensated for by strengths of other modes when combining them. These potentials are the main motive behind mixed-mode designs. Consider the limited availability of registered phone numbers, which can be compensated for by offering modes with general access to those without a registered phone number. Or consider the high costs of face-to-face interviewing, which can be compensated for by first offering cheaper modes. As a consequence, both the overall effect of modes on accuracy and the impact of single modes on total survey error components change when they are combined. Designing mixed-mode surveys is, thus, not a matter of exploring the isolated features of the modes but the combined features of the modes. This will be a returning theme throughout the book.

So, how to make decisions about mode design? Two main viewpoints exist as to how to make decisions about mode strategies. One viewpoint is the total system effect, in which no attempt is made to separate out the impact of modes on quality and costs but only the total effect of all choices (including modes) is assessed. The rationale is that there is too much within mode variation and there are too many design features that vary along with the mode to be able to isolate the impact of a single design feature. Obviously, still design choices need to be made, but these are based on scientific knowledge and empirical evidence in the literature (Biemer and Lyberg, 2003, Biemer, 2010). The other viewpoint is that of disentanglement of mode impact on different quality and cost components. Although it is recognized as a hard problem, the motivation is that, without such disentanglement, no effective choices can be made. This book tends to the second viewpoint, although Chapter 7, Field Tests and Implementation of Mixed-Mode Surveys, does present methodology to efficiently estimate total system effects.

Two important side remarks are in place before proceeding to a discussion on the types of mixed-mode surveys: mode effects and methodology to reduce mode effects.

First, although one speaks of online, paper, telephone, and face-to-face as the contemporary survey modes, these terms really represent larger classes of survey designs; within one mode, designs can be very different. As a result, there may also be large variations within one mode in quality and costs. Online devices were already mentioned as subclasses of the online survey mode, but many other features may vary as well in an online survey: the use and type of incentives; the type of invitation, e.g., a letter, email, text message, or mix; the number and type of reminders; the length and nature of the questionnaire, e.g., a single measurement versus a diary; the survey sponsor, any gamification of the questionnaire; and so on. The variety in design features is even larger for interviewer-administered surveys. The between mode variation in quality and costs is, however, the largest of all survey design features. For this reason, survey modes are a crucial design choice, regardless of the specific implementation. Nevertheless, it is important to keep the within mode variation in mind when considering and comparing different mode designs; choosing a particular combination of modes still leaves a lot of room for further refinement of the design.

Second, the differences between modes change gradually over time; they may attenuate when modes become omnipresent and/or standard tools to communicate, such as the internet, but may also become stronger when modes become outdated, such as fixed landline phones. For this reason, it is not just accuracy that is affected by modes but also comparability over time and between population subgroups, say between younger and older persons. Comparability over time is one of the quality cornerstones of repeated surveys. Comparability between population subgroups is a cornerstone to surveys in general. Mixed-mode survey designs are, thus, not fixed, but need to be reevaluated periodically.

## 2.3  How Can Modes Be Combined?

Modes can be combined in various ways. This section discusses the options and introduces some terminologies. It is important to realize that combining modes is more than just providing the mode as an option; the different data collection steps that are inherently linked to the modes need to be harmonized and aligned.

This book deals with designs in which multiple modes may be used in parallel to complete the questionnaire. In other words, the focus is on more than one mode of administration. Within the same interview, one may also shift from interviewer-administered to self-administered or vice versa, because of the nature of the survey items. Some items are always posed by the interviewer, while others are self-reporting only. This setting is not viewed as mixed-mode and is not addressed here.

Multiple modes of administration often imply multiple modes of contact. Contact, eligibility, and participation strategies cannot be seen as separate from the interview. Obviously, a face-to-face interviewer cannot conduct an interview without making contact and obtaining participation, even when an advance letter has been sent. An online response is possible only with a web URL and this has to be provided to the sample units in some form. Sending a paper letter or an email with a URL will make a big difference in whether and how units will participate. In fact, Dillman et al. (2014) propagate that the whole process from first contact to completion of the questionnaire is aligned in presentation and style. In other words, the impact of a mode on quality and costs cannot and must not be evaluated disregarding any of the previous steps in the data collection process. This dependence complicates the generalization of survey design using multiple modes, but there still are sufficient commonalities to consider survey modes as a design choice.

To start, see De Leeuw (2005). There are two basic mode options: (1) modes can be offered simultaneously, or (2) modes can be offered sequentially. The first option, termed *concurrent* mixed-mode, allows respondents to choose explicitly between modes within the same time window. The second option, termed *sequential* mixed-mode, offers only one mode at a time and only (part of the) nonrespondents to the first mode are invited for the second mode. Figure 2.1 shows the two options that can be combined into more complex mixed-mode designs. The different time windows are viewed as data collection phases.

The boundary between the two may be blurred when modes that follow sequentially are announced at first contact but are not yet open for completion of the survey. In the invitation letter, it may, for example, be announced that an interviewer will contact the sample unit when no response on paper and/or online is submitted before a specified date. The respondent then is aware of the option to provide a response through other modes. Still this is considered a sequential mixed-mode design.

The boundary is also blurred when a new mode is introduced when completion of the survey in another mode is still allowed. For example, telephone calls may start a certain week of data collection while sample units can still submit their data online or through a paper questionnaire. Or paper questionnaires may be added to a reminder letter for

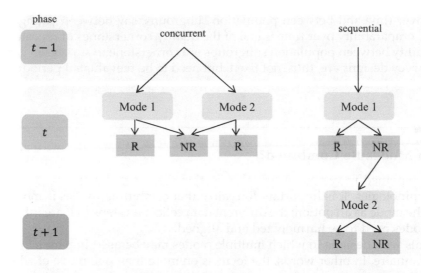

**FIGURE 2.1**
Schematic view of concurrent and sequential mixed-mode designs of a survey for which data collection starts in time period $t$ and extends to time period $t + 1$.

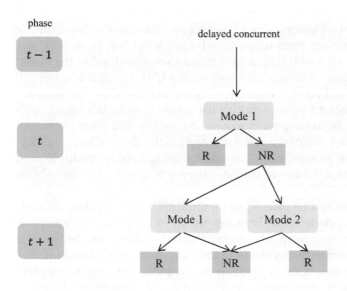

**FIGURE 2.2**
Schematic view of a delayed concurrent mixed-mode design of a survey for which data collection starts in time period $t$ and extends to time period $t + 1$.

a survey that started online. This option is depicted in Figure 2.2 and is termed *delayed concurrent* mixed-mode design.

Within sequential mixed-mode designs, only part of the nonrespondents may receive the second mode. For example, sample units refused explicitly are not eligible, while all other nonrespondents are eligible for follow-up. In Figures 2.1 and 2.2, one may distinguish different types of nonresponse, say NR1, NR2, and so on, and let only part of these nonresponse types be offered the second mode.

A special case is formed by online devices, such as desktops, laptops, smartphones, and tablets. They all can be used when a URL is available for the online survey, even though the survey institute may not make this salient to the sample unit. Hence, online surveys are concurrent mixed-device, unless one or more devices are explicitly blocked. In Chapter 11, Multi-Device Surveys, the use of different devices is discussed.

It is important to note that two or more modes that are employed simultaneously do not necessarily constitute a concurrent mixed-mode design. If sample units are called for an interview when a registered phone number is available and otherwise are visited at home, then this is not concurrent mixed-mode; the sample units never had the chance to choose between a telephone and face-to-face interview. This differentiation based on a registered phone number is but a simple form of adaptation to known characteristics of the sample unit. Such stratification may be much more elaborate and be based on all kinds of auxiliary data available through the sampling frame, linked administrative data, earlier waves prior to the survey, or paradata recorded during the survey. The adaptation itself may also be more advanced and involve multiple design features such as the type and size of an incentive, the number and type of contact attempts, and choice of the interviewer. Such designs are called adaptive or responsive (Groves and Heeringa, 2006; Schouten, Peytchev, and Wagner, 2017). Chapter 12, Adaptive Mixed-Mode Survey Designs, discusses such designs in detail. Figure 2.3 shows an adaptive mixed-mode design with two strata.

In mixed-mode designs, nonrespondents of one phase may be subsampled before entering a new data collection phase for costs or logistical reasons. Such subsampling is depicted as a dotted line instead of a straight line, see Figure 2.4.

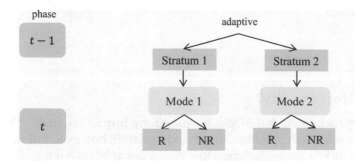

**FIGURE 2.3**
Schematic view of an adaptive mixed-mode design of a survey for which data collection runs in time period *t*.

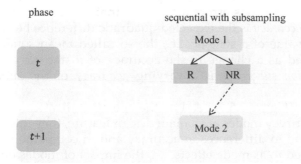

**FIGURE 2.4**
Schematic view of sequential mixed-mode design with subsampling of a survey for which data collection starts in time period *t* and extends to time period *t* + 1.

Obviously, surveys may combine concurrent, sequential, delayed concurrent, and adaptive mixed-mode within a single design. The resulting mixed-mode design then no longer is of a specific type as there are a plethora of options. Such designs are simply termed *hybrid* mixed-mode. Murgia et al. (2019) present an overview of mixed-mode designs used in the surveys of the European Statistical System (ESS) throughout Europe.

In Section 2.5, the case study surveys of this book are presented. The design of each survey is described following the terminology of this section. Below the various terms are summarized that will be used throughout the book.

*Concurrent mixed-mode* – Sample units are offered the choice between two or more modes of administration.

*Delayed concurrent mixed-mode* – Sample units are offered the choice between two or more modes of administration, but at least one mode becomes available at a later point in time.

*Sequential mixed-mode* – Sample units are offered two or more modes in sequence, allowing only one mode at a time.

*Hybrid mixed-mode* – Sample units are offered a mix of a (delayed) concurrent and a sequential design.

*Adaptive mixed-mode* – Different sample units are offered different sets of modes.

*Mixed-mode with subsampling* – Only a random subsample of (nonresponding) sample units is offered in all modes.

## 2.4 How Do Modes Affect Quality?

In the introduction, it was proclaimed that modes have a strong impact on quality. To this point, quality features of modes have not been discussed in detail, however, but they are the key factors in choosing mixed-mode designs. How do modes affect accuracy and comparability of statistics?

To discuss accuracy and comparability of statistics, the classification of errors as introduced by Groves et al. (2009) is used. Figure 2.5 shows the two main features: measurement and representation. From construct to edited response and from target population to a weighted set of respondents, various errors may occur that impact accuracy and/or comparability. In this book, *accuracy* is the expected quadratic difference between the estimated value and the true value of a statistic, i.e., the so-called mean square error. *Incomparability in time* is defined as a change in the accuracy of a statistic over time. *Incomparability between population subgroups* is a varying accuracy of the subgroups' statistics.

The steps and errors shown in Figure 2.5 depend on the operationalization and implementation of the survey. Since survey modes are a main design feature, they interact with each of the errors and, thus, lead to differences in accuracy and in comparability. Their impact is usually loosely referred to as mode effects, i.e., the impact of modes on survey errors. Later chapters discuss mode impact on single survey errors and methods to avoid or reduce the impact. Let us first make the term mode effect more rigorous.

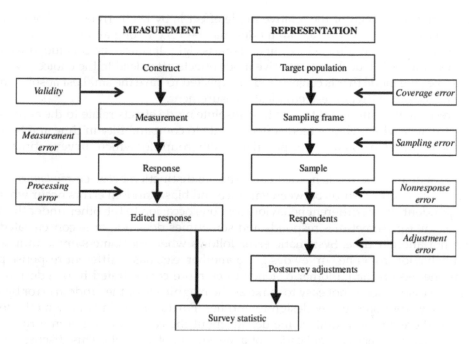

**FIGURE 2.5**
Survey errors for measurement and for representation. Taken from Groves et al. (2009).

The total impact of a single mode on a survey estimate, i.e., more formally, the mode-specific method effect, is called the *mode effect*. Following Figure 2.5, the mode effect potentially has multiple components conforming to the various steps in measurement and representation.

In survey practice, and also in survey literature, the term *mode effect* is, however, not used consistently and may represent different statistics (see also chapter 2 of Vannieuwenhuyze, 2013).

A number of remarks are needed to make the definition of a mode effect rigorous.

As a first remark, mode effects have no meaning without reference to a *population parameter* of interest, e.g., the mean income or the regression coefficient of deprivation on urbanization, and an *estimator* or statistic that employs survey data, e.g., the response mean or response regression coefficient. Consequently, within a single survey, there are as many mode effects as there are pairs of parameters and estimators. In practice, to avoid a highly multidimensional decision problem, only a few key parameters are evaluated.

The second remark is about a relative viewpoint versus an absolute viewpoint. Obviously, no method, and modes are no different, is infallible. Each mode of its own will produce a total survey error against a true population value, following Figure 2.5. One may term this the *absolute mode effect*. Even if a true measurement value is debated, still no survey mode will produce a full response, and the nonresponse will lead to varying absolute effects per mode. So absolute mode effects exist even in single mode surveys. However, in practice, it is hard, if not impossible, to estimate such absolute mode effects without validation data.

What one is stuck with in practice is a set of relative mode effects, i.e., differences between modes that are observed or estimated. The *relative mode effect* is the difference in absolute mode effects of two modes. It is, thus, not the property of a mode, but of pairs of modes. It is important to note that, without further information on the purpose of the survey estimates, the sign of the relative mode effect is by itself irrelevant. For example, when comparing survey estimates, all modes may be very wrong and one is slightly less wrong, or all modes may almost be right and one is right. It cannot be concluded which of these two scenarios is true. Hence, relative mode effects often lead to the choice of a *bench-mark mode*, i.e., the mode that is conjectured or expected to have the smallest absolute mode effect and closest to a (hypothesized) ideal measurement.

In this respect, it is important to note that absolute mode effects relate to the accuracy of survey estimates, while relative mode effects relate to comparability in time and between publication population domains (the population subgroups for which survey estimates are provided).

The third side remark that is needed concerns the distinction between *random error* and *systematic error*, in other words between variance and bias. Random errors occur when the same respondent shows different behavior from one occasion to the other under the same survey design, e.g., sometimes responds and sometimes does not, or is concentrated one time but not another time. Systematic error follows when the same sample unit shows different behavior from one survey design to another, e.g., has a different response prob-ability in one design than another or tends to be more concentrated in one design than another. The distinction is not easy to make as the magnitude of the random error by itself may also vary from one survey design to another. Furthermore, in survey methodology and theory, there has been quite some debate about the concept of random error, fueled by the fact that independent replication of a survey is not possible. This discussion goes far beyond the scope of this book. However, mode effects may, thus, refer to random error, systematic error, or both. The reference to both errors must then be done in a mean square error fashion as it is a function of both. The mean square error is the sum of the squared bias and the variance of an estimator, e.g., the estimated mean or total of a population parameter based on the survey. Random errors affect the variance term, whereas system-atic errors may affect both the bias and the variance term.

In this book, the three are labeled as *mode effects on variance, mode effects on bias*, and *mode effects on accuracy*. In literature about mixed-mode survey designs, usually, the focus is on systematic error, i.e., mode effects on bias. This will also often be the focus of this book. Mode effects refer to mode effects on bias, unless stated otherwise.

The fourth and last remark is about the survey design itself. From Section 2.3 it is clear that modes can be combined in several ways, leading to different types of mixed-mode designs. Along with these combinations the meaning and size of mode effects change. For example, the telephone mode effect for a sequential design with web followed by tele-phone in a general population survey is different from the telephone mode effect of a design with telephone as single mode. The reason is that the target population switches from the general population to the general population that does not respond to the web.

Summarizing, mode effects refer to a population parameter and estimator, depend on the target population, depend on the statistical property of the statistic that is evalu-ated, and have an absolute and relative viewpoint. All these subtleties, obviously, hamper generalization. This must be kept in mind. Nonetheless, there is general methodology to reduce the impact of mode effects, which is detailed in various chapters of the book. Before methodology is discussed, first mode effects are decomposed into their corresponding survey error components.

Let us return to Figure 2.5. The main survey errors are lack of validity, measurement error and processing error on the measurement side, and coverage error, sampling error, nonresponse error, and adjustment error on the representation side. For each survey error, there is a mode effect component that is labeled analogously, e.g., *mode-specific coverage effect*, *mode-specific nonresponse effect*, and *mode-specific measurement effect*. In this book, the specific reference to mode-specific effects is omitted, unless it leads to confusion.

Now, how do modes relate to survey errors? The various survey errors are enumerated here.

*Validity* refers to the difference between the intended (ideal) measurement and the operationalization through one or a series of questions. Mode affects the operationalization of a construct in various ways, as will be explained in detail in Chapter 3, Mode-Specific Measurement Effects. The mode itself may be restrictive to operationalization, e.g., a telephone survey is audio and a smartphone survey requires compact wording and presentation, and demand for concessions in constructing questions. The survey mode may even affect the latent construct itself when associations between the answers to questions are distorted. The latter is called *measurement inequivalence* (Klausch, Hox, and Schouten, 2013; Cernat, 2015).

Measurement error and modes are studied in depth in mixed-mode survey literature, because modes are known to have a strong impact on answering behaviors and answering conditions. Mode-specific measurement effects are also sometimes termed *pure mode effects*, or, confusingly, referred to as just mode effects. The reason is (probably) that one thinks of modes as different instruments to get answers from the same respondent, rather than as instruments to attract different respondents. See also Section 2.2. In this book, the term (mode-specific) measurement effect is used and its origins are discussed in Chapter 3, Mode-Specific Measurement Effects.

Processing error is itself a combination of two errors: *transmission error* and *editing error*. Since a mode is a communication channel, it has specific transmission features of respondent data, and, hence, its own specific transmission errors. Interviewer-assisted modes have a human intermediary in collecting survey data. Face-to-face (F2F) interviewers may use paper questionnaires, laptops, and/or mobile devices to record answers. These answers need to be transferred to the statistical office. Telephone interviewers are more likely to use fixed devices. Online surveys obviously employ web servers to which connection needs to be made. Paper questionnaires are picked up by an interviewer or returned by mail before they are transferred to digital form. In all these steps, errors may occur. Respondent data may be edited and respondent answers may be revised. Such editing is often based on edit or plausibility rules and relies on assumptions and auxiliary data. Since validity of assumptions and utility of auxiliary data may vary per mode, editing error may be mode-specific.

Coverage error refers to a difference between the target population and sampling frame. It comes in two forms: *overcoverage* and *undercoverage*. Overcoverage applies when the sampling frame contains units that are not in the target population. Conversely, undercoverage applies when the target population has units not included in the sampling frame. Overcoverage and undercoverage are properties of the sampling frame and, therefore, not directly related to modes. There are, however, settings where it plays a role, namely, when the mode determines the choice of sampling frame. When a list of email addresses or telephone numbers, motivated by the choice of mode, is used as a surrogate for the target population, coverage issues may occur. These lists may contain addresses or numbers not belonging to any unit in the target population. Survey institutes then need to assess eligibility of the sampled address or number, which may not be feasible for all units. The lists may also be incomplete or population units may not have an email or telephone.

Sampling error is the difference between a sample estimate and the corresponding population estimate. It is a direct function of the sampling design and its size is inversely proportional to the size of the sample. Sampling error depends on the mode only through the costs of conducting the survey for one sample unit. When fixing the budget, cheaper modes allow for larger samples and, consequently, smaller sampling error.

Nonresponse is perhaps the most studied of all mode effect components as response rates vary greatly between modes. Combined with coverage error, it is termed the *selection effect*. Nonresponse error has two forms: *unit-nonresponse* and *item-nonresponse*. Unit-nonresponse means that the questionnaire as a whole is not completed, whereas item-nonresponse means that only one or more answers are missing. A special case of unit-nonresponse is break-off, where a sampled unit starts the survey but does not complete it. The available data of the questions that have been answered cannot (usually) be used, so the questionnaire as a whole is missing. A refuse-to-provide-answer is a special case where the respondent explicitly expresses not to be willing to answer the question. Item-nonresponse, however, is not to be confused with a do-not-know answer, which can be a valid answer category. In this book, no separate account is given of item-nonresponse, but this type of error will show up when discussing measurement error. Selection effects are discussed in great detail in Chapter 4, Mode-Specific Selection Effects.

Although, typically, adjustment is applied to improve representation, it is based on assumptions and administrative data and paradata that are auxiliary to the survey. Adjustment error follows in the estimation stage of a survey when assumptions are made about the missing data mechanism and/or when auxiliary data are used that themselves may be subject to errors. Adjustment error does not depend explicitly on the survey mode, but it may be larger or smaller for a survey mode depending on the validity of assumptions and the reliance on auxiliary data.

How do all these errors feed back into mode effects? Total survey error is the sum of all errors, and the (total) mode effect is the compound or nett effect of all single survey error effects. This means that, in a statistical sense, the total mode effect must be written as a function of the components. Such expressions are not at all straightforward, because survey errors are nested. A population unit that is not covered by the sampling frame cannot respond to the survey. A sampled unit that does not respond to the survey cannot produce item-nonresponse or measurement error. Consequently, mode effect components have *potential outcomes*, i.e., responses or answers under the hypothetical condition that the unit was sampled or did respond. Table 2.3 contains estimates of relative mode effects, relative coverage effects, relative nonresponse effects, and relative measurement effects for the unemployment rate in the first wave of the 2012 Dutch Labour Force Survey with face-to-face as benchmark. See Calinescu and Schouten (2015). For instance, the total mode effect for being employed between web and face-to-face is estimated at 8.9% and

**TABLE 2.3**

Relative Mode Effects and Their Decomposition into Coverage Effects, Nonresponse Effects and Measurement Effects for Web and Telephone for the Unemployment Rate in the First Wave of the 2012 Dutch LFS

|            | Mode  | Mode Effect (%) | Coverage (%) | Nonresponse (%) | Measurement (%) |
|------------|-------|-----------------|--------------|-----------------|-----------------|
| Employed   | Phone | −0.4            | −1.2         | 0.1             | 1.5             |
|            | Web   | 8.9*            | 2.5*         | 5.5*            | 1.0             |
| Unemployed | Phone | −3.0*           | −0.8         | −1.1            | −1.1            |
|            | Web   | −2.6*           | 0.0          | −0.9            | −1.7            |
| Other      | Phone | 3.4*            | 2.0          | 1.0             | 0.4             |
|            | Web   | 6.3*            | 2.5*         | −4.6*           | 0.7             |

The benchmark is face-to-face. Effects significant at 5% level are marked by an asterisk

decomposes in a 2.5% coverage effect, a 5.5% nonresponse effect, and a 1.0% measurement effect. Nonresponse is, thus, the dominant effect in this case. Chapter 8, Re-interview Designs to Disentangle and Adjust for Mode Effects, returns to such decompositions and explains how relative mode effect components are defined and combined to the relative mode effect.

From mode effects of single modes to method effects of full mixed-mode designs require an extra step. Absolute method effects of mixed-mode designs are a complex interplay of the absolute mode effects of the single modes comprised in the design. Consider, for example, a sequential design with web followed by telephone. Both modes are subject to undercoverage due to unavailability of web access or telephone, but when combined, only those units without access to the internet and without telephone are still not covered. The combined nonresponse effect of the two modes may be weaker when the modes attract opposite population units but may also be larger when the modes attract similar population units.

For relative method effects, a benchmark design needs to be chosen. Given such a benchmark design, one may be satisfied with the total method effect and not consider its survey error components. Chapter 8 considers methods to efficiently assess relative total method effects for mixed-mode surveys. One may also search for estimates of survey error components in order to effectively improve accuracy and/or comparability. Section 2.5 introduces such methodology. See also the discussion in Biemer (2001).

However, mixed-mode surveys allow for more advanced benchmark designs. As an example, consider again the sequential design with web and telephone. One may judge the combined absolute selection effect, i.e., coverage and nonresponse combined, to be smaller than the single mode absolute selection effects. At the same time, one may view web measurement as superior to telephone measurement for the particular survey items. A benchmark design would then be the mixed-mode selection plus the web measurement. Such a design is called *counterfactual*, as it is not possible to actually implement it. Even if telephone interviewers, after obtaining a response, would invite sampled units to complete the survey online and these units would all do this, still the units would have been influenced by the context of the telephone invitation. Still such counterfactual benchmark designs are considered (Klausch, Schouten, Buelens, and Van den Brakel, 2017).

Mode-specific costs have not been discussed to this point, but they are often the main driver behind redesigns to mixed-mode surveys. Costs vary greatly between modes (see Table 2.1). Even if a real design is chosen as benchmark, this benchmark design may not be implementable due to its large costs. Although cheaper modes do not necessarily have larger mode effects, costs do limit the possible range of designs and the options to minimize mode effects. As a consequence, available budget is the natural counterpart of mode effects and mixed-mode method effects; they will come up in the various chapters.

Mode effects play an important role in surveys that are repeatedly conducted with the purpose to construct time series that measure period-to-period change over as long as possible periods. In mixed-mode designs, the distribution of respondents over the modes typically varies between the subsequent editions of the survey. As a result, the absolute mode effect varies between the different editions of the survey and adds additional bias to the estimates for period-to-period change. Methods to stabilize these biases in the estimation procedure are discussed in Chapter 9, Mixed-Mode Data Analysis. Another aspect is the implementation of a new data collection mode in a repeated survey. This results for similar reasons in an instantaneous difference in mode effects. To avoid confounding real period-to-period change with systematic differences in mode effects, it is important to quantify the relative mode effects. Different methods to quantify these so-called discontinuities are discussed in Chapter 8, Field Tests and Implementation of Mixed-Mode Surveys.

So far, the impression may be that mode effects and their components are a nuisance to survey design. This is not true. In fact, relative mode effects can be helpful in reducing the method effect of mixed-mode designs, as will be discussed in the next section.

An overview of terminology is given below.

*Accuracy* – Mean square error (MSE) of observed value of a statistic to its true value

*Comparability in time* – Change in accuracy between two values of a statistic in time

*Comparability between subpopulations* – Difference in accuracy between the values of a statistic for two subpopulations

*Absolute mode effect* – Accuracy of a mode relative to the true value

*Relative mode effect* – Difference in accuracy between two modes

*Mode-specific coverage effect, nonresponse effect, and measurement effect* – Components of the mode effect

*Benchmark design* – Mode or combinations of modes that are assumed to have the highest accuracy or deemed most comparable to a former mode design

*Mode effect on bias* – The impact of a mode on the bias of a statistic

*Mode effect on variance* – The impact of a mode on the precision of a statistic

*Mode effect on MSE* – The impact of a mode on the accuracy of a statistic

## 2.5 How to Reduce Mode Impact on Quality?

As explained in the previous section, survey modes, like other design features, introduce method effects.

These effects can be decomposed into corresponding survey error components. In this section, an introduction is given to methodology to reduce and adjust the various mode effect components.

Figure 2.6 depicts the plan-do-check-act cycle for surveys. The plan and act phases collide into survey design and redesign. The do phase amounts to the actual data collection. The check phase consists of fieldwork monitoring and survey data analysis. The survey data collection itself consists of a smaller plan-do-check-act cycle that is in operation during the survey fieldwork period: the fieldwork is planned and possibly adapted in the plan and act phases. The do phase conforms to data collection phases in which the design is kept fixed. These may be time periods, say a week or month, but also shifts in a design such as a change of mode in sequential mixed-mode designs. The check phase is daily monitoring of the survey fieldwork.

In the plan/act phases, mode effects can be addressed in four design stages: questionnaire, sampling, data collection, and estimation. Table 2.4 presents these four main design stages and the mode effect components they address. The four stages are used as a starting point for distinguishing methodology to account for mode effects.

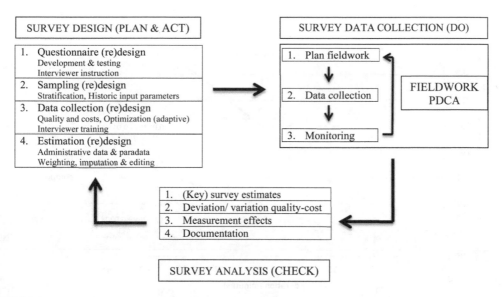

**FIGURE 2.6**
The plan-do-check-act cycle for surveys. The do phase of a survey itself has a smaller plan-do-check-act cycle linked to data collection.

**TABLE 2.4**

Methodology Design Stages and the Link to Mode Effect Components

|  | Coverage | Nonresponse | Measurement |
|---|---|---|---|
| Questionnaire |  | X | X |
| Sampling | X |  |  |
| Data collection | X | X | X |
| Estimation | X | X | X |

Questionnaire design has measurement error and item-nonresponse as its focal points but does also impact unit-nonresponse. To begin with the latter: the length, flow, general design, and layout of questionnaires affect participation and completion of the survey. Sample units considering participation may depend their decision on the appearance of the survey. This may be very salient in self-administered modes, especially paper questionnaires, and less so in interviewer-administered modes, but may always play a role. Dillman et al. (2014) have addressed these decisions in great detail. Once the interview is started, the questionnaire design also affects break-off and partial responses. Figure 2.7 shows break-off rates, defined as the proportion of the sample that answers at least the first survey question, as a function of time for different online devices for a range of surveys. The survey questionnaires have been optimized for traditional desktops and laptops but have not been adjusted for smartphones. The break-off rate for smartphones is much higher than for other devices.

Questionnaire design aims at the reduction of measurement error and item-nonresponse. It is a mix of literature review and reported best practices and guidelines, cognitive testing and usability testing, evaluation of paradata, and analysis of larger pilots and field experiments. Cognitive testing has been an element of survey design for a long time, but encountered new challenges when web became a prominent mode

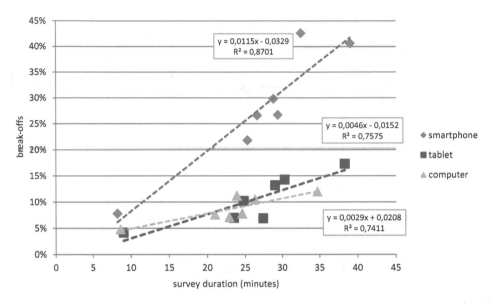

**FIGURE 2.7**
Break-off rates as a function of interview duration for desk/laptops, tablets, and smartphones averaged over a range of nonoptimized surveys.

and several online devices were introduced gradually. A crucial decision is the dominant mode in making questionnaires. This choice has long been an interviewer mode in many surveys, except for modes in which paper questionnaires are an important mode. With the rise of the web survey mode, the dominant mode tends to shift, however. And it tends to shift further with the introduction of devices. One viewpoint in multi-mode questionnaire design is that of a *unified design* or *Dillman unimode design*, in which questionnaires for different modes are as similar as possible. Given the features of modes and, especially, devices, this pursuit usually implies a dominant mode, i.e., an online first or smartphone first design. Another viewpoint is that of a *best practices approach*, in which measurement errors are reduced as much as possible within each mode. Responsive questionnaire design, where questionnaires respond to the type of device, is a good example. Chapter 6, Mixed-Mode Questionnaire Design, discusses the various steps in questionnaire design and how to combine these viewpoints. The different features of modes, with the presence of an interviewer as most influential, have challenged testing. An interviewer-administered survey can be mimicked in a cognitive test, but self-administered modes are harder and demand a stronger reliance on paradata. Examples of such data are audit trails, time stamps, audio and video recordings, and even eye-tracker data. Chapter 6, Mixed-Mode-Mode Questionnaire Design, also describes the best practices and challenges in testing over a range of modes and devices. Some varieties in questionnaire design are hard to understand without large-scale field test data. Chapter 7, Field Tests and Implementation of Mixed-Mode Surveys, discusses randomization techniques to estimate the impact of questionnaire design choices. As an example of such a randomized experiment and to fix thoughts, Table 2.5 shows associations between the size of answer category buttons on a smartphone and a range of survey variables from the health survey. The estimated association is largest for drug use and feeling nervous with values of, respectively, 0.158 and 0.151. These values still represent only a relatively modest association, however.

**TABLE 2.5**

Cramer's *V* Values for the Association between Smartphone Answer Category Button Size and 15 Key Variables from the Dutch Health Survey

| Variable | Cramer's *V* | Variable | Cramer's *V* | Variable | Cramer's *V* |
|---|---|---|---|---|---|
| Health status | 0.063 | Feeling grim | 0.139 | Eat fruit | 0.086 |
| Handicaps | 0.063 | Feeling happy | 0.050 | Alcohol use | 0.048 |
| Feeling nervous | 0.151 | Visit GP | 0.116 | Drugs use | 0.158 |
| Feeling tired | 0.090 | Eat vegetables 1 | 0.118 | Had sex | 0.029 |
| Feeling calm | 0.095 | Eat vegetables 2 | 0.175 | Injury | 0.110 |

Sampling design is oriented at the minimization of sampling error but also addresses coverage through the choice and use of the sampling frame. Overcoverage and undercoverage, essentially, depend on the mode(s) of contact and the access of a sample unit to the modes of administration. The mode(s) of contact rely on the contact information available in the sampling frame, i.e., addresses, phone numbers, emails.

Consider first overcoverage. When the same modes of contact are used for different modes of administration, then there is no mode-specific overcoverage effect. For example, when an advance letter is sent to an address announcing the visit of an interviewer, or an invitation letter is sent to the same address offering a link to an online questionnaire, they have the same (potential) overcoverage. Obviously, the mode of administration may affect the choice of modes of contact. For example, one may perform a random-digit-dialing (RDD) approach in a telephone survey, which will evidently have a different coverage than an invitation letter to an online survey. However, if RDD is merely used to invite sample units to make appointments for field interviewers or to go to an online survey, then again no mode-specific overcoverage exists. Hence, mode-specific overcoverage effects occur only implicitly through the choice of contact modes. As a result, one rarely speaks of mode-specific overcoverage.

However, detection of overcoverage does depend on the mode of administration. In the self-administered survey modes, there is less control on who is completing the survey and who is not. A sample unit not eligible for the survey may participate and overcoverage may go unnoticed. Conversely, sample units may be ineligible and (consequently) show no sign of life, which cannot be separated from a refusal or other type of nonresponse. Paper and online questionnaires may screen respondents, but ultimately it is up to the sample unit to take action. In sequential mixed-mode designs where self-administered modes are followed by interviewer modes, it occurs frequently that nonresponding sample units are concluded to be overcoverage by interviewers. This means that response rates in self-administered modes are blurred to some extent by undetected overcoverage; they tend to be underestimated.

Consider next undercoverage. Following the same reasoning, also undercoverage is merely a result of the choice of contact modes, from the sampling design point of view. But with one exception: the contact information in the sampling frame may be complete for one mode of administration but not for another. A sampling frame may, for example, contain addresses and phone numbers. The phone numbers may be incomplete or outdated, so a telephone survey may lead to undercoverage, whereas a face-to-face survey does not. Although sampling design itself cannot overcome this, it is a feature of the sampling frame. Mode-specific undercoverage may, thus, follow from faulty contact details in the sampling frame. Such undercoverage effects usually do not go unnoticed; emails will bounce, phone numbers will turn out to be not in use, and dwellings will turn out inhabited or used by others.

Data collection design mainly focuses on representation errors, but it also affects measurement error. The choice of modes, but also the implementation and combination of modes, is part of the data collection design. Obviously, per mode, data collection design can be optimized by the design of advance/invitation letters, the training of interviewers, the look and feel of online survey landing websites, the timing of calls, reminders and visits, and so on. Much of this is, essentially, independent of the combination of modes within a mixed-mode design. In this book, the focus is on data collection design specific to mixed-mode surveys. Chapter 5, Mixed-Mode Data Collection Design, discusses methods to reduce representation errors.

Choices made in data collection design may be adaptive when different sample subgroups get different selections and sequences of modes. Such adaptive mixed-mode designs employ auxiliary data that are available at the onset of a survey or that become available during data collection. These designs are described in detail in Chapter 11, Adaptive Mixed-Mode Survey Designs.

Finally, estimation design may consider all survey errors. The estimation design concerns all processing and corrections that are made after the survey is completed. Traditionally, there has been little interest in estimation designs that explicitly account for mode-specific survey errors. The last ten years, following the introduction of the web mode and the gradual transformation of surveys to mixed-mode including web, have seen an increased interest in such designs. Without exception such methods rely on additional information, either through auxiliary data or through assumptions. A distinction can be made between designs that employ experimental designs collecting auxiliary data, and designs that do not. Since undercoverage and unit-nonresponse are often hard to separate, they are typically adjusted for simultaneously in the estimation. However, estimation methods do attempt to separate representation error from measurement error. Chapter 8, Re-interview Designs to Disentangle and Adjust for Mode Effects, presents various approaches to separate mode-specific representation effects from mode-specific measurement effects. Chapter 9, Mixed-Mode Data Analysis, more generally, considers inference from mixed-mode survey designs.

## 2.6 Case Studies

In this book, three case studies are employed as running examples: the Crime Victimization Survey (CVS), the LFS, and the European Health Interview Survey (EHIS). All three surveys are official surveys that are conducted in many countries with a relatively similar set of topics and questionnaires. Within the EU, the LFS is a mandatory survey for the member states. This holds as well for the EHIS, which is a mandatory five-yearly extension of country health surveys. The CVS is not a mandatory EU survey but is conducted in many countries.

The three surveys differ in their characteristics such as number of items, complexity of required information, need for recall, and sensitivity of required information, and, thus, are expected to be affected differently by the choice of survey modes. The surveys also differ greatly in available budget, which implies that decisions to allocate efforts to estimate, reduce, and adjust mode effects are made differently. This makes them interesting and realistic examples to illustrate the challenges and solutions presented in the previous sections. Each survey is briefly presented.

### 2.6.1 Dutch Crime Victimization Survey

A Crime Victimization Survey may have a broad set of topics and may contain questions about the perceived quality and safety of the neighborhood, feelings of unsafety, victimizations for a range of crimes, contacts with police departments, and perceived performance of the police municipality and cybercrime.

The Dutch CVS has as target population all persons registered in the Netherlands of 16 years and older. The sampling design is a stratified simple random sample with regions as strata. The survey consists of a base sample that is self-weighting, and additional samples on request by specific municipalities, usually the bigger cities. The CVS is a repeated survey with monthly samples. Statistics are published on an annual basis.

Figure 2.8 shows the former and current mixed-mode designs of the CVS. The former CVS mixed-mode design is a hybrid design with two data collection phases in which the first is single mode and the second is adaptive on availability of a telephone number, i.e., represented as listed versus nonlisted. In the first data collection phase, sampled persons were also given the opportunity to apply for a paper questionnaire but few persons used this option. This hybrid design was used between 2006 and 2011. In 2011, a large-scale experiment was launched to estimate relative mode effect components. This experiment led to a redesign to a delayed concurrent mixed-mode design with web and paper as survey modes. This design is still used. The delayed concurrent design is implemented as follows: An invitation letter is sent with a link to the CVS online questionnaire. After one week and after two weeks a reminder letter is sent, including a paper questionnaire. It is an example of a web-push design where respondents are stimulated to fill in the online questionnaire for reasons of cost and timeliness. If a person does not respond after three weeks, then interviewers make telephone reminders. However, the questionnaire cannot

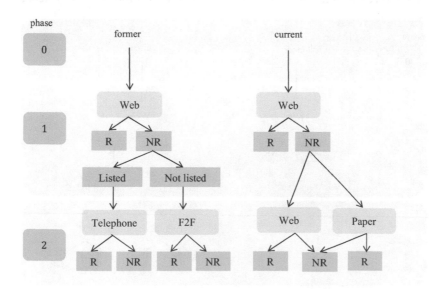

**FIGURE 2.8**
The CVS design up to 2011 and the CVS design after 2011.

be completed over the phone in order to avoid measurement effects. A conditional incentive is used in the form of a lottery.

### 2.6.2 Dutch Labour Force Survey

The Labor Force Survey is an ESS survey that is conducted in a relatively similar form throughout Europe. It contains questionnaire modules on employments, unemployment, working hours, working overtime hours, highest attained and followed educational levels, and current professional courses and education. The LFS is a household survey with all persons of 16 years and older living in the country as target population. The LFS is a rotating panel with multiple waves and a three-month time lag. The first wave is most extensive and longest. In this book, the focus is on the first wave. The LFS produces both monthly and annual statistics.

In Eurostat projects, DCSS (Data Collection for Social Surveys) and MIMOD (Mixed-Mode Survey Designs), inventories have been made of the modes that are used in the LFS in the ESS. See Murgia, Lo Conte, and Gravem (2018) for an overview of the 31 ESS countries. For the LFS recruitment wave, 13 countries use 1 survey mode, 7 countries mix survey data with administrative data, 8 countries use a mixed-mode design, and 3 countries use a mixed-mode survey design plus administrative data. The mixed-mode designs are web-telephone (2x), mail-F2F (4x), telephone-F2F (2x), web-telephone-F2F (2x), and mail-telephone-F2F (1x). So, to date, the LFS consists of a mix of designs across countries.

In the Netherlands, the LFS is based on a two-stage sampling design with municipalities and addresses as stages. In the second stage, addresses are stratified based on age, registered unemployment, and ethnicity. Addresses with persons below 25, with persons registered unemployed and with persons of non-Western nonnative ethnicity, are oversampled, whereas addresses with persons above 65 years are undersampled.

Figure 2.9 depicts the former single mode and current mixed-mode design of the LFS first wave. Up to 2010 the LFS had a single mode face-to-face design. Within a time span of three years, the survey was transformed into a hybrid mixed-mode survey design. During

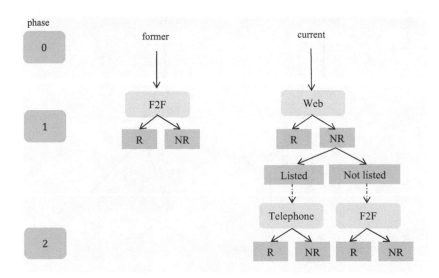

**FIGURE 2.9**
The LFS first wave design up to 2010 and the LFS design after 2012.

these years, first an adaptive mixed-mode design was used with the telephone and face-to-face modes. The adaptation was based on the availability of a telephone number, i.e., again represented as listed versus nonlisted. As a second stage, the web mode was added in sequence before the adaptive telephone and face-to-face modes. This design has been used since 2012 and is still in use. The web mode is implemented as an invitation letter and two reminder letters, all containing a link to the online questionnaire. The telephone mode is implemented as three call days of three calls each. The face-to-face mode consists of a monthly data collection with at most six visits that are spread over different week days and times of the day. The eligible sample for follow-up in CATI and CAPI is subsampled in order to ensure a pre-fixed interviewer workload.

### 2.6.3 European Health Interview Survey

The European Health Interview Survey is a diverse survey asking questions about self-perceived health, use of various medical and health care facilities, diseases, handicaps and injuries, medicine use, nutrition, and daily activity and leisure. The population comprises all registered persons in the country. For persons below 16 years, parental consent is usually necessary. For persons below 12, years parents need to be present and/or report for their children.

As for the LFS, in Eurostat projects, country mode designs have been collected with the ESS. Murgia, Lo Conte, and Gravem (2018) report that 12 out of the 31 countries use a mixed-mode design. The designs are mail-F2F (1x), web-F2F (6x), mail-web (2x), web-telephone (1x), mail-telephone-F2F (1x), and web-telephone-F2F (1x). Five out of these 12 countries in addition employ administrative data. Like the LFS, the EHIS is based on a mix of designs.

In the Netherlands, the sampling design is two-stage with municipalities and persons as stages. The sample is self-weighting. Official statistics are published annually.

Figure 2.10 presents the former single and current mixed-mode design of the HS. As for the LFS, the HS had a single mode face-to-face design up to 2010. It then went through

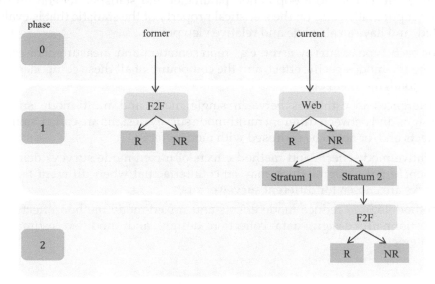

**FIGURE 2.10**
The EHIS design up to 2010 and the HS design after 2017.

a series of changes. First a hybrid mixed-mode survey design was introduced with two data collection phases similar to the LFS current design and the CVS former design. Shortly after the introduction of web and telephone, it was concluded that the telephone mode was not suited for the HS, because of the length of the survey and the sensitivity of some of the HS survey items. The telephone mode was omitted and the HS became a sequential mixed-mode design with web and face-to-face modes. This design was used between 2012 and 2017. In 2018, the face-to-face second data collection phase was made adaptive, in which some strata are allocated to face-to-face and others are not. The latter stratum, thus, receives no follow-up. The two strata are formed based on age, urbanization degree of the area of residence, ethnicity, and household income.

## 2.7 Summary

The main takeaway messages from this chapter are:

- Mixed-mode designs are divided into concurrent, sequential, delayed concurrent, and hybrid designs depending on the timing and order of the survey modes;
- Within the same survey mode, there may be considerable variation in survey errors because of varying design features other than the mode and the contact strategy that precedes the interview;
- Mode effects are the net differences in an estimated population parameter between pairs of modes;
- Mode effects refer to a population parameter and statistic, depend on the target population and on the statistical property of the statistic that is evaluated, and have an absolute and relative viewpoint;
- For each type of survey error, e.g., representation and measurement error, there is a mode-specific effect, and the compound of all these components is the (total) mode effect;
- Differences in estimates between single mode and multi-mode survey designs and between different multi-mode survey designs are called method effects and/or not to be confused with mode effects;
- Relative mode effects and method effects of mixed-mode surveys demand a benchmark design, which may be counterfactual when different benchmarks are chosen for different survey errors;
- Methodology to reduce mode effects and mixed-mode method effects are questionnaire design, data collection design, and modified estimation strategies.

# Part II

# Mode Effects

# 3

## Mode-Specific Measurement Effects

## 3.1 Introduction

This chapter has three objectives: to discuss the (1) occurrence, (2) impact, and (3) detection of *mode-specific measurement effects*. Mode-specific measurement effects refer to the differences in answers that the same respondent provides when questions are posed in different modes. They are also termed pure mode effects and have been at the core of multi-mode survey research from the very start. Mode-specific measurement effects are one component of the total mode effect (as discussed in Section 2.4). In this chapter, the focus is on the measurement part of total survey errors, see Figure 3.1.

These three objectives of the chapter will lead from the cognitive answering process to the analysis of respondent survey data and paradata. The occurrence of measurement effects is usually related to the various steps in answering a survey item and the properties of the survey item. Both will be described and elaborated. Here, a survey item refers to the combination of the introduction text, the question, and the answer options or categories. The *impact* of measurement effects may be complex to interpret, as explained in Section 2.4, since they may concern both systematic and random error. Moreover, a true value is often not available, and so a benchmark needs to be chosen against which differences are compared. To date, still, no general guidelines exist as to how to choose such a benchmark. However, some general approaches will be given. Finally, the *detection* of measurement effects relies heavily on auxiliary data and/or on experimental designs. In this chapter, the restriction is to the collection of auxiliary data that facilitates the detection of measurement effects. Experimental designs are discussed in Chapter 8, Re-Interview Designs to Disentangle and Adjust for Mode Effects.

A crucial limitation and barrier in the analysis of mode-specific measurement bias is the confounding of measurement error with representation. Naturally, measurement follows selection, and, as a consequence, differences between modes or mixed-mode designs can be the result of both differences in selection and differences in measurement. In the extreme scenario, two designs attract disjoint, complimentary respondent populations, and measurement effects become hypothetical rather than real. Nonetheless, one may argue that potential outcomes exist and a respondent would have provided a potentially different answer had he or she been interviewed in another mode. This confounding becomes even stronger when sample units are offered a mode choice, as respondents may choose the mode in which they feel most comfortable and/or that requires the least effort. The latter means that a benchmark value is individual rather than global. There is not much evidence from the literature, however, that respondents self-select into the optimal mode, but studies have demonstrated the association between unit-nonresponse and measurement error (Olson, 2007; Olson and Parkhust, 2013; Kreuter, Müller, and Trappmann, 2010). In order to

DOI: 10.1201/9780429461156-3

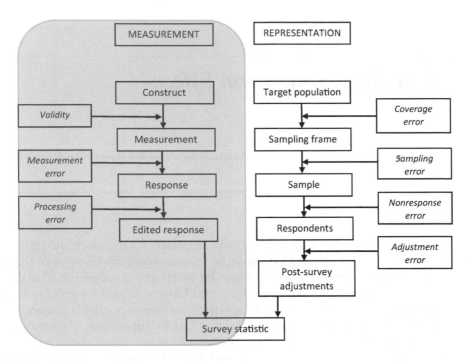

**FIGURE 3.1**
Survey errors for measurement and for representation. *Source*: Groves et al. (2009).

not further complicate the study of measurement effects, this chapter focuses on general population surveys.

Mode-specific measurement effects have been studied extensively from both statistical, socio-metrical, and psychometrical perspectives. Key references are Dillman et al. (2009); Jäckle, Roberts, and Lynn (2010); Krumpal (2013); Vannieuwenhuyze (2013); and Klausch (2014, chapter 2).

Mode-specific measurement effects result from the interplay of mode features, characteristics of the survey and its items, and the characteristics of the respondent.

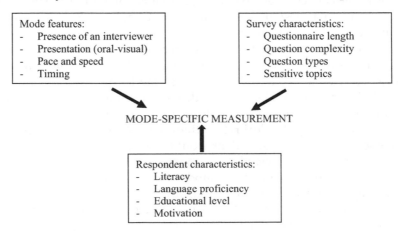

In this chapter, Section 3.2 elaborates on the basic features of modes. Section 3.3 introduces mode-specific answering behaviors and the interaction between mode features, survey characteristics, and respondent characteristics. Section 3.4 discusses measurement effect detection. Takeaway messages are given in Section 3.5.

## 3.2 Measurement Features of Modes

To understand the origin of measurement effects due to modes, one needs to dive into the specific characteristics of the modes. How do the modes differ from each other, and what features do they offer? This section discusses the origin of measurement effects from the perspective of the mode features. In the next section, this perspective is combined with the other actor in measurement effects: the respondent.

In this chapter, the focus is on the modes that are used to complete the survey and not so much on the modes that are used to make the initial contact with the respondent. This does not at all mean that the contact modes do not impact measurement. On the contrary, a respondent may, for example, be more or less motivated depending on how intrusive the contact and recruitment was perceived, on whether and how trust of the respondent in the survey institute was established, and/or on how recruitment and instruction material was designed and aligned with the questionnaire. Careful design of contact modes and material is paramount. Here, it is assumed that such material was designed according to guidelines and recommendations in the literature and that they have no strong interaction with measurement error in the modes of administration. See Section 5.6 for communication strategies. The reader is also referred to Schaeffer and Dillman (1998); De Leeuw et al. (2007); Millar and Dillman (2011); Bandilla, Couper, and Kaczmirek (2014); and Vogl, Parsons, Owens, and Lavrakas (2020).

Modes are, in essence, communication channels and their introduction and form follow trends in societies, but with some time lag. If a communication channel is adopted on a large scale by individual population units, becomes prominent in more formal communications, and has personal identifiers that can be listed in some way, then a channel is a candidate for a survey. Using different tools and arousing different senses, such modes, are, however, disparate on various features. In Section 3.3, it is discussed how features of survey modes may interact with steps in answering survey questions. Here, features are discussed first.

Survey mode features have been discussed in the literature. Couper (2011) distinguishes the degree of interviewer involvement, the degree of contact, the channel of communication, the locus of control, the degree of privacy, and the degree of computer technology as dimensions. Pierzchala (2006) lists eight features: the type of presentation, the manner of responding, the segmentation of the questionnaire, the dynamic versus passive nature of the questionnaire, the type of administration (self-administered versus interviewer-assisted), the pace of the interview, the medium of the interview, and the training of the person recording the responses. Based on these ideas, six main features are distinguished here:

1. Intimacy: The extent to which the communication is entering personal circumstances and daily life;
2. Interaction: The extent to which communications are interactive and resemble human conversation;
3. Assistance: The extent to which respondents can seek/ask for help in navigation and in interpretation;

4. Speed and pace: The speed at which communications using the mode tend to take place and that respondents may conform to;

5. Presentation (aural or visual): The form in which survey questions and answer options are presented to respondents;

6. Timing: The extent to which respondents can choose the place and time to do the survey.

Table 3.1 describes the four contemporary modes on the six features. The average durations for phone and face-to-face interviews were taken from surveys conducted at Statistics Netherlands. The rating of the modes is to some extent subjective.

The first three main features all relate to the presence of an interviewer. The interviewer has been a traditional actor in surveys for a long time. Communications have been nondigital in the past, thus limiting the number of potential channels and making personal communication much more common than today. Interviewers act as representatives of the survey institute, give it a face, and make communication more personal and intimate. Interviewers are spokespersons of the survey institute and can convey the message, relevance, and urgency of the survey. Finally, interviewers are experts and can perform the survey task more easily than respondents; they can assist respondents and navigate through the questionnaire. Being from a remote location, telephone scores weaker on intimacy and interaction than face-to-face. Interviewer modes lead to a form of intimacy that may be prohibitive to sensitive survey topics. Certain survey topics, such as health, victimization, political, and religious values, are normally not discussed with strangers. Although interviewers will avoid performing judgments and conclusions and may mirror respondents to some extent, they are strangers and bring with them social and behavioral norms. This may lead to so-called social desirability in answering questions. See Blom and West (2017) for an extensive overview of the role of interviewers in all phases of a survey. Self-administered surveys have been possible since the existence of paper questionnaires either as drop-off questionnaires or via mail. The absence of personal communication makes mail and online surveys less intimate, reduces interaction, and formalizes contact to helpdesk messages and phone calls. Online surveys may be perceived as slightly more intimate as they are conducted on personal devices. Being computer-assisted, online surveys allow for some flexibility in assistance through extensive help buttons and frequently asked questions, which are too extensive to provide on paper. Paper questionnaires can be complex when surveys contain filter questions and different routing and then demand more assistance than computer-assisted modes.

**TABLE 3.1**

Main Features of the Four Survey Modes, Web, Mail, Phone, and Face-to-Face

|              |             |             |                   |                   |
| ------------ | ----------- | ----------- | ----------------- | ----------------- |
| Intimacy     | Some        | Little      | Moderate          | Strong            |
| Interaction  | No          | No          | Some              | Strong            |
| Assistance   | Some        | No          | Yes               | Yes               |
| Speed-pace   | Independent | Independent | 15 sec on average | 18 sec on average |
| Presentation | Visual      | Visual      | Aural             | Mostly aural      |
| Timing       | Any time    | Any time    | Day & early evening | Day & early evening |

The fourth main feature is the speed and pace of the interview. These essentially correspond to two subfeatures: the average speed of the interview and the variation in speed during the interview. Personal interaction determines to a large extent the average speed of a survey but also the potential variation. Phone interviews tend to go faster than in-person interviews, as silences over the phone are less common. Interviewers, obviously, notice this when respondents increase their speed during the interview and are trained to moderate the pace. Self-administered modes tend to show more variation in speed. See, for example, Fricker et al. (2005) and Tourangeau and Yan (2007). Self-administered modes are free of such conversational norms and respondents may pause and regain the survey.

The fifth main feature is the general presentation of the questionnaire, which ranges from fully aural for phone to fully visual for paper questionnaires. The form of the presentation determines the cognitive processing of information by the respondent. Aural presentation prevents the use of supporting material that may assist the survey. This is especially strong for survey questions with many possible answer categories. Face-to-face interviews still allow for some visualization as opposed to phone interviews but are still primarily an aural mode. The presentation also includes navigation as this is an integral part of the questionnaire. An important tool in presentation is the computer, which organizes design, lay-out, and navigation. Most telephone and face-to-face surveys are computer-assisted.

The sixth main feature is the timing of the interview. For interviewer-assisted surveys, the timing is limited to the working hours of the interviewers: often day time and early evening. Self-administered surveys can, in theory, be completed at any time but may depend on other factors, e.g., being somewhere (e.g., at home) where one can use a desktop computer

Figure 3.2 gives a rough and subjective rating of the four modes plus smartphones on each of the six dimensions associated with the mode features. Smartphones are included

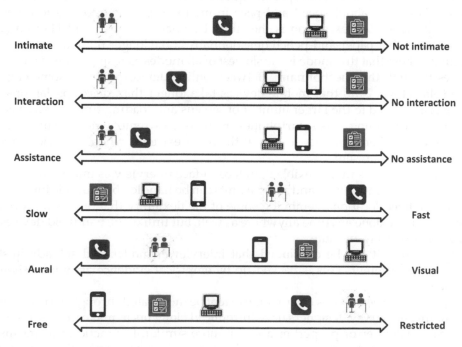

**FIGURE 3.2**
The four modes rated on the six features. Smartphones are added as a separate mode.

separately, because they are potential devices in online surveys, but they differ from fixed devices such as desktops and laptops.

Face-to-face interview is the most intimate mode as interviews tend to take place in the respondents' homes. Mail is the least intimate as there is no recording of the living conditions of a respondent, nor is it performed on a personal device. Online surveys are more intimate than mail surveys as personal devices are used; for smartphones, this holds stronger than for fixed devices that are often used by the whole household.

Face-to-face also is closest to a personal conversation and as such supports full verbal and nonverbal interaction. Telephone conversation, being distant and audio, has weak nonverbal interaction. Self-administered modes lack interaction almost completely. Online surveys do have some interaction as the user interfaces respond to answers through routing. This is a bit more limited in smartphone surveys because of the screen size.

Assistance from the survey institution is stronger for interviewer-assisted modes as interviewers represent the institution and are trained to answer general and specific questions. Mail scores the weakest on interaction as it lacks both computer assistance and human assistance. Through help buttons and search options in helpdesk libraries, online surveys provide some interaction. Smartphones score a bit weaker through the limitations of the screen size and navigation.

When it comes to speed, the picture is less clear. Based on recent years of paradata, telephone surveys are rated at 15 seconds per question and face-to-face surveys at 18 seconds per question at Statistics Netherlands. Self-administered surveys tend to take longer, up to 30%, but between-person variations are stronger than for the interviewer-assisted surveys. Also, within-person variation in speed during the questionnaire is stronger for self-administered surveys. The latter conforms to the norm that personal communications do not allow for long silences. The within-person variation also hints at problems that respondents may encounter in interpreting questions or navigating through the questionnaire. Little is known about speed and pace in mail surveys, as opposed to other self-administered modes. An exception is the study by Fuchs, Couper, and Hansen (2000), which concludes that paper-and-pencil questionnaires take longer than face-to-face ones. It is speculated here that this mode is the slowest of all modes as respondents have to route themselves through the questionnaire. Between online devices, there is some empirical evidence (Mavletova, 2015) that mobile devices take longer than desktops/laptops. This difference is attributed to the larger number of screens and loading times of screens.

The last feature, timing, has smartphones in the one extreme, where a survey can be done anytime anywhere and face-to-face in the other extreme, where interviews require the interviewer to be present.

Telephone interview is more flexible than face-to-face interview as interviewers usually are not assigned to sample units and sample units can be handled by multiple interviewers. Desktops have fixed locations, removing some of the flexibility that online surveys have. Mail surveys can be done anytime anywhere as well, but unlike the handheld devices cannot be performed easily en route.

From the ratings, it can be concluded that interviewer modes and self-administered modes are most different. This is backed up by empirical evidence as will be shown in later chapters.

The disparity between modes can to some extent be moderated through added elements. Questionnaire and survey designers have introduced elements to make modes more alike, so that measurement error properties are also more similar. In practice, this means that features on which a mode scores very strongly are transferred to modes that score weakly on the features.

On the intimacy dimension, added features go both ways. Self-administered modules are added in face-to-face for (very) sensitive topics, which is called self-interviewing. For part of the survey, the respondent is left alone with a questionnaire on a laptop or on paper. The interviewer will wait until the completion of these modules and remain present, or pick up the completed questionnaire later.

Attempts have been made to computerize telephone surveys by introducing interactive voice response (IVR); a human interviewer is then replaced by automated cues and questions and speech recognition. Conversely, the web mode has been made more intimate by exploring virtual interviewers that read out introductions and questions, while the answering process may still be conducted through the keyboard. To date, IVR and virtual interviewers still are marginally used. See, e.g., Fuchs and Funke (2007), Bloom (2008), Fuchs (2009), and Conrad et al. (2015), and Figure 3.3 for an example taken from Fuchs (2009).

On the interaction dimension, the contrast is between interviewer-assisted and self-administered modes.

Web surveys can be given more interaction by introducing a landing page providing information, background, and frequently asked questions. In the online survey itself, various help buttons are added to assist potential questions that respondents may have. The recent emergence of chatbots has stimulated the investigation of more advanced interaction in surveys, i.e., in the form of chats rather than plain questions on a screen. See Figure 3.4 for an example of a chat-like survey. All questions are posed in the form of a chat textbox, and respondents also type in answers in similar textboxes.

Conversely, the interaction in interviewer-assisted modes is often reduced or limited. Interviewers are instructed to adhere to the exact question and answer texts and not to use their own wording or mirror respondents in the choice of wording. If a respondent asks

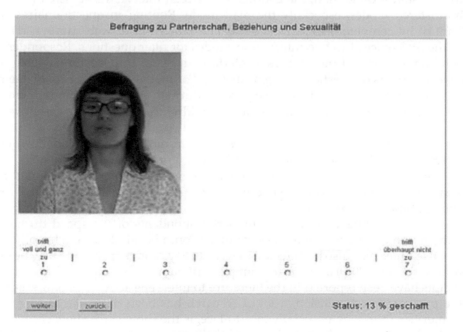

**FIGURE 3.3**
Video-enhanced web survey. *Source*: Fuchs (2009).

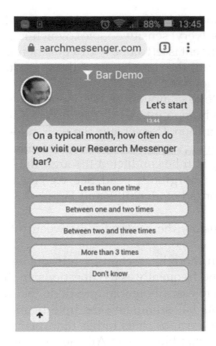

**FIGURE 3.4**
Example of an interactive survey.

for clarification of a term or definition, then the interviewer is instructed to answer that the term or definition is what the respondent thinks it is.

On the assistance dimension, the contrast is between interviewer-assisted modes and self-administered modes, especially the mail mode. Routing in questionnaires can be taken care of by computer-assisted decisions and edit rules in web surveys, removing much of the difference of web to interviewer modes for filter questions. Respondents still do depend on a clear and intuitive user interface and cues in order to navigate within and between screens of questions in web surveys. For mail surveys, the challenge is bigger as respondents need to find the correct route through the questionnaire themselves. Chapter 6, Mixed-Mode Questionnaire Design, addresses these contrasts between the modes.

On the speed-space dimension, differences between modes can be moderated only to some extent through interviewer training and through prompts to respondents.

In face-to-face and telephone interviews, interviewers are instructed and trained to read and repeat all answer options and to not guess respondents' answers, i.e., to give respondents enough time to process questions.

Some investigations have been made to alert respondents of the speed through web questionnaires and to prompt them to slow down (Conrad et al., 2017). On the other hand, long response times may also be detected in web surveys and respondents may be offered help. Kunz and Fuchs (2019) provide an example of dynamic instructions. However, to date few attempts have been reported in the literature to intervene actively in online surveys.

On the presentation dimension, a lot of research has been done. An element that is used in interviewer modes, especially face-to-face, is the show card. Show cards provide a visual display of part of the questions and their answer categories. They are used for questions with a relatively large number of answer categories, and/or answer categories with a lot of words/text. Show cards thus mimic the visual element of the paper and web modes. See Dijkstra and Ongena (2006); Jäckle, Roberts, and Lynn (2010) for discussions, and Figure 3.5 for an example taken from Lund et al. (2013).

Finally, the timing dimension can be addressed in two ways. To become less restrictive in the timing, interviewers are usually instructed to search for a good time to do the interview rather than to force a decision directly. Making appointments strikes a fine balance, as sample units may also use lack of time as an excuse for a direct refusal.

Conversely, smartphones may be discouraged or even blocked when the survey is considered too demanding to be performed anytime anywhere. Reminders are used to let the sample unit know that a response through web and/or paper has a finite time horizon.

The three case study surveys of this book, in Section 2.4, use subsets of the four modes. The Crime Victimization Survey (CVS) uses self-administered modes: web and paper. During the 2011 redesign, the interviewer modes and self-administered modes were deemed too different to employ them simultaneously. The Health Survey uses web and face-to-face and allows for smartphone interviews. In an earlier redesign, phone was discarded as a mode, as it was considered too disparate from the other modes. The Labor Force Survey is the only survey that uses three modes: web, phone, and face-to-face. The questionnaire

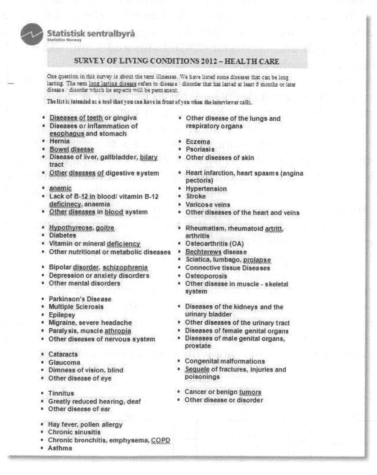

**FIGURE 3.5**
Example of show card used in the Survey on Income and Living Conditions 2012.

is simply too long (and, thus, complex) due to many filter questions and routing to use paper questionnaires. The Dutch LFS questionnaire when printed has hundreds of pages. However, the disparity between the other modes is deemed sufficiently small.

So how come these surveys use different sets of modes? The answer is the characteristics of the survey items contained in the questionnaire and how they interact with respondent characteristics. This is the topic of the next section.

## 3.3 Mode-Specific Answering Behaviors and Response Styles

This section combines mode features, as introduced in Section 3.2, with survey item characteristics and respondent characteristics. Here, and in the following, a *survey item* is a survey question including its answer categories and its introduction and instruction, if present.

When the combination of the features and characteristics leads to recurrent measurement error, it is termed an *answering behavior*. When answering behavior is recurrent for the same respondent, i.e., affects multiple survey items, it is often called a *response style*. See Tourangeau and Rasinski (1988); Krosnick (1991); Greenleaf (1992); Billiet and McGlendon (2000); Baumgartner and Steenkamp (2001); Harzing (2006); Weijters, Schillewaert, and Geuens (2008); Heerwegh and Loosveldt (2011); Lynn and Kaminska (2012); Aichholzer (2013); and Geisen and Romano Bergstrom (2017).

The cognitive process for answering a survey question is usually described using a four-step model from Tourangeau and Rasinski (1988), as depicted in Figure 3.6: Interpretation, Information retrieval, Judgement, and Reporting. In the interpretation step, the respondent reads the question or listens to the question and tries to understand the information that is asked for. Then, in the information retrieval step, the respondent searches for the requested information either in his/her own memory or through external administration or archives. The next step is the judgement step: the respondent combines the collected information and tries to transform it into an answer. Finally, in the reporting step, the respondent compares his own answer to the answer category options. This process may show deficiencies. When a step is problematic, the respondent may return to earlier steps. In questionnaires, survey items for which steps are similar are usually grouped within modules or within grids of questions. Geisen and Romano Bergstrom (2017) break the four steps down into substeps, but this is beyond the scope here.

Measurement error results from a deficiency in one or more of the four steps. This deficiency may be consciously made but also unconsciously. For example, a respondent may think he/she understands the question but in fact misinterprets, or, a respondent may realize he/she does not know the terminology used in a survey item. The deficiency may be conscious when the respondent is not willing to perform one or more steps or may be accidental.

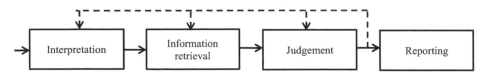

**FIGURE 3.6**
Cognitive answering process as proposed by Tourangeau et al. (1988).

Well-known answering behaviors leading to measurement error are socially desirable answering; i.e., the respondent misreports the answer on purpose or provides a noninformative response. Another well-known behavior is satisficing, which implies that one or more steps are circumvented by the respondent. Satisficing comes in a wide array of forms, from primacy (reporting one of the earlier answer options) and recency (reporting one of the later answer options) to acquiescence (agreeing with the tendency of the question) and do-not-know answers. A vast, and ever-expanding, literature has investigated these fallacies in the answering process (Baumgartner and Steenkamp, 2001; Aichholzer, 2013). Bais et al. (2017) provide a literature overview of the most common response styles.

The important question in this chapter is whether such answering behaviors depend on the survey mode. Literature has shown this is indeed the case. Specific discussions of the impact of the survey mode on the answering process can be found in De Leeuw (1992); Holbrook, Green, and Krosnick (2003); Fricker, Galesic, Tourangeau, and Yan (2005); Greene, Speizer, and Wiitala (2008); Chang and Krosnick (2009); and Klausch, Hox, and Schouten (2013). In light of this, the mode features of Section 3.2 are considered again. The mode feature dimensions are briefly considered.

The intimacy, interaction, and assistance dimensions show their greatest contrasts between self-administered and interviewer-assisted modes.

An interviewer may mediate the answering process by motivating the respondent, explaining the survey item, and assisting in one or more answering steps. Interviewers bring social communication norms. Respondent motivation plays a gradually more and more important role as the interview progresses. For this reason, the longest of surveys are face-to-face; personal communication is considered less burdensome than other forms of communication. Although interviewers are often instructed not to explain survey items and to ignore respondent questions (in a friendly manner), they often do assist in explaining terminology. The interviewer, as a representative of the survey institute, may, however, lead respondents to mask sensitive information. Although the interviewer will attempt to mirror a respondent and copy communication style to some extent, some sensitive information is not revealed. Summarizing, interviewer modes, through their higher intimacy, interaction, and assistance, may help avoid misinterpretation and satisficing behavior but may arouse more socially desirable answering.

Speed and pace, being affected by the mode, may lead to more or less hastiness in performing the answering process steps. This may lead to misinterpretation, insufficient information, such as recall error, errors in combining information, or a careless choice of an answer category. This impact may be strongest over the phone.

The presentation varies greatly between modes and, essentially, implies different sensory stimuli and, as a consequence, different cognitive processes. The aural presentation may be less clear and respondents have to memorize questions and answer categories. This may lead to satisficing and/or reporting errors. Well-known satisficing examples are primacy and recency, where, respectively, earlier and later answer categories are chosen more often. The first occurs more in visual modes and the latter more in aural modes.

The timing dimension may affect answering behavior either by imposing time constraints or by evoking external stimuli. When respondents can choose their own timing, then chances are that they are less in a hurry and can devote more time to each of the answering process steps. This is beneficial and leads to less measurement error. However, the increased freedom also allows for completing questionnaires at settings or occasions that are distracting attention away from the survey and, thus, from performing the steps in good order. Telephone interviewers will check the circumstances under which an interview is conducted but this cannot be done for the self-administered modes.

Survey modes may, thus, arouse answering behaviors that give a variable risk of measurement error. The discussion of the various mode features highlights that such impact is complex and entails both positive and negative influences. For some survey items some influences may be stronger than for others. Does this impact then depend on the type of survey item and the type of respondent?

Literature has devoted some attention to the classification of survey items with the aim of predicting measurement error risk. A recent account is given by Bais et al. (2017), which focused on mode as a source of differential measurement error. They use an item coding scheme that is based on the SQP (Survey Quality Predictor) typology of Saris and Gallhofer (2007) and Gallhofer et al. (2007), and on the typology of Campanelli et al. (2011). Table 3.2 presents the item characteristics that are proposed and that have been applied to a range of different surveys. Four types of characteristics are distinguished: types that are related to the introduction, the question, the answer options, and the context. The various characteristics will be briefly described below:

Introduction: Differences may occur between modes if instructions are not consistently given to all respondents or do not contain exactly the same information. For example, interviewers are instructed to read instructions if they think it is necessary. If the same instruction is available in the web or mail versions of the questionnaire, then it is up to the respondents to read them or not.

**TABLE 3.2**

Examples of Survey Item Characteristics

| Element | Characteristic |
| --- | --- |
| Introduction | Is an instruction provided? |
| Question | Concept of question: opinion, knowledge, or fact |
| | Complexity: Length of question in words |
| | Complexity: Does question use difficult language? |
| | Complexity: Are there conditions or exceptions in question? |
| | Complexity: Does question require recall? |
| | Complexity: Does question contain a hypothetical setting? |
| | Complexity: Does question require calculations? |
| | May questions arouse strong emotions? |
| | May respondents perceive question as a filter question? |
| | Is question multidimensional? |
| | Is question formulated as a statement? |
| | Time reference of question: past, present, or future |
| Answer | Number of answer categories |
| | Is there a mismatch between question and answer options? |
| | Is scale ordinal? |
| | If ordinal, are categories presented as a grade? |
| | If ordinal, is range of scale bipolar? |
| | If ordinal, is direction of scale positive or negative? |
| | Is DK explicitly offered as answer category? |
| Context | Questionnaire section to which an item belongs |
| | Is the item an element of a battery of items? |
| | If the item in battery: Relative position in battery |
| | Absolute position in questionnaire |

*Source:* Bais et al. (2017).

Question:

- The concept of the question is a choice between attitude, knowledge, and fact. Questions about attitudes, and to a lesser extent about knowledge, are supposed to be more sensitive to the presence of an interviewer than factual questions.

- The complexity of a question may influence the extent to which respondents need assistance but may also determine the amount of time they need to process the question. For these reasons it is suspected that the mode has a stronger impact on complex questions than on easy questions. Defining what is a complex question is, however, not straightforward, and for this reason a number of subcharacteristics are proposed: the length of a sentence (counted in the number of words), the use of difficult words, the use of conditions or exceptions, the use of a hypothetical setting, the requirement to recall past events, and the need to perform calculations.

- When a question has an emotional content, respondents may be reluctant or, on the contrary, be eager to answer it. Interviewers may mitigate this effect by motivating the respondents to answer, or checking if the answer is correct. An example is forward or backward telescoping: when people are victims of a crime, they may place this victimization further in the past than the reference period or, on the contrary, report it as happened in the reference period while it actually happened before.

- Respondents may suspect that a question is a filter question and that some answers may lead to more questions, some of which may be sensitive. Respondents may avoid such answers in order to reduce burden and/or to not having to answer sensitive questions.

- Multidimensional questions are questions that consist of two or more subquestions. For example, do you like the house and the neighborhood you are living in? These types of questions are confusing for the respondent because he/she does not know which subquestion to answer. This confusion may again be mitigated by the amount of time the respondent has to answer the question and also by the interviewer who may be aware of this from experience.

- In order to assess a respondent's opinion, researchers often use statements to which the respondent can either agree or disagree. There has been a strong debate about the use of questions posed as statements in the literature and the answering process of statements is conjectured to be affected by the mode (Fowler, 1995; Saris, Revilla, Krosnick, and Shaeffer, 2010; Ye, Fulton, and Tourangeau, 2011). The answering process is considered to be more complex for statements, which is moderated by the mode.

- The time reference of the question is a choice between past, present, and future. It overlaps partially with the complexity dimension on the need to recall past events. However, it is introduced as a separate dimension as also expectations about the future may complicate the answering process.

Answer:

- The number of answer categories complicates the reporting of an answer. The greater the number of categories, the more likely it is that a respondent will pick the first category that seems to apply when reading the answer categories (a primacy effect), or, conversely, to report one of the last categories that seems to apply when being read the answer categories (a recency effect).

- A mismatch between a question and the answer options occurs when the response options do not correspond to the question. The impact of a mismatch is conjectured to be stronger when the answer options are visually presented (web/paper) or are being read out aloud to the respondent (telephone/face-to-face). When the answer categories are not read aloud, then the impact is thought to be less strong.

- Several answer characteristics are linked to the measurement level of the survey item, i.e., nominal versus ordinal. When the level is ordinal, then three more indicators are constructed. The first is whether categories do not have clear labels and are presented as marks. This type of labelling prevents recency and primacy effects and is more adapted to multi-mode surveys. Similarly, the visual presentation of the categories (horizontal or vertical) may play a role. The second indicator is whether the range of the scale has a single pole (unipolar) or has two poles (bipolar). The third indicator is whether the answer categories are ordered from negative to positive or from positive to negative. The direction of the scale may interact with primacy and recency effects and some surveys use randomization of the direction of items for this reason.

- The availability of a "do not know" (DK) and/or "refuse to answer" (RF) answer category is considered influential in the literature. Various mechanisms may enforce or weaken each other. First, respondents may feel pressured to give a substantive answer and tend to give such an answer, even if they do not have an opinion (Beatty and Hermann, 2002). This tendency will be strong when an interviewer administers the questionnaire. In modes without interviewers, respondents might feel less pressure to give substantive answers and will admit more frankly that they have no opinion. In this respect, mode effects are expected for questions on topics that many respondents have no opinion about. Second, respondents may feel reluctant to give a truthful substantive answer. This holds especially if an interviewer is present and the answer may be socially undesirable. Under such circumstances a respondent can easily revert to a DK (Beatty and Hermann, 2002). This mechanism operates to a lesser extent in modes without interviewers. Third, a lack of motivation is suspected to have a greater effect when no interviewer is present to encourage a substantive answer. These mechanisms are expected to create differences in mode effects, if the DK answer is presented in an identical way in all modes. The presentation of DK and RF answer categories has received a lot of interest for that reason (Couper, 2008).

Context: With regard to the context, three characteristics are introduced: is the item part of a grid/battery or a group of items, the topic of the questionnaire section, and the position of the item (begin, middle, or end) in the whole questionnaire? In case the item is an element of a grid or battery, the exact position in the grid/battery may play a role as well (e.g., is it the first item or the last). The first two characteristics form the local context of the item, whereas the position in the whole questionnaire is part of the global context. Clearly, the global context of the item is also formed by the content of all preceding questionnaire sections, but it is hard or impossible to translate that full context to simple indicators. The context of the item has a complex impact on the answering process. The most obvious impact is that on motivation and concentration, which is conjectured to be mitigated by the interviewer and by the possibility to complete the questionnaire in steps and at a self-selected point in time.

Table 3.3 shows the scores on the survey items contained in the Dutch CVS based on the classification scheme. Remarkable are the large proportions of survey items with long question texts that are coded as difficult, that require memory or calculation, and that are included in a grid (or battery). In other words, the CVS scores high on characteristics that are related to answering behaviors and, consequently, is at risk of mode-specific measurement bias. This risk was deemed very real, and a large-scale experiment was set up to estimate mode-specific measurement biases relative to face-to-face. The experimental design is discussed in Chapter 8, Re-interview Designs to Disentangle and Adjust for Mode Effects.

Table 3.4 shows estimated measurement biases for web, mail, and telephone relative to face-to-face for four survey statistics of the CVS. The biases are large, especially for the victimizations and for feeling unsafe. A closer look at the question texts and evaluations based on cognitive interviews revealed that respondents indeed have difficulty interpreting the exact definitions of victimization.

**TABLE 3.3**

Frequency Distributions for CVS Survey Item Characteristics

| Property | Frequency | Property | Frequency |
|---|---|---|---|
| Concept | 27% fact and 73% opinion | Formulation | 33% are a statement |
| Length | 24% have >25 words | Mismatch | 8% have a mismatch |
| Language | 29% are difficult | Response scale | 79% are ordinal |
| Conditional | 25% with the condition | Labels as marks | 2% as marks |
| Memory | 17% require memory | Position | 52% are unipolar |
| Hypothetical | 10% hypothetical setting | Direction | 62% are negative to positive |
| Calculation | 16% need calculation | Battery | 71% in battery |
| Time ref | 83% present and 17% past | Instruction | 15% modes differ |
| Emotion | 2% evoke emotion | DK available | 79% modes differ |
| Dimensional | 2% are multidimensional | | |

Properties position and direction apply to ordinal items only. Table taken from Beukenhorst et al. (2014).

**TABLE 3.4**

Estimated Mode-Specific Measurement Biases in the CVS for Web, Mail, and Telephone Relative to Face-to-Face for Four Survey Statistics: Number of Victimizations per 100 Persons, Being Victimized, Feeling Unsafe at Times and Satisfaction with the Neighborhood on 5-Point Scale

| | Mode | | |
|---|---|---|---|
| | Mail | Web | Telephone |
| Number of victimizations | 12.5 | 15.3 | 4.7 |
| Being victimized | +3.3% | +3.9% | 4.0% |
| Feeling unsafe | +1.2% | +6.3% | 2.8% |
| Satisfaction with neighborhood | 0.06 | 0.21 | 0.16 |

So, what about the respondent characteristics, the third element in the mode-specific measurement effects? In order to go through the four steps of a survey item, the respondent needs motivation and skill; the answering process steps need to be doable for him/her and he/she needs to be sufficiently motivated to repeat the steps for different items. Between respondents both the ability and motivation vary.

Motivation and ability mark the main respondent characteristics related to answering behavior. Since these are rarely measured or available prior to the survey, proxy variables such as educational level are used. Educational level, literacy and language proficiency, and, indirectly, ethnicity are associated with the ability to perform the survey task. Bais, Schouten, and Toepoel (2019) provide an overview of literature into the relation between various respondent characteristics and answering behavior. Table 3.5 is taken from their paper.

Motivation is much harder to grasp as it is strongly dependent on the topics of the survey. There are general differences in attitudes toward the utility, enjoyment, and burden of surveys (Tourangeau, Groves, Kennedy, and Yan, 2009), that have been shown to be related to nonresponse and panel attrition (Lugtig, 2017). Given the common causes between nonresponse and measurement error, they predict answering behaviors in any survey; i.e., some respondents are less motivated in general than other respondents. However, survey-specific topics affect motivation more strongly than do general attitudes toward surveys.

What is the role of modes in the interplay between survey characteristics and respondent characteristics? The modes have different potential to address the inability of respondents and lack of motivation. Let us consider again the mode feature dimensions.

With the emergence of new communication instruments, those not familiar or new to the instruments may have more difficulty performing the survey task. Interviewers interact

**TABLE 3.5**

Literature Overview

| Respondent Characteristics | Relevant Literature |
| --- | --- |
| Gender | Bernardi (2006); Hox et al. (1991); Marshall and Lee (1998); O'Muircheartaigh et al. (2000); Pickery and Loosveldt (1998); Zhang and Conrad (2014) |
| Age | Alwin and Krosnick (1991); Andrews and Herzog (1986); Greenleaf (1992); He et al. (2014); Hox et al. (1991); Kieruj and Moors (2013); Meisenberg and Williams (2008); O'Muircheartaigh et al. (2000); Pickery and Loosveldt (1998); Schonlau and Toepoel (2015); Zhang and Conrad (2014) |
| Education | Aichholzer (2013); Alwin and Krosnick (1991); Greenleaf (1992); He et al. (2014); Krosnick (1991); Krosnick and Alwin (1987); Krosnick et al. (2002); Marin et al. (1992); McClendon (1986, 1991); Narayan and Krosnick (1996); O'Muircheartaigh et al. (2000); Pickery and Loosveldt (1998); Schuman and Presser (1981); Zhang and Conrad (2014) |
| Domestic situation | Alwin and Krosnick (1991); Holbrook et al. (2003); Kellogg (2007); Lavrakas (2010); Lavrakas et al. (2010); Lynn and Kaminska (2012); Olson et al. (2019); Schwarz et al. (1991) |
| Primary occupation | Butler and MacDonald (1987); Lynn and Kaminska (2012); McClendon (1991); Schräpler (2004) |
| Income | Greenleaf (1992); Lynn and Kaminska (2012); McClendon (1991); Schräpler (2004) |
| Origin | Bachman and O'Malley (1984a,b); Baumgartner and Steenkamp (2001); Bernardi (2006); Chen et al. (1995); Chun et al. (1974); Cheung and Rensvold (2000); Dolnicar and Grun (2007); Harzing (2006); He and Van de Vijver (2013); Hui and Triandis (1989); Johnson and Van de Vijver (2002); Marin et al. (1992); Marshall and Lee (1998); Si and Cullen (1998); Smith (2004); Stening and Everett (1984); Van Herk et al. (2004); Watkins and Cheung (1995); Zax and Takahashi (1967) |
| Received a PC? | Schonlau and Toepoel (2015); Zhang (2013); Zhang and Conrad (2014) |

*Source:* Bais et al. (2019).

with respondents and can assist respondents when the survey task is complex. Doing so, they may reduce the potential impact of less able respondents. They can do so only when respondents admit they have problems processing the survey questions, and to the extent they are allowed to by interviewer protocols. The higher intimacy of telephone and, especially, face-to-face may counteract this; respondents may not admit to the problems they have and may mask their inability.

The self-controlled speed and pace of the interview in self-administered modes can provide more time for respondents to interpret questions and to retrieve the requested information. Respondents requiring more processing time may benefit from this more than others.

Visual presentation and user interface may help respondents to understand questions by appealing to presentations they have seen in other communications.

Like speed and pace, the self-controlled timing of self-administered modes allows less able respondents to seek a time and place that is optimal to them for performing the cognitive task.

Lack of motivation may be mitigated by interviewers, especially when the interview progresses. This holds the strongest for face-to-face due to the interaction. For this reason, interviewer-assisted surveys support longer questionnaires. Face-to-face is deemed the only survey mode that allows for questionnaires that take longer than 45 minutes.

The presentation dimension is also important for motivation in the self-administered modes. In recent years, there has been an increased interest in the introduction of gamification elements in order to make surveys more enjoyable and, consequently, increase motivation (Keusch and Zhang, 2017). These elements have been not (yet) been embraced widely by national statistical institutes. Nonetheless, more attention to visual elements is deemed important. For more on this, see Chapter 10, Multi-Device Surveys.

Self-controlled place and time may enhance motivation. The self-administered modes may, thus, perform better on motivation.

All in all, for both ability and motivation, the picture is very mixed and modes may interact positively and negatively with respondent characteristics. The net influence of survey modes is, therefore, hard to isolate across surveys. The only consistent finding is that long surveys are predominantly done face-to-face in order to keep all respondents motivated and concentrated.

Returning to the CVS example, respondent characteristics are important in two ways. First, the CVS is a relatively long survey with various modules on victimization, feelings of unsafety, satisfaction with the neighborhood, satisfaction with police performance and municipality, and cybercrime. Keeping respondents motivated through an attractive and intuitive presentation of questions is considered crucial. Second, there is a risk that respondents who have strong opinions about safety and/or may have experienced one or more crimes in their family or work networks are more motivated to go through the questionnaire. Section 3.4 returns to this risk in the CVS.

Chapter 6, Mixed-Mode Questionnaire Design, discusses how mode features, survey item characteristics, and respondent characteristics can be accounted for in the design of questionnaires.

## 3.4 Detection of Mode-Specific Measurement Effects

This final section discusses the options to detect measurement effects. How can measurement error risk be evaluated without going through large quantitative (and, thus, costly) studies?

A lot of literature has been devoted to the detection of answering behaviors that affect multiple survey items, so-called *response styles*. There are three main options:

1. Substantive knowledge about the (likely) associations between survey items;
2. Paradata about answering behavior;
3. Gold standard data.

Each of the three options usually requires expert knowledge. General introductions to the detection of measurement error are given by Alwin (2007), Millsap (2011), and Klausch (2014). The three options are discussed first. They are then applied to the CVS example. At the end of the section, recommended steps in mode-specific measurement bias are presented.

The first option is substantive knowledge about the survey items themselves, especially the anticipated relations between answers to the questions.

In psychometrical literature, it is common to assume a latent structure or configuration to the survey items. For a discussion on this, see, for instance, Weijters et al. (2008); and Klausch, Hox, and Schouten (2013). When acting on batteries of survey items with the same answer categories, response styles may be detected through structural equation models or multi-trait-multi-method models (Heerwegh and Loosveldt, 2011; Saris and Gallhofer, 2007; Van Rosmalen, Van Herk, and Groenen, 2010). In these settings, it is usually assumed that the survey items form scales with some known structure. The detection of the response styles is then dependent on the assumed associations; for cross-national and mixed-mode surveys, it is sometimes heavily debated whether such a structure even exists. If one believes in the imposed latent variable configuration, the analysis reveals direct measures of error over items. A natural example is a factor representing straightlining. When respondents tend to choose the same answer categories in a battery of questions, the items load strongly on one latent variable, i.e., the straightlining factor.

Paradata about answering behavior are a second option. For a general introduction, see Kreuter (2013). Examples are the average administration time per question, or the average length of blocks of questions; the frequency of failures to edit rules, inconsistencies, or lack of coherence in answers; the proportion of missing items; break-offs during the questionnaire; device switches; coded recordings of sections of the questionnaire, e.g., via Computer-Assisted Recording of Interviews (CARI); audit trails containing all keystrokes; eye movements; or an interviewer assessment of the pace of the interview.

Some paradata may, however, not be suitable in a regular survey environment for practical reasons or for reasons of performance of the survey. Examples are CARI recordings or eye movements that are demanding in analysis but also intrusive for respondents. Furthermore, some paradata depend heavily on the survey mode itself. The mail survey mode supports little paradata observations.

Gold standard data linked to the sample from administrative data or other sources are the third option to detect and identify response styles. There are two options: validation data and record check data. Validation data concern administrative data or other external data containing part of the variables in the survey. The answers to the survey items are compared to these validation data, and they provide a direct measure of error on the items. See, for example, Bakker (2012). The validation data may be the closest option to true measurement error, but they are rarely available on multiple survey items. Furthermore, in some countries, and/or for some survey organizations, it may not be possible to link administrative data, or respondents may need to be asked for consent.

Record check data are a second form. A more expensive instrument is employed to check the correctness of data. Dutch municipalities are, for example, forced to check the correctness of household dwelling data. The record check data may again be compared to the survey data.

Table 3.6 presents indicators for each option that may be estimated during or after data collection. Some of the indicators require coding of all the survey items. Bais et al. (2017) describe coding of approximately 2400 survey items that were asked in ten surveys in the Dutch LISS panel. Bais, Schouten, and Toepoel (2019) provide the estimates for the paradata indicators of these items.

Now, how do the survey modes fit in? In general, this is a complicated question to answer due to the confounding of selection into the mode and measurement within a mode; observed differences may be due to a different subset of respondents from one mode to the other. Another complication is the mode-specific options to record paradata. Paper questionnaires carry only weak paradata as opposed to web audit trails.

One option to overcome these complications is qualitative research and testing of questionnaires for different modes. Chapter 6, Mixed-Mode Questionnaire Design, discusses how mode features, survey item characteristics, and respondent characteristics can be evaluated in cognitive questionnaire lab tests.

Another option involves the adjustment for mode-specific selection biases, which is described in Chapter 8, Re-interview Designs to Disentangle and Adjust for Mode Effects.

The three options gold standard data, latent constructs, and paradata are now discussed for the CVS and HS case studies (see Section 2.4).

Gold standard data may be extracted from linked administrative data. Statistics Netherlands has access to a range of administrative data sources, including police records about various forms of reported crimes. Access to these data is (obviously) restricted. Nevertheless, through linking survey data with register data, it can be derived whether a sample person was the victim of a combination of crimes. In 2011, a multi-mode experiment

**TABLE 3.6**

Examples of Response Quality Indicators Based on Gold Standard Data, Latent Variable Models, and Paradata on Answering Behavior

| Type | Response Quality Indicator |
|---|---|
| Gold standard data | Difference to validation data |
| | Difference to audit or record check data |
| Latent variable models | Amount of random measurement error in scale items (reliability) |
| | Loading on common factors/classes representing response styles or latencies |
| Paradata on answering behaviors | Average duration per completed item |
| | Variance in durations over completed items |
| | Average decrease in duration per completed item over course of interview |
| | Percentage of items with missing data |
| | Percentage of items with do-not-know answers |
| | Percentage of items with order effects |
| | Percentage of items with agree answers |
| | Occurrence of rounding of answers (continuous measurement levels) |
| | Percentage of items with answers in nonsensitive categories |
| | Percentage of items with answers that skip filter questions |
| | Variance of responses to batteries of items |

was performed within the CVS, in which a sample of 8800 persons was randomly allocated to web, paper, telephone, and face-to-face. Afterward in 2012, police records containing information about reported crime were linked to the sample. Table 3.7 shows the proportion of respondents per mode that reported the crimes found in the police records. The number of respondents that appeared in the police records varies greatly due to the varying mode response rates (with web lowest and face-to-face highest). The results show that for paper respondents less than half of the crimes are reported in the CVS, while for web this is close to 100%. Table 3.8 presents the proportion of CVS-reported crimes that were correctly found in the police records for online and in-person submissions at police departments. The proportions of correctly traced crimes in police records are relatively low, which may be due to administrative errors as well as underreporting of crimes to the police. The differences between modes are relatively modest. Numbers are small, so

**TABLE 3.7**

Proportion of Police Reported Crimes That Were Reported Also in the CVS 2011

| Mode | Number of Respondents Occurring as Victim in Police Records | Proportion Reporting Crime in CVS |
|---|---|---|
| Web | 27 | 93% |
| Paper | 41 | 46% |
| Telephone | 42 | 67% |
| Face-to-face | 54 | 76% |
| Total | 164 | 69% |

*Source:* Reep (2013).

**TABLE 3.8**

Proportion of CVS 2011 Reported Crimes That Were Found in Police Records By Type of Submission to Police Departments

| Type of Submission | Mode | Number of CVS Victims | Proportion Traced in Police Records |
|---|---|---|---|
| Total | Web | 35 | 43% |
| | Paper | 18 | 22% |
| | Telephone | 64 | 25% |
| | Face-to-face | 75 | 31% |
| | Total | 192 | 30% |
| In-person | Web | 17 | 41% |
| | Paper | 14 | 7% |
| | Telephone | 41 | 29% |
| | Face-to-face | 55 | 29% |
| | Total | 127 | 26% |
| Online | Web | 18 | 44% |
| | Paper | 4 | 75% |
| | Telephone | 23 | 17% |
| | Face-to-face | 20 | 35% |
| | Total | 65 | 34% |

*Source:* Reep (2013).

strong conclusions cannot be drawn. Surprisingly, in-person reported crimes show lower proportions than online submissions.

Regarding latent constructs, socio-metrical literature defines measurement equivalence as a property that allows for comparisons of statistics across different methods, i.e., across modes in this book. Measurement equivalence is generally considered (Vandenberg and Lance, 2000; Van de Schoot et al., 2015; Hox, De Leeuw, and Zijlmans, 2015), to have a hierarchy consisting of three nested levels. These levels are linked to a factor model in which survey items load on a smaller number of independent latent, i.e., unobserved, variables. Typically, at least three survey items load on every factor. It is possible that some survey items load on more than one factor, but, usually, there is just one underlying factor by the construction of the questionnaire. The exact configuration, i.e., the number of factors and the relations to the survey items, may be based on a theoretical model or may be determined through an explanatory analysis. The weakest measurement equivalence level is termed 'configural'. Configural equivalence holds when the same survey items load on the assumed factors under different methods. In terms of survey modes, this means that respondents may provide different answers, but their answers relate to the same underlying concepts. The second level is metric equivalence, which is true when the loadings are the same for different methods. For survey modes, this amounts to the same regression coefficients and relationships between the survey items. However, the means of the survey items may differ between modes. The strongest level is scalar equivalence and holds when also the intercepts are the same over different methods. Scalar equivalence implies that modes give the same means.

Table 3.9 presents the synthesis of results from Klausch, Hox, and Schouten (2013), in which measurement equivalence was tested for a series of 11 survey terms taken from the CVS 2011. The items belong to three scales: a neighborhood traffic problem scale (NTP, four items), a police visibility scale (PV, four items), and a duty to obey the police scale (DTO, three items). The items all are ordinal five-point scale questions. For this reason, the authors used a multi-group confirmatory factor analysis with ordinal variables instead of continuous variables. In the ordinal setting, thresholds are used rather than an intercept, and metric and scalar equivalence are replaced by scale equivalence on the loadings and on the thresholds. Klausch, Hox, and Schouten (2013) also evaluated the equivalence of random error and systematic error. They conclude that configural equivalence holds for all scales, but for two scales (NTP and PV) scale equivalence does not hold, and thresholds are affected by the choice of mode. Furthermore, they found mode effects on random error and systematic error.

The paradata option is complicated for survey modes as different modes bring different forms of paradata. Paper has very little paradata, while face-to-face can be very rich

**TABLE 3.9**

Comparison of Scale Equivalence, Random Error, and Systematic Bias in the Three CVS Scales

|  | NTP | PV | DTO |
|---|---|---|---|
| Configural equivalence | Yes | Yes | Yes |
| Scale equivalence | Threshold bias | Threshold bias | Full equivalence |
| Random error | Web/paper lower errors than F2F/tel | Web/paper lower errors than F2F/tel | Web/paper lower errors than F2F/tel |
| Systematic bias | Yes | Yes | Yes |

Results modified from Klausch et al. (2013).

in paradata. For the Health Survey (HS) example, interview duration is derived from time stamps in the questionnaire and monitored for both web and face-to-face. Figure 3.7 shows the average duration per mode for three auxiliary variables: age in classes, ethnicity, and degree of urbanization. From the results, it is clear that web surveys take longer on average than face-to-face interviews. The durations also show that considerable variation exists between age classes. The older the respondent, the longer the interview. Mode variation between age, ethnicity, or urbanization classes points at an increased risk of mode-specific measurement effects. Slow responding may point at complex questions and/or respondent fatigue, while fast responding may point at satisficing. Such variation seems to exist for age and ethnicity. Non-Western respondents are relatively faster in face-to-face than other ethnic groups. Older and younger respondents are relatively faster in face-to-face than in the web. These findings alerted designers of the HS questionnaire about the differences between these groups that need to be accounted for.

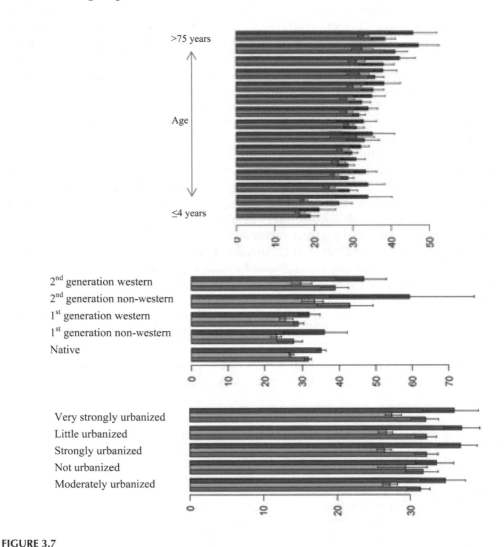

**FIGURE 3.7**
Average interview duration in the Health Survey 2016 for web, face-to-face, and combined for age classes, ethnicity classes, and degree of urbanization. Web durations are in blue, face-to-face in green, and combined in red. 95% confidence intervals are provided based on normal approximation.

All options presented in this section provide a partial view of the risk of mode-specific measurement effects. Ideally, a combination of these evaluations and comparisons is applied in practice. Furthermore, the options may go hand in hand with mixed-mode questionnaire design and cognitive testing. This section is ended with a general set of recommended steps in the exploration of mode-specific measurement effects. The following steps are recommended:

1. Score the survey items on characteristics that interact with mode features, such as the sensitivity of the topics and complexity of the questions and answer categories. An extensive list of characteristics is given in Section 3.3.

2. Summarize the characteristics into an overall inventory for the survey as a whole, including the anticipated length of the survey.

3. Determine whether less/more able and less/more motivated respondents are relevant subgroups in the survey.

4. Review the survey characteristics and respondent characteristics on their interplay with the six mode dimensions.

5. Determine whether the survey is at risk of mode-specific measurement effects.

6. If the risk is deemed high, then
   a. Explore whether gold standard data is available and can be linked with or without consent;
   b. Use substantive knowledge to define what associations between survey questions are to be expected and decide to include extra questions that allow for the estimation of latent constructs;
   c. Enable paradata observation and estimate quality indicators such as presented in this section;
   d. Use the three answering behavior detection options to guide questionnaire design and testing.

7. If the risk is deemed very high, decide to conduct experimental designs aimed at disentangling mode effect components to separate measurement from selection. Methodology to do this is discussed in Chapter 8, Re-interview Designs to Disentangle and Adjust for Mode Effects.

## 3.5 Summary

The main takeaway messages from this chapter are:

- Mode features that impact measurement are intimacy, interaction, assistance, presentation (visual/aural), timing of the interview, and pace/speed of the interview.
- The largest difference comes from interviewer versus self-administered modes, which show disparity on all mode features.

- The impact of survey modes is a complex interplay between mode features, survey item characteristics, and respondent characteristics.
- Various typologies exist for the classification of survey item characteristics that are deemed related to the impact of the survey mode.
- Survey modes to a larger or lesser extent mitigate inability and/or lack of motivation to perform the cognitive answering process steps Interpretation, Information retrieval, Judgement, and Reporting.
- Options to detect measurement effects are gold standard or validation data, latent construct models, and paradata on answering behaviors.
- It is recommended to apply a combination of all three as each option has its strengths and limitations.
- The disparity of smartphones relative to other devices and modes has yet to be explored (see Chapter 11, Multi-Device Surveys).

# 4

## Mode-Specific Selection Effects

### 4.1 Introduction

In Chapter 3, Mode-Specific Measurement Effects, the focus was on the measurement side of survey errors. This chapter considers the representation side of survey errors (see Figure 4.1).

In most situations, there is no single mode that covers the entire population. As we will see in this chapter, not everyone has a listed telephone, not everyone is connected to the internet, CAPI is sometimes made impossible by gated communities or other barriers, and mail surveys may not reach the sample household. Mixing modes may circumvent coverage problems in one mode by offering one or more other modes.

Undercoverage occurs if a survey's sampling frame does not contain all elements of the target population. Consequently, there will be elements that have a zero chance of being selected in the sample. If these elements differ from those in the sampling frame, there is a risk of estimators being biased (Bethlehem, 2015). As will be discussed in this chapter, there is ample evidence that there are differences between those with and those without internet access, those with and those without a listed telephone, and very probably between those who live in gated communities and those who do not.

People who are selected in the sample (and are eligible) but do not take part in the survey, either because they are not willing to or are not able to, are nonrespondents. The boundary between coverage and nonresponse is however fluid (Eckman and Kreuter, 2017). For example, the fact that people do not have internet access is mostly considered to be a coverage problem. However, as is explained in Chapter 2, undercoverage is the result of the mode of contact: if a statistical office sends sample persons a letter to inform them of their selection in a sample and invites them to take part via a web questionnaire, they are covered. Even if people are not able to comply with the request, for example because they do not have a computer, they are part of the sample. People without personal access to the internet could go to a library or ask a friend, for example. Not filling in the questionnaire logically makes these persons nonrespondents. Another example where the boundary between coverage and nonresponse is not hard is the case of illiteracy. Literacy has many levels and a sample person may be fluent in one or more languages but not the language of the survey. Chapter 2 defines the boundary between coverage and nonresponse as having personal access to a mode and the ability to read the invitation letter and questionnaire. Unless one does a re-approach study, as explained in Chapter 8, Re-interview Designs to Disentangle and Adjust for Mode Effects, or asks about the reasons for nonresponse in a sequential mode that does not suffer from coverage problems, undercoverage is in practice not always distinguishable from nonresponse, however.

DOI: 10.1201/9780429461156-4

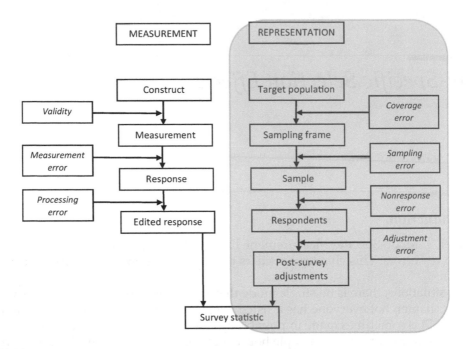

**FIGURE 4.1**
Survey errors for measurement and for representation. *Source*: Taken from Groves et al. (2010).

In Section 4.2, an overview is given of trends in coverage issues in CAPI, CATI, mail, and web surveys. Especially for CATI and web, we will touch on demographic differences between those who are and those who are not covered. In Section 4.3, response and subgroup response in a number of concurrent and sequential designs are discussed. Illustrations come from various mixed-mode designs used by Statistics Netherlands that use either a sequential or a hybrid mixed-mode design starting with web, where persons are invited by letter to participate. Dillman et al. (2014) calls these web-push designs. In a web-push design, respondents are approached (with a letter) to fill in a questionnaire on the internet. Only after several reminders will other modes then be introduced. The section discusses questions like what is the effect of increasing web response on total response, and response distribution in a survey; do we find other people in web mode than we would have with other modes; do different surveys with similar designs but with different response rates display different response distributions? Section 4.4 summarizes the takeaway messages in this chapter.

## 4.2 Coverage Issues in CAPI, CATI, Mail, and Web Surveys

### 4.2.1 CAPI Coverage

Some countries have access to population registers, whereas others use multistage area sampling to build a housing frame. In the latter case, field staff travel through the sampled areas and generate lists of addresses of housing units. In some countries, (commercial) address lists that are updated via postal services are used as a frame for address-based sampling (AAPOR, 2016). When sampling frames that contain addresses are used, an additional step is necessary to select a person within the household as the sample person.

Potentially, face-to-face surveys have the least coverage problems, because they are less dependent on the existence of population registers that can be used as sampling frame (de Leeuw, 2008; Groves, Fowler, Couper, Lepkowski, Singer, and Tourangeau, 2004). Interviewers can be trained to implement within address and within household selection methods to secure random selection. However, whether interviewers are provided with a list of registry addresses or whether enumerators are listing housing units for an area frame, errors are made. Housing units may be excluded from the frame because they are difficult to find, or are located above a business (Lohr, 2008). Eckman (2010) shows that even well-trained and experienced listers may produce substantially different frames, making errors of both under and over coverage. Dillman (2017) describes that coverage problems are increasing, as a result of locked multi-unit buildings and gated-residential communities that make it impossible to reach households in-person.

### 4.2.2 CATI Coverage

Mohorko, de Leeuw, and Hox (2013b) sketch the European situation, wherein the ten years from 2000 to 2009 dramatic changes were observed both in the coverage of landlines or fixed telephone lines and mobile phones. For example, while 16% of the population of Hungary had a fixed landline in 1988, the percentage had grown to 64 in 2005. But while 94% of the population in Finland had a fixed landline in 1988, the percentage dropped to 52% in 2005. And while 47% of the Finnish population had only a mobile phone, and no landline anymore in 2005, this percentage was 1% in Sweden, where 100% of the population had a fixed landline. In 2009, the percentage mobile-only in Finland had increased to 74%. Other European countries likewise show a very high percentage of mobile-only population: Slovakia 51%, Latvia 54%, Lithuania 59%, the Czech Republic 74%. In some western European countries, people abandon their landline connection in favor of mobile phone, while in some former Eastern European countries, no-phone households skip the landline phase and choose a mobile phone instead.

Mohorko et al. (2013b) summarizes a number of studies that show differences in key demographics like age and education between those with and those without a mobile phone. Differences were also shown in a number of substantive findings: mobile phone owners are more liberal or democratic and differ on a variety of health and lifestyle variables. If mobile phones are excluded in telephone surveys, there will be an increasing overrepresentation of female and older respondents, an overestimation of right-wing respondents and of people who are more satisfied with life. The trend in this bias over time varies over countries, however. After including the mobile phone-only population in their study, the bias reduced and was significant only for estimates of age and education.

If and how mobile phones should be included in the sample depends on the situation in the country of interest. In Europe, the number assignment for mobile phones varies between countries and also between telephone providers within countries. In the United States, the mobile phone numbering system follows the landline telephones. Good list-based sampling frames with mobile numbers are not always available. Including mobile phones complicates the survey process. More elaborate screening is necessary: is the respondent safe, not abroad, of eligible age (AAPOR, 2010)? Weighting is more complicated to compensate for differential inclusion probabilities, and calculating inclusion probabilities can be difficult. As a result, and also because calling cell phones is more expensive, the costs of including cell phones in the survey can be substantial.

Table 4.1 shows how the trends described in Mohorko et al. (2013b) continued unto 2017. The data come from the Eurobarometer, a face-to-face survey on European sentiments, that is held several times each year among 1000 persons in each European country. Questions are also asked about telephone and internet possession and use. The table shows the percentage of people in Europe who have a landline available, a mobile available, who have no telephone and who have a mobile telephone only. Almost everyone in Europe had a

**TABLE 4.1**

Availability of Landlines and Mobile Phones across Europe, 2017

| 2017 | Landline Available % | Mobile Available % | No Telephone % | Mobile-Only % |
|---|---|---|---|---|
| Czech Republic | 7 | 98 | 2 | 92 |
| Finland | 12 | 99 | 0 | 87 |
| Latvia | 12 | 98 | 1 | 86 |
| Turkey | 18 | 98 | 1 | 81 |
| Slovakia | 13 | 86 | 12 | 75 |
| Romania | 17 | 90 | 8 | 74 |
| Poland | 23 | 92 | 4 | 73 |
| Cyprus TCC | 26 | 96 | 1 | 73 |
| Lithuania | 25 | 93 | 3 | 72 |
| Bulgaria | 23 | 90 | 6 | 72 |
| Montenegro | 30 | 97 | 1 | 68 |
| Albania | 31 | 98 | 1 | 68 |
| Austria | 31 | 93 | 2 | 68 |
| Estonia | 36 | 96 | 1 | 63 |
| Denmark | 38 | 97 | 0 | 62 |
| Hungary | 35 | 90 | 5 | 60 |
| Ireland | 41 | 95 | 2 | 57 |
| Makedonia | 41 | 93 | 3 | 56 |
| Sweden | 49 | 99 | 0 | 51 |
| Belgium | 54 | 95 | 1 | 45 |
| Italy | 52 | 89 | 4 | 44 |
| Slovenia | 63 | 96 | 1 | 36 |
| Cyprus (Republic) | 65 | 95 | 1 | 34 |
| Portugal | 63 | 92 | 4 | 33 |
| Croatia | 74 | 92 | 1 | 25 |
| East Germany | 76 | 92 | 1 | 23 |
| Spain | 77 | 92 | 1 | 22 |
| France | 78 | 91 | 1 | 21 |
| United Kingdom | 80 | 94 | 1 | 19 |
| Luxembourg | 82 | 97 | 1 | 17 |
| Serbia | 82 | 93 | 1 | 17 |
| The Netherlands | 86 | 98 | 0 | 14 |
| Greece | 86 | 92 | 0 | 13 |
| West Germany | 90 | 94 | 0 | 10 |
| Malta | 93 | 93 | 1 | 6 |

*Source:* Eurobarometer table ZA6921.

telephone available 2017, with very small percentages in some countries without one. The only exception is Slovakia, with 12% without phone. A very high percentage of people in almost all countries has a mobile phone: the lowest percentage is 86% in Slovakia. In almost half of the countries, the percentage is over 95%. However, the distribution of availability of landline or mobile-only varies immensely between countries. Landline availability ranges from 7% in the Czech Republic to 93% in Malta, with relatively low landline penetration in Eastern Europe, and high landline penetration still in Western Europe: 90% in West Germany, 86% in the Netherlands and Greece, and around 80% in the United Kingdom, France, and Spain. Mobile-only mirrors this distribution, with a high of 92% in the Czech Republic and a low of 6% in Malta.

Referring back to the examples in Mohorko et al. (2013b), Finland had a mobile-only population of 74% in 2009, which had increased to 87% in 2017. Where in Sweden 1% of the population was mobile-only, this was 51% in 2017. In Hungary, the increase in landlines reported in 2005 has decreased again, in favor of an increase of mobile-only to 60%.

Although most people have a telephone in one form or another, it is not always straightforward for the statistical office to match a sample of addresses or persons with telephone numbers. Sampling from frames with telephone numbers will likewise result in undercoverage, as a large and increasing number of people do not register their telephone numbers (Lipps, Pekari, and Roberts, 2015). Luiten, Hox, and de Leeuw (2020) describe the International Questionnaire on Nonresponse, an initiative to inventory response trends and fieldwork practices in European, American, Canadian, and Australian statistical offices. One of the questions was of what percentage of the population telephone numbers could be linked for CATI surveys, in the period from 1998 to 2015. Not all countries reported on this question, and not all countries could report on the entire period. Table 4.2 shows the percentages over the periods that were reported.

There are large differences between countries: from around 96% reported by the United States and some northern European countries to around 12% by Canada. Most countries also report the diminishing number of linked telephone numbers. In Italy, the number diminished from 46% in 2006 to 25% in 2015. In the United Kingdom, the number diminished from 32% in 2011 to 13% in 2015; in the Netherlands from 57% in 2009 to 51% in 2015. This trend is not visible in all countries. In Latvia, the number of linked telephone numbers increased substantially over the research period: from 30% in 2010 to 65% in 2015. In the

**TABLE 4.2**

Percentage of Telephone Numbers Linked to Sample for CATI Surveys (2006–2015)

|  | 2006 | 2007 | 2008 | 2009 | 2010 | 2011 | 2012 | 2013 | 2014 | 2015 |
|---|---|---|---|---|---|---|---|---|---|---|
| Canada |  |  |  |  |  |  |  | 13.0 | 12.0 | 11.0 |
| Finland |  |  |  |  |  |  |  |  |  | 70.0 |
| Iceland | 95.6 | 94.9 | 94.2 | 92.5 | 93.7 | 97.0 | 97.3 | 97.0 | 95.6 | 94.7 |
| Italy | 46.0 | 48.0 | 39.0 | 34.0 | 36.0 | 34.0 | 31.0 | 30.0 | 27.0 | 25.0 |
| Latvia |  |  |  |  | 30.5 | 37.0 | 42.8 | 51.1 | 55.0 | 65.1 |
| The Netherlands |  |  |  | 57.3 | 58.1 | 58.2 | 52.6 | 52.3 | 53.6 | 51.2 |
| Norway | 96.7 | 95.8 | 95.7 | 97.1 | 96.4 | 94.7 | 93.5 | 93.5 | 92.8 | 94.3 |
| Sweden | 90.0 | 90.0 | 90.0 | 90.0 | 90.0 | 90.0 | 90.0 | 90.0 | 90.0 | 90.0 |
| Switzerland |  |  |  |  | 75.0 | 75.0 | 74.0 | 74.0 | 73.0 | 73.0 |
| United Kingdom |  |  |  |  |  | 32.2 | 30.0 | 20.8 | 15.0 | 12.9 |
| United States |  | 94.7 | 95.8 | 95.5 | 96.3 | 96.4 | 96.4 | 96.5 | 96.7 | 96.9 |

countries where a high percentage of telephone numbers can be linked, there is no downward trend. These countries have either access to telephone registries (Statistics Iceland, Norway, Sweden) or are otherwise not dependent on commercial vendors to secure telephone numbers (US Census Bureau).

A number of authors showed that telephone registration is related to various demographic characteristics. Cobben and Bethlehem (2005) and Cobben (2009) showed that registration was related to ethnic background, family composition, age, and urbanicity. People from non-Western ethnic background have far less listed telephone numbers as do divorced people and people who live in urban environments. People of 55 years and older were overrepresented on the other hand. Roberts, Groffen, and Paulissen (2017) replicated these findings and showed that unlisted persons were to a far larger extent young or single, lived in urban areas, had a minimum income, were more often on social support, and had less often work.

Lipps et al. (2015) found that in Switzerland younger people, singles or divorced, French or Italian speakers, and those living in larger municipalities are underrepresented, both as a result of having a lower proportion of landlines and a lower probability of being listed in the public directory. Additional fieldwork effort could diminish, not remove the effect. Sala and Lillini (2015) were able to differentiate bias as a result of not having a landline, and not having a listed telephone number for the 1997 to 2012 data from the Italian Multipurpose Survey, Aspects of Everyday Living. Of the 50% Italians who would not have been contacted, had the multipurpose survey been a CATI survey, 32% would have been excluded because they did not own a landline, while 18% would be excluded because they were not listed. There were a large number of differences between persons who would have been available for CATI, and those who were not. The biases caused by not having a landline were more severe than those by not being listed. The combined influence of the bias on substantive measures was significant, however, and could not be adjusted by standard weighting procedures. Joye, Pollien, Sapin, and Stähli (2012) also come to the conclusion that the coverage bias as a result of not having a landline is larger than that of not being listed. The authors warn, however, that the latter group grows more rapidly than the former.

Blumberg and Luke (2016) show that nearly half (49.3%) of US households had only cell phone service. This percentage varied by demographic characteristics: more than 70% of persons aged 25–34 years had only a cell phone; 63% of adults living in poverty; and 64% of Hispanic adults. Seven percent of US households had only landline telephone service.

The unavailability of frames of telephone numbers has been circumvented for a number of decades by RDD sampling. However, RDD telephone interviewing has lost its appeal. Dutwin and Lavrakas (2016) analyzed telephone response rates for nine organizations. They found that response rates for landlines declined from 16% in 2008 to 9% in 2015 (a relative decline of 41%). Cell phone response rates declined from 12% to 7% (a relative decline of 40%). This decline was not the result of increasing refusal rates but of increasing no answers and answering machines. The increase was 10 percentage points for landlines and 24 for cell phones. In addition, an increasing proportion of landline telephone numbers appeared to be not working. This trend is one of the reasons that research on web-push methodology is blooming (see Section 4.2). However, extensive cell phone RDD frames are being built and are expected to continue to grow (AAPOR, 2017). Peytchev and Neely (2013) have provided evidence to suggest that RDD using only cell phone numbers may be a viable alternative for traditional RDD telephone surveys in the United States. They have shown negligible coverage bias for a number of studies. For specific samples, like 65 years and older, cell phone-only sampling is not feasible yet, but for most surveys they conclude that adequate coverage of the general population and most subgroups of the general population can be obtained with only the cell phone RDD frame.

### 4.2.3 Mail Coverage

Mail surveys can have a very high coverage if adequate sample frames are available for addresses or persons. Those can be population registers like in the Netherlands, Italy, or the Scandinavian countries; address lists based on census information; commercial address list; and address lists that are constructed from postal services. If such a detailed sampling frame does not exist, lists like the telephone directory may be used (De Leeuw, 2008). Obviously, the coverage problems that are described in Section CATI Coverage are relevant for this method as well.

Even in the most up-to-date registers, there are bound to be small differences between the situation at the time of the interview and the time of sampling. It may happen that mail is returned to the statistical office as 'addressee unknown', 'no such street'. Dillman, Smyth, and Christian (2014) give guidelines to minimize coverage errors as a result of undelivered mail. Mail not reaching the right person may be the result of people moving out to other places, clerical errors resulting in wrong addresses, and refused and unclaimed letters that people do not take out of their mailboxes. The first two categories can be resolved to some extent by additional efforts to locate sample persons, e.g., by consulting other registries or directories, or trying to call the person. The third category can either be the result of extended absences or implicit refusals, both of which would end up as nonresponse.

A subject that does not receive much attention in the literature on mail questionnaires is that of literacy. In the same vein as people without a computer receiving a letter to participate in a web survey, people without the skills to understand the written questionnaire exist in the gray area between nonresponse and undercoverage. Where people without computers could circumvent the problem by going to a library, people who cannot read will not be able to participate, unless they ask for help. Although the number of completely illiterate persons in Europe is virtually none, a substantial number of people is what is called 'functionally illiterate'. People with literacy difficulties can read at best simple texts. They are not able to deal with longer or more complex texts and to interpret beyond what is explicitly stated in the text. This makes it difficult to find or keep a job and increases the risk of poverty and social exclusion (ELINET, 2015). People with literacy difficulties have a 1.5 to 2 times higher mortality rate than people without literacy difficulties, for example because they are less able to understand medicine leaflets or to read or interpret prescriptions correctly (Twickler et al., 2009).

Low literacy makes it extremely difficult to fill in paper questionnaires. First, a large number of words that statistical offices use in their questionnaires are too difficult, and second, following routing instructions is very complicated for low literate people (Gerber and Wellens, 1995). According to the Program for the International Assessment of Adult Competencies (PIAAC) (2015), 16% of the adult international population had literacy difficulties, ranging from 11% in Finland to 28% in Italy and Spain. The mean level in the United States is 18% (NCES, 2015), on a level with the United Kingdom, Canada, and Germany, for example.

### 4.2.4 Web Coverage

Undercoverage is one of the main concerns for the validity of conclusions based on internet surveys (Couper, 2000). Although internet access is growing, there are still many individuals not covered. This differential coverage would not be a problem if there would be no difference between those with and those without internet with respect to important survey variables. However, even in countries with a high coverage, internet access is unevenly distributed over the population. Both in the United States and in Europe, persons with internet connection are more often highly educated, younger, have a high income and higher labor force participation, and live in urban areas (Bethlehem and Biffignandi, 2012; Couper, 2000; Mohorko, Hox, and de Leeuw, 2013a; Mamedova and Pawlowski, 2018).

In the remainder of this section, the recent situation and the development since 2010 are described, with a focus on the European situation.

In 2007, a majority of households in the EU had access to the internet (55%). In 2016, this had increased to 85%[1] (Eurostat, 2017a). The highest proportion (97%) of households with internet access in 2016 was recorded in Luxembourg and in the Netherlands, while Denmark, Sweden, the United Kingdom, Germany, and Finland also reported that more than 90% of households had internet access in 2016. The lowest rate of internet access in Europe was observed in Bulgaria (64%). However, Bulgaria, together with Spain and Greece, observed an increase of 19 percentage points between 2011 and 2016. Table 4.3 gives an overview of the percentage of people using the internet in 2016, and the increase since 2010 across the world.

Almost all countries and parts of the world show a mean increase of 15 percentage points more internet use between 2010 and 2016. The parts of the world with lower internet use are catching up. The largest increase is seen in the Middle East and North Africa, and the United Arab Emirates. India, Latin America, and the Caribbean show large increases as well. Table 4.4 sketches the situation in Europe, between 2010 and 2017.

In several European countries, internet coverage approaches 100%. As will be shown below, only the eldest age groups in these countries do not all use the internet. The mean increase of 6 percentage points is smaller than the world increase described above, which

**TABLE 4.3**

Percentage of Individuals Using the Internet, 2010 and 2016

| Country Name | 2010 | 2016 |
| --- | --- | --- |
| World | 29 | 46 |
| Japan | 78 | 93 |
| United Arab Emirates | 68 | 91 |
| Canada | 80 | 90 |
| New Zealand | 80 | 88 |
| Australia | 76 | 88 |
| European Union | 71 | 81 |
| United States | 72 | 76 |
| Latin America and Caribbean | 35 | 57 |
| China | 34 | 53 |
| East Asia and Pacific | 34 | 53 |
| Middle East and North Africa | 25 | 48 |
| Arab World | 25 | 43 |
| India | 8 | 30 |
| Sub-Saharan Africa | 7 | 20 |
| Eritrea | 1 | 1 |

*Source:*   Worldbank, 2018 (online data codes API_IT.MET.USER.
        ZS_DS2).

**TABLE 4.4**

Percentage of Individuals Using the Internet, 2010 and 2017

|  | 2010 | 2017 |
|---|---|---|
| Iceland | 96 | 98 |
| Norway | 96 | 98 |
| Denmark | 93 | 97 |
| Luxembourg | 94 | 97 |
| Sweden | 94 | 96 |
| The Netherlands | 93 | 95 |
| United Kingdom | 89 | 95 |
| Finland | 90 | 94 |
| Switzerland | 94 | 94 |
| Germany | 85 | 90 |
| Austria | 81 | 88 |
| Belgium | 83 | 88 |
| Estonia | 81 | 88 |
| France | 81 | 87 |
| Czech Republic | 76 | 85 |
| Spain | 75 | 85 |
| Slovakia | 79 | 82 |
| Cyprus | 67 | 81 |
| Ireland | 74 | 81 |
| Latvia | 74 | 81 |
| Malta | 71 | 80 |
| Slovenia | 74 | 79 |
| Lithuania | 69 | 78 |
| Hungary | 69 | 77 |
| Poland | 68 | 76 |
| Republic of Macedonia | 64 | 75 |
| Portugal | 63 | 74 |
| Italy | 61 | 71 |
| Montenegro | 71 | 71 |
| Greece | 57 | 70 |
| Serbia | 70 | 70 |
| Croatia | 61 | 67 |
| Romania | 50 | 64 |
| Bulgaria | 53 | 63 |

*Source:* Eurostat 2018 (online data code isoc_ci_ifp_iu).

has to do with the ceiling effect of the countries that had already a very high internet coverage in 2010. The largest increase is shown in Rumania, Cyprus, and Greece.

Mohorko, de Leeuw, and Hox (2013a) compare the availability of the internet at home in Europe in 2006 and 2009 for a number of demographic variables. The general pattern in this period is that those with the internet at home are more often male, younger, and highly educated. However, the gap for sex was rapidly closing and not existent anymore in countries like Sweden, Slovenia, Ireland, and the Netherlands in 2009. Also, the age difference is becoming smaller over time and hardly exists anymore in Sweden and the Netherlands, although it is still high in other countries, like Bulgaria.

Mohorko et al. (2013a) conclude that differences in age, sex, education, and life satisfaction between those with and those without internet access are diminishing or even disappearing across countries in Europe. To analyze to what extent this trend has continued in later years, recent Eurostat data were studied (Eurostat, 2018). Table 4.5 shows Eurostat figures about the percentage of persons in 2010 and 2017 who used the internet in the last three months (a different operationalization than used by Mohorko et al. (2013a), by age, sex by age, education, employment status, income, and urbanicity).

In the age groups to 54 years old, internet use is very high all over Europe and is reaching total coverage in several countries. This trend is even more visible in the youngest age groups. In the age groups over 54, internet use is notably lower, although not in all countries. The difference between men and women of 16–24 was already nonexistent in 2010

**TABLE 4.5**

Internet Use (Last Three months) in Europe by Age, Sex, Education, Employment, Income, and Urbanicity, 2010 and 2017

|  | 2010 | | 2017 | |
|---|---|---|---|---|
|  | % | Range | % | Range |
| 16–24 | 89 | 63–100 | 95 | 87–100 |
| 25–34 | 85 | 51–100 | 96 | 85–100 |
| 35–44 | 78 | 35–99 | 93 | 74–100 |
| 45–54 | 64 | 22–97 | 84 | 52–100 |
| 55–64 | 46 | 8–87 | 69 | 27–97 |
| 65–74 | 25 | 3–64 | 43 | 11–93 |
| 75+* | 13 | 2–41 | 20 | 0–65 |
| Women 16–24 | 93 | 63–100 | 98 | 88–100 |
| Men 16–24 | 93 | 63–100 | 98 | 88–100 |
| Women 25–54 | 76 | 39–98 | 92 | 75–100 |
| Men 25–54 | 78 | 41–99 | 91 | 75–100 |
| Women 55–74 | 34 | 5–72 | 59 | 28–94 |
| Men 55–74 | 43 | 11–84 | 63 | 32–95 |
| No or low formal education | 48 | 15–87 | 64 | 20–97 |
| Medium formal education | 74 | 37–95 | 85 | 65–99 |
| High formal education | 91 | 76–99 | 96 | 34–100 |
| Employed, aged 25–64 | 78 | 43–98 | 92 | 74–100 |
| Unemployed, aged 25–64 | 60 | 24–100 | 78 | 46–100 |
| Inactive, aged 25–64 | 44 | 10–81 | 67 | 33–97 |
| Lowest household income quartile | 44 | 10–90 | 67 | 24–97 |
| Second household income quartile | 60 | 20–94 | 78 | 49–93 |
| Third household income quartile | 74 | 38–97 | 88 | 69–100 |
| Highest household income quartile | 86 | 53–100 | 94 | 77–100 |
| Densely populated | 74 | 53–96 | 88 | 74–99 |
| Intermediate urbanized area | 71 | 43–95 | 84 | 62–93 |
| Sparsely populated | 63 | 24–91 | 79 | 48–98 |

*Figure not provided by all countries.
*Source:* Eurostat (online data code is oc_ci_ifp_iu).

and remains so in 2017. Internet use is almost 100% all over Europe in this group. Likewise, there is no more difference between men and women from 25 to 54 years old. There is even a slightly higher percentage of women using the internet in 2017. Only in the age group of 55–74 is the number of men using the internet still slightly higher, although the increase in this time period was larger for women (25 percentage points) than for men (20 percentage points). So, in all likelihood, the internet divide between the sexes that Mohorko et al. (2013a) remark on will be history in a few short years.

The social economic aspects of education, employment, and income are very determinant of internet use as well. While 64% of Europeans with a low formal education used the internet in 2017, this was 96% for Europeans with a high educational level. Because of a ceiling effect in the high education group, the increase in use between 2010 and 2017 was larger for the people with low education. So, most probably, this divide will also disappear with time. Employment status too has a large influence on internet use. While 92% of employed people in Europe used the internet in 2017, this was only 67% in the inactive population of the same age. Again, however, the increase between the two periods was largest for those with the lowest educational attainment. Not surprisingly, the same pattern is shown for income: the higher the income, the higher the internet use. Again, the increase was largest, the lower the income. Table 4.6 illustrates these patterns with three countries that symbolize trends in high, low, and intermediate internet penetration: Iceland, Bulgaria, and Spain.

**TABLE 4.6**

Internet Use (Last Three Months) in Iceland, Spain, and Bulgaria by Age, Sex, Education, Employment and Income 2010 and 2017

|  | Iceland | | Spain | | Bulgaria | |
|---|---|---|---|---|---|---|
|  | **2010** | **2017** | **2010** | **2017** | **2010** | **2017** |
| 16–24 years old | 99 | 100 | 93 | 98 | 78 | 88 |
| 25–34 years old | 99 | 100 | 84 | 96 | 62 | 85 |
| 35–44 years old | 99 | 100 | 73 | 96 | 52 | 81 |
| 45–54 years old | 95 | 100 | 60 | 90 | 39 | 69 |
| 55–64 years old | 87 | 97 | 55 | 74 | 19 | 44 |
| 65–74 years old | 64 | 90 | 14 | 44 | 3 | 17 |
| Women 16–24 | 99 | 100 | 95 | 98 | 78 | 88 |
| Men 16–24 | 99 | 100 | 92 | 98 | 77 | 88 |
| Women 25–54 | 97 | 100 | 71 | 94 | 52 | 78 |
| Men 25–54 | 99 | 100 | 74 | 94 | 50 | 78 |
| Women 55–74 | 72 | 94 | 21 | 59 | 12 | 31 |
| Men 55–74 | 84 | 94 | 31 | 63 | 13 | 33 |
| No or low formal education | 87 | 96 | 39 | 68 | 16 | 33 |
| Medium formal education | 94 | 99 | 79 | 95 | 45 | 65 |
| High formal education | 99 | 100 | 92 | 98 | 81 | 91 |
| Income in first quartile | 75 | 97 | 37 | 68 | 10 | 24 |
| Income in second quartile | 90 | 98 | 59 | 82 | 20 | 49 |
| Income in third quartile | 97 | 99 | 81 | 92 | 44 | 72 |
| Income in fourth quartile | 99 | 99 | 92 | 98 | 66 | 84 |

*Source:* Eurostat (online data code isoc_ci_ifp_iu).

In Iceland, internet use is almost 100%, Only some of the oldest inhabitants lag behind a little. The small difference between the sexes that still existed in 2010 has disappeared in 2017. Likewise, the lag for people with low education and low income has all but disappeared in 2017. In Spain, the lower internet use is caused by the people of 55 and older. In the youngest population, internet use is nearing 100%, and in the other age groups below 55, more than 90% use the internet. The educational gap is still relatively large but diminishing. In 2010, there was a 54 percentage point gap between the highest education and the lowest. In 2017, this gap had diminished to 30 percentage points. The same pattern is noticeable for income. Bulgaria has the lowest mean internet use of Europe, 63%, but there too, younger people use the internet to a large extent: 88% in the youngest age group and 81% in the 35–44 age group. The middle group lags behind some 20 percentage points but has shown a large increase since 2010. In the 65–74 age group, internet use remains very rare. There is no sex difference in Bulgaria for internet use. The highly educated population in Bulgaria shows a high internet use of 91%, but the difference with the low educated group is large (58%) and has not diminished in the last seven years. Likewise, the income divide in internet use is large and has not diminished. Summarizing, the relatively lower use of the internet in both Spain and Bulgaria is seen predominantly for older, poorer, and lower-educated persons. Other groups use the internet to a very high extent as well. The divide shows signs of diminishing in Spain, but not (yet) in Bulgaria.

Where for CATI coverage a distinction was made between having a telephone and a survey organization's ability to retrieve the telephone number, a similar distinction exists between a person having internet access and the survey organization's ability to access that. Except for specific populations, like customers in a customer satisfaction survey or students in education surveys, registries of email addresses are rarely available. There is also no sampling algorithm for email addresses that will provide a known nonzero chance of being selected, like RDD sampling did for telephone surveys (Dillman, 2017). One way around that problem is sending sampled persons a letter. Having a sampling frame of addresses of persons is a prerequisite for this option, however.

This section has described how all modes suffer to some extent from coverage issues. In the next section, it will be studied to what extent mixing modes can overcome these issues.

## 4.3 Response and Subgroup Response Rates in Mixed-Mode Designs

In this section, response rates in a number of different mixed-mode designs are studied. Obviously, response rates in any given mode will be different when the mode is part of a mixed-mode design compared to when the mode is used as a single mode. Also, the distribution of response rates in subgroups may change as a result of the specific mixed-mode design. For example, CATI response rates will be different when CATI is the only mode offered compared with CATI as the follow-up mode of a web survey, or CATI in a CAPI/CATI mixed-mode design. Web response rates will be different when the web is the only mode offered versus when the web is offered alongside another mode, like mail.

In a recent assessment of mixed-mode practices across the 31 countries of the European Statistical System (Lo Conte, Murgia, Coppola, Frattarola, Fratoni, Luiten, and Schouten, 2019), it was shown that in the majority of European NSIs, some form of mixed-mode design is used for one or more of the official social surveys. There are differences between surveys in design: some surveys are generally mixed-mode (e.g., the second wave of the Labor Force Survey), while others are mostly single mode (e.g., the European Time Use

Survey). Of the mixed-mode surveys, a slight majority offers a form of concurrent design, where several modes are in the field at the same time. A large part of the concurrent designs is used in the second and later waves of the Labor Force Survey and the Survey of Income and Living Conditions (SILC), where respondents in the first wave can indicate a preference for the mode of the later waves. But concurrent designs in cross-sectional surveys are offered as well. A number of NSIs target a specific mode to groups of respondents (e.g., telephone interviewing to persons with a matched telephone number) but offer the alternative mode as well. These may be all kinds of mode combinations (CATI-CAPI, web-CATI, web-CAPI). Four NSIs send out a paper questionnaire for some of their surveys while informing the respondent that the questionnaire is also available on the web. Chapter 5, Mixed-Mode Data Collection Design, will go into more detail on the kind of designs and design choices that are made.

Giving people a choice between modes is funded in the hope that this may reduce nonresponse (de Leeuw, 2005). De Leeuw summarizes research to show that up to 2005 this hope was not backed by empirical evidence. In a number of studies, the choice between web and mail, mail and IVR, and phone versus web did not lead to increased response rates compared to a single-mode control survey. Mauz et al. (2018a) compared response rates and substantive findings in a study about self- and parent-reported health for German children and adolescents aged 0–17 years. They compared a paper questionnaire with three mixed-mode designs: a concurrent paper-web design, where both modes are offered; a concurrent design, where respondents have to send back a postcard indicting their mode of choice; and a sequential web-paper design. They found that response rates were comparable for the mail-only, the concurrent design where the two modes were made available, and the sequential design while the version where respondents had to indicate their preferred mode first had a significantly lower response rate. Male adolescents and young male fathers, persons with higher income and higher education, chose the web questionnaire over the paper version in all designs. Differences were small and not significant, however. In the design with the lower response rate, sample distribution was nevertheless comparable to both the single mode, the concurrent design, and the web- paper sequential design. In a second study among German adults (Mauz et al., 2018b) a concurrent mail-web-telephone design was compared with a sequential design where a web round was followed up with a mail questionnaire with a telephone option. Again, no differences were found with regard to socio-demographic characteristics of the achieved samples, or the prevalence rates of the studied health indicators. The authors concluded that the sequential design was more cost- and time-effective.

While the above studies did not find any difference in response rates in a concurrent design and a single-mode control group, there is evidence that giving people a choice may actually lead to lower response rates. In a meta-analysis of 19 concurrent designs, Medway and Fulton (2012) showed that giving people a choice between mail and web actually significantly *reduced* response rates by three to four percentage points. In addition, when offering web in a concurrent design with a paper questionnaire, only about 20% of respondents will actually fill in the questionnaire online, thereby greatly diminishing the potential for costs saving (Dillman, 2017; Luiten and Schouten, 2013; Millar and Dillman, 2011; Medway and Fulton, 2012). The European assessment described above (Lo Conte et al., 2019) also indicated that NSIs expect to reduce response burden and increase response rates by offering a choice, even though giving respondents a choice is harder to manage than making that choice for them in the office.

These findings support De Leeuw's (2005) conclusion that it is more efficient to use a sequential mixed-mode design, starting with the cheapest mode. Dillman, Smyth, and Christian (2014) and Dillman (2017) summarize a host of research showing that stimulating respondents to fill in the questionnaire online before other, more costly alternatives are offered may lead to response rates and response distributions that are comparable to single-mode research. The experiments showed that respondents in web mode were different from the mail respondents: they were younger, had higher educational attainment, had higher income, and were less likely to be single. The combined web and mail data showed that the demographic composition was very similar to the mail-only data.

Sakshaug, Cernat, and Raghhunatan (2019) found in a panel study among young drivers that a second mode can reduce both nonresponse and measurement bias, but the sequence makes a difference: the mail-telephone sequence minimized bias to a greater extent than the telephone-mail sequence. The CATI starting mode especially minimized nonresponse bias, while the mail starting mode did a better job of minimizing measurement error bias. An interesting observation in this study was that a large nonresponse bias in a mixed-mode design can sometimes be offset by an opposite measurement effect, with the result that the total bias becomes negligible.

Bianchi, Biffignandi, and Lynn (2017) studied the effect of introducing a sequential web-face-to-face mixed-mode design over three waves of the longitudinal UK Household Study, in which members were previously interviewed face-to-face. No differences were found between the mixed-mode design and face-to-face design in terms of cumulative response rates and only minimal differences in terms of sample composition. They conclude that the study paints a rather positive picture of the potential for mixed-mode data collection in panel surveys.

Suzer-Gurtekin, Elkasabi, Lepkowski, Liu, and Curtin (2019) performed a series of experiments in the University of Michigan's Surveys of Consumers. Groups either received a request to complete the survey by mail or web or a request to complete the survey by web before offering a mail alternative. These approaches were compared in terms of response rates, process measures, sample composition, and key substantive measures. The studies showed no clear differences between response rates or substantive findings on economic attitudes. It was hoped that the web survey would attract different subgroups to participate who would not have participated in the mail survey, but this was not the case. Sample composition was quite similar in the two designs.

Cornesse and Bosjnak (2018) concluded on the basis of a meta-analysis of over 100 studies that mixed-mode surveys attain better sample representativeness compared to single-mode surveys. They also found that web surveys are on the other hand less representative than other modes. The meta-analysis also showed that higher levels of response are associated with better sample representativeness.

In the following section, a number of mixed-mode designs as used by Statistics Netherlands are described. These designs all start with the web and may be followed up by telephone and/or face-to-face interviewing. It will be shown that web response rates fluctuate as a result of offering incentives, the subject of the survey, or the population. It will be determined what the effect of differences in response rate signify for distribution of response over a number of demographic and socioeconomic variables. In addition, the effect of the fluctuations in web mode on response rates and response distribution in subsequent modes is described.

### 4.3.1 Mixed-Mode Designs at Statistics Netherlands

Most Statistics Netherlands' social surveys employ web interviewing as the first mode. In a limited number of surveys, the web is the only mode, but more generally the nonrespondents of the web round are reapproached for a CATI or CAPI interview. Nonrespondents in web mode are called if a telephone number can be linked to a sample address. Otherwise, the follow-up is in CAPI. In general, nonresponse in CATI is not followed up in CAPI. Other designs are possible as well: web-CATI is fairly common, but web-CAPI is used as well. All social surveys are voluntary. Statistics Netherlands has invested in a large number of experiments with the aim of increasing web response. These are described in Chapter 5, Mixed-Mode Data Collection Design.

Table 4.7 shows for a number of SN surveys on what topic they are, what the population is, whether a person or household sample is drawn, what the length of the questionnaire is, which modes are used, whether or not an incentive is used, and what kind of incentive: an unconditional €5 voucher included in the advance letter, or a lottery of iPads. The last columns show web response rates and the total response rate.

As can be seen, there are large differences between web response rates, from 18% for the Travel Survey to 47% for the Survey of Social Cohesion. The high response rate in the latter survey is due to the unconditional incentive in this survey. But even without incentive there may be large differences: for example, the Health Survey has a web response of 34% even though the questionnaire length is almost double that of the Travel Survey. The increase in response as a result of the follow-up modes differs as well. In the four surveys with a complete CATI and CAPI follow-up, the increase in response is 30 and 35 percentage points for the two surveys without incentive and 14 and 17 percentage points for the surveys with unconditional incentive.[2] The three surveys with a one-mode follow-up all show an increase in response rates of about 20 percentage points. The one survey where a paper questionnaire could be requested showed an increase of 2 percentage points.

The sampling frame for persons and households in the Netherlands is the (municipal) population register. The register contains information on a number of demographic characteristics of each person: age, sex, and ethnic background. As a result, it is possible to study the characteristics of both respondents and nonrespondents. For the analyses in this section, we also linked information on neighborhood income and urbanicity. These are derived from neighborhood information on postcode 6 level (PC6), a fine-grained mapping on street level, or even finer. This information on PC6 level has in previous analyses shown to be highly related to response behavior.

From the overview of surveys in Table 4.7, we selected five surveys to study the impact of a number of measures: ICT2016, with a web-CATI design and no incentive; ICT2017 with a web-CATI design and a lottery incentive; the 2017 Health Survey, with a web-CAPI design and no incentive; the 2017 Survey of Social Cohesion (SSC), with a web-CATI-CAPI design and an unconditional incentive; and the 2017 Crime Victimization Survey (CVS), with a web-mail design without incentive. The CVS and the Health Survey are described in greater detail in Section 2.6. The other surveys are added because of the possibilities they offer to illustrate the effect of design differences on response and mode distribution. A short illustration of their features is included in the appendix.

- The comparison between the ICT surveys of 2016 and 2017 shows the impact of a lottery incentive on web response, and subsequent follow-up response.
- Comparing ICT2017 with the Survey of Social Cohesion shows the relative impact on response and response distribution of a lottery incentive (ICT2017) versus an unconditional incentive (SSC).

**TABLE 4.7**

Web Response and Overall Response in Several Statistics Netherlands' Surveys

| Name | Topic | Population | Sample | Length of Survey | Modes | Web response | Total response | Incentive |
|------|-------|-----------|--------|------------------|-------|--------------|----------------|-----------|
| Labor Force Survey | Work status present or past, education present and past | Dutch population not in institutions, 15 year and older, oversampling of registered unemployment, undersampling of 65 years and older, oversampling of ethnic minorities | Households, all persons of 15 years and older in households; proxy is permitted | 27 min for web, 16 for CATI, 20 for CAPI | Web with CATI and CAPI follow-up | 24.4 | 54 | Gradual introduction of lottery incentive (iPad) in 2017 |
| Health Survey | General health, illness, contact with various specialists, etc. | Dutch population 0–125, not in institutions | Persons | 33 minutes | Web with CAPI follow-up | 34.4 | 58 | No incentive |
| Life style monitor | Alcohol and drugs use, sexual habits (topics vary somewhat over the years) | Dutch population 0–125, not in institutions | Persons | 27 minutes | Web with CATI and CAPI follow-up | 44.8 | 58 | Unconditional incentive of €5 voucher |
| Working conditions | Working conditions and work-related accidents | Employees, between 15 and 74 years old with oversampling of persons < 24 and non-Western nonnative employees | Persons | 20 minutes | Web with paper on request | 29.5 | 32 | Lottery incentive (€250 vouchers) |
| Travel survey | Travels during one day, mode of travel and reason for travel | Dutch population not in institutions | Persons 0–125 yo | 16 minutes | Web with CATI and CAPI follow-up | 18.6 | 54 | No incentive |
| Social cohesion | Happiness, life satisfaction, contacts with family and friends, volunteer work, political and social participation, trust in institutions, environment | Dutch population not in institutions | Persons 0 to 125 yo | 24 minutes | Web with CATI and CAPI follow-up | 46.5 | 63 | Unconditional incentive of €5 voucher |

*(Continued)*

**TABLE 4.7 (CONTINUED)**

Web Response and Overall Response in Several Statistics Netherlands' Surveys

| Name | Topic | Population | Sample | Length of Survey | Modes | Web response | Total response | Incentive |
|------|-------|-----------|--------|-----------------|-------|-------------|---------------|-----------|
| Crime Victimization Survey | Being victim of various crimes, feelings of unsafety, judgment of police | Persons of 15 years or older; complicated oversampling strategy | Persons | 20 minutes | Web invitation with paper questionnaires sent at second and third reminders | 20.9 | 41 | No incentive |
| Consumer confidence | Confidence in economy, and own financial situation; willingness to buy large items | Persons of 15 years or older | Persons | 8 minutes | Web invitation, cati follow-up of nonrespondents | 40.9 | 53 | Lottery incentive iPads |
| School leavers | Satisfaction with education, connection of education with subsequent work | School leavers with or without diploma of secondary schools and vocational training | Persons | 17 minutes | Web only; not possible on smartphone | 21 (range 15–45 per school type) | 21 | Lottery incentive iPads |
| ICT use | ICT use in work and private situation | Persons of 15 years or older persons | Persons | 24 minutes | Web with cati follow-up | 33.1 | 45 | Lottery incentive iPads |

*All response rates are unweighted for unequal inclusion probabilities, with the exception of the overall response of the LFS.

- Comparing ICT2016 with the Health Survey may show what the impact is of relatively high versus low response rates: is the response rate in the Health Survey higher for all groups of people, or are the respondents to the two surveys different?
- Comparing ICT2016 with the Crime Victimization Survey shows the effect on response and response distribution of a follow-up with a paper questionnaire to all nonrespondents versus follow-up with CATI for the 50% nonrespondents with a listed telephone.
- Finally, the comparison of surveys with different modes sheds light on the question of how the various modes influence response and response distribution.

First, some general observations about these surveys. The highest overall response rate among these three is found in the Survey of Social Cohesion. The SSC offers an unconditional incentive of €5 with the advance letter and has a three-mode design. The web response rate is 13 percentage points higher than that of the Health Survey, but the overall response is only 8 percentage points higher. That means that relatively more response is generated in the second round of the Health Survey. The higher response of CAPI in the Health Survey compared to the SSC is the result of the fact that a large portion of the 'easier' respondents goes to CATI in the SSC, i.e., the ones with a listed telephone.

The web response of the Crime Victimization Survey is relatively low. The oversampling of difficult groups (see Table 4.7) may play a role here. The total response of the CVS after mail follow-up is comparable to the response of the ICT2016 after CATI follow-up. Although the CATI response is higher than the mail response in the CVS, the paper questionnaire can be sent to all nonrespondents. CATI can only be offered to the 50% nonrespondents with a listed telephone. Although the overall response is similar in the two designs, there are subgroup differences in the relative success of CATI versus mail as a follow-up mode.

Table 4.8 shows the response per mode and the total response for these five surveys by sex, age, neighborhood income in quartiles, urbanicity, and ethnic background.

High web response rates can be attained in some groups under some circumstances. In the Social Cohesion survey, response rates of more than 50% are attained with the 45–64-year olds, persons with the highest income, persons living in more rural areas, and for native Dutch persons. Although the absolute level of the response is lower in the other surveys, these are consistently the groups with the highest response web rates. In the ICT 2017 and the Health Survey, there is no difference between the response of the 45–64-year olds and that of the 65 and older group. The lowest web response rates are consistently found for the youngest two age groups (16–29 and 30–44), the lowest-income groups and people with a non-Western migration background. There are no large differences between the sexes in web response rates.

In general, the groups with the highest response rates in web are also the groups with the highest overall response rate. And vice versa, the groups with the lowest response in web are the groups with the lowest response overall. Neither the use of incentives nor the use of different follow-up modes changes that pattern. This is, on the one hand, because a large part of the total response comes from web (65%, 74%, 59%, 74%, and 51% for the ICT2016, ICT2017, Health Survey, Social Cohesion Survey, and Crime Victimization, respectively). On the other hand, the follow-up modes do not always succeed in diminishing the contrast

**TABLE 4.8**

Response Rates by Mode and Survey for Sex, Age, Neighborhood Income, Urbanicity, and Ethnic

| | | ICT 2016 | | | ICT 2017 | | | Health Survey | | | Social Cohesion | | | | Crime Victimization Survey | | |
|---|---|---|---|---|---|---|---|---|---|---|---|---|---|---|---|---|---|
| | | Web | CATI | Total | Web | CATI | Total | Web | CAPI | Total | Web | CATI | CAPI | Total | Web | Mail | Total |
| Gender | M | 29 | 34 | 43 | 34 | 27 | 44 | 34 | 41 | 55 | 48 | 32 | 35 | 64 | 23 | 20 | 40 |
| | F | 28 | 38 | 44 | 31 | 31 | 44 | 34 | 43 | 56 | 48 | 38 | 38 | 66 | 19 | 26 | 43 |
| Age | 16–29 | 23 | 30 | 35 | 29 | 22 | 37 | 29 | 39 | 52 | 42 | 32 | 37 | 61 | 17 | 11 | 28 |
| | 30–44 | 24 | 24 | 33 | 28 | 23 | 37 | 29 | 40 | 52 | 46 | 25 | 34 | 61 | 18 | 14 | 31 |
| | 45–64 | 34 | 38 | 50 | 36 | 28 | 47 | 38 | 39 | 57 | 54 | 35 | 39 | 70 | 24 | 25 | 46 |
| | >=65 | 30 | 55 | 55 | 37 | 47 | 56 | 37 | 51 | 59 | 45 | 43 | 38 | 64 | 22 | 44 | 59 |
| Neighborhood monthly income in € | <1900 | 21 | 34 | 34 | 23 | 26 | 33 | 25 | 42 | 49 | 36 | 31 | 37 | 55 | 12 | 21 | 32 |
| | 1900–2300 | 24 | 37 | 38 | 31 | 28 | 42 | 31 | 43 | 54 | 45 | 35 | 36 | 63 | 18 | 24 | 40 |
| | 2300–2800 | 30 | 37 | 47 | 35 | 30 | 47 | 36 | 42 | 57 | 51 | 35 | 38 | 67 | 23 | 24 | 44 |
| | >=2800 | 34 | 36 | 50 | 38 | 31 | 50 | 41 | 39 | 59 | 55 | 37 | 36 | 71 | 28 | 24 | 48 |
| Urbanicity | Very urban >=2500 addresses/km² | 25 | 28 | 35 | 30 | 19 | 36 | 28 | 34 | 48 | 40 | 28 | 27 | 55 | 17 | 17 | 33 |
| | 1500–2500 addresses/km² | 28 | 35 | 42 | 31 | 26 | 42 | 33 | 42 | 55 | 47 | 36 | 40 | 65 | 20 | 24 | 41 |
| | 1000–1500 addresses/km² | 26 | 37 | 41 | 34 | 34 | 47 | 36 | 44 | 58 | 50 | 33 | 39 | 66 | 23 | 26 | 45 |
| | 500–1000 addresses/km² | 32 | 39 | 49 | 35 | 33 | 48 | 38 | 42 | 58 | 54 | 36 | 44 | 71 | 25 | 28 | 48 |
| | Rural <500 addresses/km² | 31 | 41 | 51 | 34 | 34 | 48 | 37 | 49 | 60 | 49 | 39 | 45 | 69 | 23 | 28 | 48 |
| Ethnic background | Native Dutch | 29 | 38 | 46 | 35 | 31 | 47 | 37 | 44 | 59 | 51 | 37 | 42 | 68 | 23 | 25 | 45 |
| | Non-Western migrant | 15 | 18 | 21 | 16 | 15 | 21 | 16 | 32 | 38 | 24 | 18 | 28 | 41 | 8 | 13 | 21 |
| | Western migrant | 27 | 26 | 36 | 23 | 24 | 31 | 29 | 35 | 48 | 35 | 23 | 19 | 47 | 15 | 16 | 29 |

between the highest and lowest response groups but do in some cases even increase the contrast.

For a clearer interpretation of the effect of the various designs on response representativeness, it is more informative to look at the difference between how much of the total response comes from that group, and the size of that group in the sample. This so-called relative bias is the difference in the proportion of respondents in a particular group compared to the proportion of the sample units in that group. The relative bias is used for descriptive purposes, as the sign of this estimate indicates the over- or underrepresentation of specific groups. The absolute relative bias is the sum of the absolute values of these biases. The absolute relative bias gives a measure of overall discrepancies related to specific categories (e.g., age, income) or a complete design.

Table 4.9 shows relative and absolute bias calculations for the response rates of the same five surveys. The relative bias is defined as the percentage point difference between the rate of a subcategory in the sample and in the response. For example, the web response in the ICT 2016 shows a 1.4% overrepresentation of men and a 1.4% underrepresentation of women. In the sample, 50% of persons are men, while in the web response 51.4% comes from men.

As mentioned above, the contrast in response propensities between groups in web mode is not always counterbalanced in the follow-up modes. Both ICT surveys (with a follow-up of CATI) illustrate this phenomenon: generally, the same groups are over- and underrepresented in CATI as in web and to a larger extent. The result is that the total response shows larger relative biases than the initial web response. In the Health Survey (with a CAPI follow-up) and the Survey of Social Cohesion (with CATI and CAPI follow-up), on the other hand, the final response is far more balanced than the initial web response, even though here too the same patterns of over- and underrepresentation continue to exist. The Crime Victimization study shows the largest contrasts in response propensities between groups, both in web mode and especially in the mail follow-up mode. Especially the large contrast for age as a result of the mail mode is striking.

A mail questionnaire can be sent to all nonrespondents of the web round, while a CATI follow-up can only be offered to those persons with a listed telephone (in this case, about 50% of the sample). Theoretically, one could assume therefore that the representativeness of the web-mail design would be higher than that of a web- CATI design. That appeared not to be the case however. Whether this is generally the case can only be determined by experimental research. The surveys described here differ in sample and topic, which may influence generalizability.

It is often hoped that introducing web mode will bring in responses of groups that would otherwise not participate. If that is indeed the case depends on the mode of the survey before moving to the mixed-mode design. In a telephone survey in countries where a complete sampling frame of telephone numbers does not exist (in most cases, as we have seen in the previous section), it does: it brings in people without listed telephone, who would otherwise have been under covered. For example, in the ICT2017, the web response of people without listed telephone was 25%. On the other hand, if the single-mode before the introduction of web was CAPI, or perhaps CATI/CAPI mixed-mode, introducing web will not bring in other people. Response rates in web mode consistently mirror those in the other modes: the highest response in web mode comes from groups with the highest response in other modes: the middle aged and elderly, the people with the higher incomes, natives, and people who live in relatively rural areas. The fact that the youngest

**TABLE 4.9**

Relative Bias and Absolute Relative Bias by Mode and Survey for Sex, Age, Income, Urbanicity, and Ethnic Background

| | | ICT 2016 | | | ICT 2017 | | | Health survey | | | Social Cohesion | | | | Crime Victimization Survey | | |
|---|---|---|---|---|---|---|---|---|---|---|---|---|---|---|---|---|---|
| | | Web | CATI | Total | Web | CATI | Total | Web | CAPI | Total | Web | CATI | CAPI | Total | Web | Mail | Total |
| gender | M | 1,4 | -2,4 | -0,5 | 2,3 | -4,0 | 0,2 | -0,4 | -1,2 | -0,5 | -0,2 | -4,3 | -2,0 | -0,8 | 4,0 | -6,9 | -1,7 |
| | F | -1,4 | 2,4 | 0,5 | -2,3 | 4,0 | -0,2 | 0,4 | 1,2 | 0,5 | 0,2 | 4,3 | 2,0 | 0,8 | -4,0 | 6,9 | 1,7 |
| | abs rel bias | 2,8 | 4,8 | 0,9 | 4,5 | 8,0 | 0,5 | 0,8 | 2,4 | 0,9 | 0,5 | 8,5 | 4,1 | 1,6 | 8,0 | 13,8 | 3,3 |
| age | 16–29 | -3,8 | -3,6 | -4,1 | -2,6 | -5,6 | -3,6 | -2,8 | -1,5 | -1,3 | -2,4 | -1,1 | -0,1 | -1,1 | -3,9 | -11,9 | -7,4 |
| | 30–44 | -3,9 | -7,3 | -5,8 | -3,0 | -4,6 | -3,4 | -3,4 | -1,1 | -1,3 | -0,9 | 0,4 | -2,3 | -1,3 | -3,0 | -9,4 | -5,6 |
| | 45–64 | 7,0 | 2,0 | 5,7 | 3,4 | -0,9 | 2,4 | 4,3 | -1,6 | 1,0 | 4,7 | 3,7 | 1,9 | 2,7 | 5,5 | 3,1 | 4,0 |
| | >=65 | 0,7 | 9,0 | 4,2 | 2,2 | 11,1 | 4,7 | 1,9 | 4,2 | 1,6 | -1,3 | -2,9 | 0,5 | -0,4 | 1,4 | 18,2 | 9,0 |
| | abs rel bias | 15,4 | 21,9 | 19,8 | 11,2 | 22,1 | 14,2 | 12,4 | 8,4 | 5,2 | 9,3 | 8,0 | 4,7 | 5,5 | 13,8 | 42,5 | 26,0 |
| neighborhood monthly income in € | <1900 | -4,8 | -1,2 | -4,0 | -6,2 | -1,8 | -4,9 | -5,7 | 0,2 | -2,5 | -5,1 | -2,1 | 0,2 | -3,0 | -9,4 | -1,8 | -5,1 |
| | 1900–2300 | -3,5 | 0,3 | -2,8 | -1,4 | -0,6 | -1,1 | -2,2 | 1,1 | -0,4 | -1,2 | 0,4 | -0,6 | -0,5 | -2,6 | 0,4 | -1,0 |
| | 2300–2800 | 1,9 | 1,0 | 2,2 | 2,4 | 1,0 | 2,0 | 1,7 | 0,1 | 0,8 | 1,5 | 0,1 | 0,8 | 0,9 | 2,5 | 1,0 | 1,6 |
| | >=2800 | 6,4 | -0,1 | 4,6 | 5,2 | 1,5 | 3,9 | 6,2 | -1,3 | 2,1 | 4,7 | 1,6 | -0,4 | 2,6 | 9,5 | 0,4 | 4,5 |
| | abs rel bias | 16,5 | 2,6 | 13,6 | 15,1 | 3,4 | 11,9 | 15,7 | 2,7 | 5,8 | 12,6 | 4,3 | 2,1 | 7,0 | 24,0 | 3,6 | 12,1 |
| urbanicity | very urban >=2500 addresses/km2 | -2,4 | -3,8 | -4,2 | -1,8 | -6,6 | -3,7 | -3,8 | -4,2 | -2,9 | -3,4 | -3,4 | -8,4 | -3,0 | -5,3 | -8,6 | -6,5 |
| | 1500 to 2500 addresses/km2 | -0,4 | -0,9 | -0,6 | -0,9 | -2,4 | -1,2 | -0,7 | -0,1 | -0,3 | -0,2 | 0,6 | 1,9 | 0,1 | -0,6 | 0,4 | -0,1 |
| | 1000 to 1500 addresses/km2 | -1,3 | 0,3 | -1,1 | 0,5 | 3,2 | 1,4 | 0,9 | 1,0 | 0,7 | 0,7 | -1,0 | 1,2 | 0,3 | 1,5 | 2,0 | 1,6 |
| | 500 to 1000 addresses/km2 | 2,1 | 1,6 | 2,4 | 1,4 | 2,4 | 1,6 | 2,1 | 0,2 | 0,8 | 2,4 | 0,8 | 2,6 | 1,6 | 2,8 | 3,0 | 2,7 |
| | rural <500 addresses/km2 | 2,0 | 2,8 | 3,4 | 0,7 | 3,4 | 1,9 | 1,5 | 3,1 | 1,6 | 0,6 | 3,0 | 2,6 | 1,0 | 1,7 | 3,2 | 2,3 |
| | abs rel bias | 8,2 | 9,5 | 11,7 | 5,3 | 17,9 | 9,9 | 9,0 | 8,5 | 6,3 | 7,3 | 8,9 | 16,7 | 6,0 | 11,8 | 17,1 | 13,2 |
| ethnic background | native Dutch | 3,6 | 3,9 | 4,8 | 6,0 | 4,7 | 5,9 | 7,6 | 5,0 | 4,9 | 5,7 | 5,4 | 9,4 | 4,6 | 9,0 | 7,6 | 7,6 |
| | non-western migrant | -3,4 | -2,6 | -3,9 | -4,2 | -3,8 | -4,2 | -5,9 | -3,2 | -3,5 | -4,1 | -3,6 | -4,7 | -3,0 | -7,0 | -5,4 | -5,7 |
| | western migrant | -0,2 | -1,3 | -0,9 | -1,8 | -0,9 | -1,7 | -1,6 | -1,8 | -1,3 | -1,6 | -1,8 | -4,7 | -1,6 | -2,0 | -2,2 | -2,0 |
| | abs rel | 7,3 | 7,8 | 9,7 | 12,0 | 9,4 | 11,8 | 15,1 | 10,0 | 9,7 | 11,4 | 10,7 | 18,9 | 9,2 | 18,0 | 15,2 | 15,3 |

persons in practically all countries have practically all embraced the internet does not mean that they will therefore start filling in our questionnaires. And, on the other hand, the fact that internet coverage in the older age groups is relatively low does not mean that web response rates for this group need to be low as well. We find this pattern not only in the Netherlands. In the 2013–2014 ESSnet[3] on Mixed-Mode Data Collection (Luiten et al., 2015), it was concluded likewise for web response patterns in Finland and in the United Kingdom.

The surveys in these examples were chosen because they can shed light on a number of issues. The findings in Tables 4.8 and 4.9 illustrate these issues in a descriptive way. In Chapter 5, Mixed-Mode Data Collection Design, a number of controlled experiments are described that give more rigorous answers to some of these questions.

The first question was what the impact is of the introduction of a lottery incentive (of iPads) on the web response, the follow-up response, the overall response, and the response distribution in the ICT survey. The lottery incentive increased the web response, but not the overall response. Without the lottery, 65% of responses came from web mode. The lottery increased that amount to 74%. The lottery somewhat decreased the relative bias in the web mode. The incentive brought in more men, more young people, and more urban people, the groups that are usually underrepresented. But it also discouraged the nonnative inhabitants and the persons in the lowest-income group. All in all, however, the sum total of the absolute biases showed that the introduction of the incentive led to a more representative response distribution. We will see in Chapter 5, Mixed-Mode Data Collection Design, that this result is also found in experimental settings. Because of the larger amount of web response, the relative bias in the ICT2017 total response was somewhat lower than in ICT2016. The principal effect of the introduction of a lottery incentive in the ICT survey was that less CATI follow-up was necessary, with costs saving as a result.

Comparing ICT2017 with the Survey of Social Cohesion shows the relative impact on response and response distribution of a lottery incentive (ICT2017) versus an unconditional incentive (SSC). The unconditional incentive results in a far higher web response (46.5%) than the lottery incentive (33.1%), although the difference in survey topic may play a role here as well. Overall response rates cannot be compared, as the follow-up modes are different (SSC employs both CATI and CAPI). Another observation is that the relative bias in the web results of the two surveys is rather similar. The same groups are over and underrepresented in the two surveys, even though the absolute response rate of the surveys differs.

The subsequent question, what the differences are between a survey without incentive and a low response rate (ICT2016) and a survey without incentive and a high response rate (the Health Survey), shows results in the same vein. The web response rates of the HS are about 5 percentage points higher than those of the ICT. The Health Survey shows better representativeness of sex, age, and income, but especially that for ethnicity is substantially worse. The overall absolute bias for the web in the two surveys is as a result actually worse for the survey with the higher response. The overall response in the HS is 15 percentage points higher than that of the ICT. We have seen earlier that the relative absolute bias of the total response of the Health Survey is substantially lower than that of the ICT. Especially the distribution of age, income, and urbanicity is better in the HS. It is probable that the inclusion of CAPI in the design is responsible for this result, rather than the higher response rate.

## 4.4 Summary

The main takeaway messages from this chapter are:

- There are large differences in coverage of telephone in the world. Whether these patterns lead to coverage bias depends on whether or not mobile phones are called, if telephone numbers can be linked to sample units, and if RDD is a viable alternative.

- Web coverage is increasing very rapidly around the world. Younger people and people with high education have access to the internet to a very high degree, also in countries with a lower overall level of internet access.

- People can be asked to go on the internet by letter. These web-push designs are used increasingly and successfully in various parts of the world.

- High internet penetration does not automatically imply high web response and vice versa.

- Mixed-mode designs including CAPI lead to less response and coverage bias than mixed-mode designs that do not include CAPI, at least in the five Statistics Netherlands' case studies.

- None of the five designs described in the case studies performed uniformly better for all socio-demographic groups. This illustrates the need for adaptive survey designs, where strengths of specific designs for specific populations can be optimized (e.g., Groves and Heeringa, 2006; Schouten, Peytchev, and Wagner, 2017).

## Notes

1. The figure differs slightly from the figure derived from the Worldbank (see Table 4.3).
2. The response rates mentioned for total sample are the cumulative number of responses, divided by the total sample, unadjusted for ineligibility in CATI and CAPI. Incorrect telephone numbers are thus considered as nonresponse.
3. An ESSnet is a network, commissioned by Eurostat, of several European Statistical organizations, aimed at providing results that will be beneficial to the whole European Statistical System (ESS).

# Part III

# Design

# 5

## Mixed-Mode Data Collection Design

### 5.1 Introduction

This chapter discusses the first option to reduce mode effects, namely through the data collection design (see Figure 5.1). The focus will be on the reduction of mode-specific selection effects as discussed in Chapter 4, Mode-Specific Selection Effects. However, as will be discussed in later chapters, the data collection design affects also the mode-specific measurement effects.

The choice of a mixed-mode survey design, which modes to use and in which sequence, depends heavily on the motives for mixing them: minimizing total survey error, reducing costs, or increasing the speed of data collection. In addition, the availability of sampling frames for certain modes, the wish to minimize disruptions in time series, survey features, and survey tradition all play a role in the decision of which modes to use. For example, in the gradual introduction of mixed-mode data collection in European Statistical Institutes, the traditional mode of ongoing surveys was always the starting position to which one or more modes were gradually added. Nevertheless, mode choice may change over time as a result of gradual changes in population coverage, response rates and representativeness, and survey costs. Examples are the emergence of the web as a feasible survey mode, the introduction of devices that allow easy access to the web, and the decreasing coverage of landline phones.

These different conditions may be one of the reasons why literature shows varying results when comparing modes and mode sequences, as will be illustrated in this chapter. For the same reason, it is impossible to describe 'the best' mode combination or 'the best' sequence that meets all the requirements, national backgrounds, and specific survey features. This chapter examines how NSIs around the world have operationalized these constraints in the choices that were made for their fieldwork designs. Section 5.2 gives a summary of the literature on mode combinations. Although most literature focuses on maximizing response rates and minimizing costs (Sakshaug, Cernat, and Raghunathan, 2019), attention to nonresponse bias and measurement error has been given too. In Section 5.2 illustrations are given as to how these dimensions are operationalized in different countries. For these illustrations use is made of the data collected in two Eurostat-funded projects: the ESSnet on Data Collection in Social Surveys using Multiple Modes (DCSS) (Blanke and Luiten, 2014) and the ESSnet on Mixed Mode Designs for Social Surveys (MIMOD) (Murgia, LoConte, Coppola, Frattarola, Fratoni, Luiten, and Schouten, 2019). In both projects, the countries in the European Statistical System (ESS) filled in a questionnaire on mode use in various ESS surveys. The questionnaire also covered relevant aspects related to the design of mixed-mode surveys: whether modes are used sequentially or

DOI: 10.1201/9780429461156-5

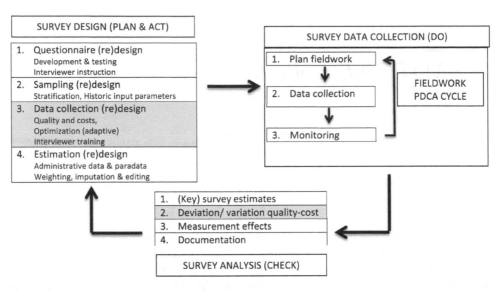

**FIGURE 5.1**
The plan-do-act cycle for surveys.

simultaneously, about the communication strategy, when it is decided whether a next mode will be used, what measure are taken to increase web response rates, the implications of the introduction for response rates and fieldwork and operational costs, what, if any, changes were made to existing questionnaires to cater for the new modes, and various other topics relevant for the introduction of mixed-mode data collection. In the DCSS project additional information was gathered from other than ESS countries.

Apart from the choice of mode, other design decisions need to be made. For instance, whether to offer the modes concurrently or sequentially and in which order and how to time the mode switch. These are the topics of Section 5.3, where again a discussion of the literature is followed by illustrations from the DCSS and MIMOD projects.

In a mixed-mode design where web is one of the modes, considerable costs savings can be attained by increasing the web response. Section 5.4 discusses the literature and illustrates the various practices in the ESS. Section 5.5 describes how mixed-mode designs impact sample composition and data quality. In Section 5.6 communication strategies for increasing web response are discussed, focusing on contacting the household or person by letter, email, or text message. The second part of this section focuses on the use of incentives for increasing web response rates. Literature overviews and illustrations from DCSS and MIMOD on these topics are followed by the description of two series of experiments by Statistics Netherlands on respondent communication (Section 5.6.3) and the use of incentives in web-push designs (Section 5.6.4.6). The chapter concludes in Section 5.7 with a summary of findings and illustrations, culminating in a number of preliminary guidelines.

In this chapter, the choice of design is mostly uniform for the entire sample, although several countries describe that some modes may be offered to part of the sample only. While mixing modes may put additional strain on survey logistics and infrastructure, it becomes even more complicated when additional constraints are put on the conditions under which certain modes are offered to certain people. This is the topic of Chapter 11, Adaptive Mixed-Mode Survey Designs.

## 5.2 Mode and Mode Combinations

When introducing a mixed-mode design, a careful balance must be found between costs and survey error. Figure 5.2 shows an overview of the relative position of the four modes used by Statistics Netherlands on the four dimensions of costs, coverage, response, and representativeness. For example, the web mode is by far the cheapest mode, coverage of internet is close to 100%, but the response rates in web surveys may be quite low, resulting in overall low representativeness. Where each mode would position itself in various countries may, however, be very different. In countries with registries of telephone numbers, CATI would slide to the right on the coverage and representativeness dimension. In countries without population register, web would slide to the right on the costs dimension, and so on. The combination of modes may therefore likewise have very different characteristics and implications in various countries.

Figure 5.2 depicts a question mark for the dimension 'measurement error'. While self-administered modes are known to be less prone to socially desirable answering (e.g., Tourangeau, Rips, and Rasinski, 2000; Hope, Campanelli, Nicolaas, Lynn, and Jäckle, 2014; Heerwegh and Loosveldt, 2011), they may on the other hand be more prone to satisficing (non-differentiation, acquiescence, selecting middle categories, primacy and recency effects), or suffer from larger amounts of item missing values. The danger of socially desirable answers in CATI and CAPI may be offset by the ability of interviewers to help respondents with complex questions or concepts. Some modes are more disparate from others. The more modes differ, the greater the risk of measurement effects (de Leeuw, 2005, 2018). Measurement errors in modes and mode combinations is the topic of Chapters 3, Mode-Specific Measurement Effects, and 6, Mixed-Mode Questionnaire Design.

Cultural differences in the position of the modes on the dimensions depicted in Figure 5.2 are reflected in the differences in the modes and mode combinations in the surveys in Europe, North America, and Eurasia. Table 5.1 shows the modes and mode combinations that were used in Europe in 2018 for a number of key surveys: the Labour Force Survey (LFS, first wave (LFS1) and later waves (LFS2+)), the Survey of Income and Living Conditions (SILC), the European Health Interview Survey (EHIS), the Adult Education Survey (AES),

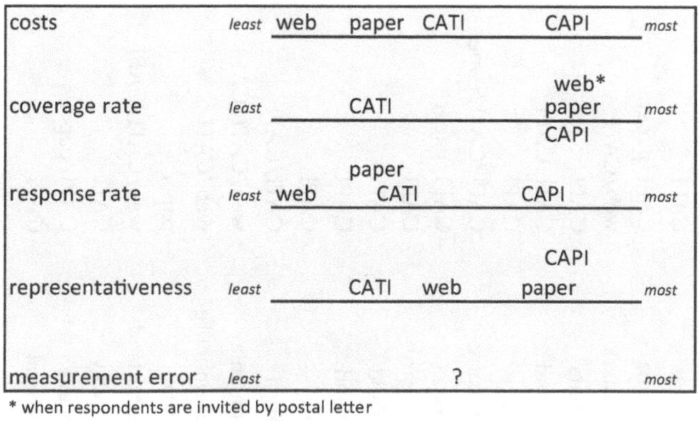

* when respondents are invited by postal letter

**FIGURE 5.2**
Relative position of modes on costs and error dimensions.

**TABLE 5.1**

Modes and Mode Combinations Used in Key European Surveys (2019)

| | LFS1 | LFS2+ | SILC1 | EH IS | AES | ICT |
|---|---|---|---|---|---|---|
| Austria | CAPI | CATI/CAPI | CAPI | web/CAPI | web/CAPI | CATI |
| Belgium | CAPI | web/CATI | CAPI | CAPI | web/paper | paper/web |
| Bulgaria | paper | paper | paper | paper | paper | paper |
| Croatia | CAPI | CATI/CAPI | CAPI | CAPI | CAPI | web/CATI/CAPI |
| Cyprus | CAPI | CATI | CAPI | CAPI | CAPI | CAPI |
| Czechia | CAPI/paper | CATI/CAPI/paper | CAPI/paper | CAPI/paper | CAPI/paper | CAPI/paper |
| Denmark | web/CATI | web/CATI | web/CATI | - | - | - |
| Estonia | CAPI | CATI/CAPI | CAPI | web/CAPI | CATI/CAPI | web/CATI |
| Finland | CATI/CAPI | CATI | CATI | - | web/CAPI | web/CATI |
| France | CAPI | CATI | CAPI | CAPI | CAPI | web/CATI/paper |
| Germany | CAPI/CATI/paper | CAPI/CATI/paper | paper | web/paper | CAPI | paper |
| Greece | CAPI/paper | paper | web/paper | paper | paper | paper |
| Hungary | CAPI | CATI/CAPI | web/CAPI | CAPI | web/CAPI | CAPI |
| Iceland | CATI | CATI | CATI | - | - | CATI |
| Ireland | CAPI | CATI/CAPI | CAPI | paper | CAPI | CAPI |
| Italy | CAPI | CATI/CAPI | CATI/CAPI | paper | CATI | web/paper |
| Latvia | CATI/CAPI | web/CATI/CAPI | CATI/CAPI | web/CATI/CAPI | web/CATI/CAPI | web/CATI/CAPI |
| Lithuania | web/CATI/CAPI | CATI/CAPI | web/CATI/CAPI | web/CAPI | web/CAPI | web/CATI/CAPI |
| Luxembourg | web/CATI | web/CATI | CAPI | web/paper | web | CATI |
| Malta | paper | CATI | CATI/CAPI | CAPI | CAPI | paper |
| Netherlands | web/CATI/CAPI | CATI | web/CATI | web/CAPI | web/CATI | web/CATI |
| Norway | CATI | CATI | CATI | CATI | web/CATI | CATI |
| Poland | CAPI/paper | CATI/CAPI/paper | CAPI/paper | CAPI/paper | CAPI/paper | web/CAPI |
| Portugal | CAPI | CATI/CAPI | CAPI | web/CAPI | CAPI | web/CATI/CAPI |
| Romania | paper | paper | paper | paper | paper | paper |
| Slovakia | CAPI/paper | CATI/CAPI/paper | paper | paper | paper | paper |
| Slovenia | CAPI | CATI/CAPI | CAPI | web/CAPI | web/CATI/CAPI | web/CAPI |
| Spain | CAPI | - | CAPI | CAPI | web/CATI/CAPI | web/CATI/CAPI |
| Sweden | CATI | CATI | CATI | web/CATI/paper | CATI | web/CATI |
| Switzerland | CATI | CATI | CATI/CAPI | - | CATI | web/CATI |
| United Kingdom | CAPI | CATI/CAPI | CAPI | CATI | CATI | CATI/CAPI |

*(Continued)*

**TABLE 5.1 (CONTINUED)**

Modes and Mode Combinations Used in Key European Surveys (2019)

| | LFS1 | LFS2+ | SILC1 | EHIS | AES | ICT |
|---|---|---|---|---|---|---|
| Single mode | 20 | 12 | 20 | 15 | 16 | 13 |
| Mixed-mode | 11 | 18 | 11 | 12 | 13 | 17 |
| Web | 4 | 4 | 5 | 5 | 7 | 10 |
| CATI | 11 | 15 | 11 | 11 | 11 | 15 |
| CAPI | 22 | 27 | 20 | 15 | 17 | 16 |
| Paper | 8 | 6 | 7 | 11 | 10 | 12 |

*Source:* Gravem et al. (2018a).

and the survey on Information and Communication Technologies (ICT) (Murgia et al., 2019; Gravem et al., 2018c).

A number of things stand out in this overview. First, the large number of different modes and mode combinations used for one and the same survey; second, there are differences between surveys in a number of countries that use a mixed-mode design. While the first wave of the LFS and the SILC is still primarily unimode, the second wave of the LFS and also the ICT mostly have a mixed-mode design. For the LFS, CATI is often added in the second wave because of the possibility of getting a phone number from the household in the first wave. Internet questionnaires are a significant part of the ICT survey (used by ten countries), much more than in the LFS and SILC. This may be related to the household component in the latter questionnaires, but also to the fact that the ICT survey is a relative newcomer in the ESS, only introduced in 2010.

A third observation is that within one country completely different modes and mode mixes may be used for different surveys. A number of countries, such as Estonia, Italy, and Luxembourg, use five different modes and mode combinations for the six surveys depicted in this chapter. This is a strong indication that circumstances like tradition, the complexity of the subject matter, survey length, the kind of sample, or the presence of a panel structure influence what could be the optimal design for any given survey.

In 2013, mixed-mode data collection was more common in the non-European respondents to the questionnaire: Australia gave information on the LFS, Canada on the LFS, the United States on the American Community Survey, the Current Population Survey and the National Cancer Institute, Turkey on the LFS, and New Zealand on the General Social Survey and Household Economics Survey. Of the 15 observations this yielded, 75% were mixed mode. The most common mix was an interviewer mode plus self-completion paper questionnaire, although even then five of the mixes contained an internet questionnaire.

The illustrations discussed support what is shown in the literature on mixed-mode designs: a varied picture emerges on designs and the effect of the introduction of additional modes. Nevertheless, it is evident that, especially in cross-sectional surveys, starting with a cheaper (self-administered) mode (mail or web) and subsequently using more expensive modes (telephone or face-to-face) for nonresponse follow-up will result in greater cost savings than other sequences (Patrick, Couper, Laetz, Schulenberg, O'Malley, Johnston, and Miech, 2018; Wagner, Arrieta, Guyer, and Ofstedal, 2014; Sakshaug, Cernat, and Raghunathan, 2019). This strategy is currently used in several large-scale surveys. The Australian LFS, for example, has a web-then-CAPI design for the first wave and a web-then-CATI design for later waves. Their Housing Survey has a web-then-CATI design. The Canadian LFS has a web-then-CATI-CAPI design, and the American Community Survey from the US Census Bureau has a web-CATI-CAPI-paper self-completion design. The ACS first sends sample dwellings a mail questionnaire and an invitation to complete the survey online. Nonrespondents are followed up by telephone if a telephone number can be found. Finally, a subsample of the remaining nonrespondents is contacted face-to-face.

In panel surveys it may make sense to start with more expensive modes of data collection for the first wave and use less expensive modes in later waves. The American Current Population Survey (CPS) uses this strategy. The first wave of data collection is done face-to-face and the next three waves are done by telephone. After eight months, the final four waves use the same pattern. A number of European NSI's expressed the concern that self-administration in the first wave of the Labor Force Survey and the Survey of Income and Living Conditions could backfire by increasing the costs of the follow-up waves. Interviewers are deemed necessary to secure high response rates and good quality data in the first wave (Murgia et al., 2019).

The potential for cost savings in cross-sectional surveys is larger when more expensive modes are initially withheld and not offered concurrently. If respondents are given a choice, e.g., by sending them a paper questionnaire and at the same offering the option of filling in by web, by far most of them will choose the paper version. Medway and Fulton (2012) show that fewer than 20% of the respondents who had the option to complete the survey by web did so when a mail survey was offered at the same time.

However, there is ample evidence that this sequence of starting with less expensive modes may lead to lower response rates. This patterns is seen both in combinations of self-administered modes (Beebe, Locke, Barnes, Davern, and Anderson, 2007; Israel, 2009; Friese, Lee, O'Brien, and Crawford, 2010; Smyth, Dillman, Christian, and O'Neill, 2010; Messer and Dillman, 2011; Medway and Fulton, 2012; Lesser, Newton, Yang, and Sifneos, 2016), in combinations of self-administered and interviewer modes (Harris, Weinberger, and Tierney, 1997; Lagerstrøm, 2008; Lynn, 2020 for individual responses in a household survey), and also in combinations of interviewer modes (Janssen, 2006; Lynn, 2012). Other research shows that there may be circumstances in which no (large) differences in response rates as a result of mode order are observed, neither in combinations of self-completion surveys (Olson, Smyth, and Wood, 2012; Patrick, Couper, Laetz, Schulenberg, O'Malley, Johnston, and Miech, 2018) nor in combinations with interviewer modes (Dillman, Phelps, Tortora, Swift, Kohrell, Berck, and Messer, 2009; Bianchi, Biffignandi, and Lynn 2017; Lynn, 2020 for household response in a household survey; Kappelhof, 2015).

Studying the design of these papers, it is evident that there are large differences between studies, in survey population, contents, use of incentives, number of reminders, additional email support, etc., that could influence comparability. These differences hinder the interpretation of findings. For example, the web-push approach that works with students (e.g., Millar and Dillman, 2011) may not generalize to other populations. Meta-analyses (e.g., Manfreda, Bosnjak, Berzelak, Haas, and Vehovar, 2008; Shih and Fan, 2008; Medway and Fulton, 2012; Weigold, Weigold, and Natera, 2018) show that study elements like the year of publication, the continent of the study, the population (e.g., business professionals or college students or medical patients), the type of measures (e.g., attitudinal or behavioral questions), the number of contacts, incentives, the sensitivity of the questions, whether or not a survey is mandatory, sponsorship, salience of topic, the mode of invitation, all influence the choice for a mail or web questionnaire. Similar meta-analyses are necessary for other mixed-mode combinations to understand the influence of these various study conditions on findings on mixed-mode designs.

Not all countries and not all surveys consider the introduction of mixed mode at the present time. Villar and Fitzgerald (2017) summarize six experiments on mixed-mode data collection in the European Social Survey.[1] The ESS is traditionally a face-to-face survey with a duration of about one hour. Three of the experiments focused on assessing the effect of mode on measurement. Three others on the feasibility and practical challenges of implementing different mode designs in cross-national surveys. The first of these assessed the possibility of using telephone interviewing, either with the original 60-minute questionnaire, a shorter 45-minute version and a split questionnaire version of two times 30 minutes. The experiment was performed in four countries, in a sample of people with fixed landlines. Results showed large response differences between countries and interview length, but generally the response in the face-to-face survey was 2–3 times higher than for the telephone design, although invitation protocol differences may have played a minor role as well.

The second experiment was conducted in the Netherlands, where sample persons with a known telephone number were called and asked to participate. In one condition they were offered the choice between three modes (web, telephone, and face-to-face). In the other conditions they were asked to participate in a web survey. Nonrespondents were called back for a telephone interview or subsequently a face-to-face interview. Sample persons with whom telephone contact was not possible were visited. Those who refused to participate were offered the web survey. The response rates in either mixed-mode design were lower than the face-to-face control group: 43% for the sequential design, 44% for the concurrent design, and 52% for the face-to-face design.

In the third experiment, countries chose the design that best fitted their survey environment. Estonia used a web–face-to-face sequential design, where sample persons received a named invitation to participate in the web survey, plus two reminders. Nonrespondents were approached for a face-to-face interview. In the UK a similar design was used. However, because the UK lacks a sample frame of individuals, the letters were addressed to 'the resident', and additional instructions were included to select a random household member. An incentive of £15 or £35 was promised upon completion. Sweden compared a response enhancing face-to-face then telephone design and a cost-saving telephone then face-to-face design. The results showed that the web then face-to-face design in Estonia can save costs without overly harming the survey response rates: the response rates were 68% for the control group, and 64% for the mixed-mode group. In the UK, the mixed-mode design led to 17 percentage points response loss, in spite of the offered incentive. Surprisingly, the response rate in the Swedish response-enhancing design was much lower than in the standard ESS design (38% vs 50%). However, differences in fieldwork protocol may partly have contributed to this result.

The conclusion from these series of experiments in nine different countries was that in the context of the ESS at the time, the cost savings were too limited to justify the switch to mixed-mode data collection. It was expected that the reduction of quality in most countries and on many key quality dimensions would far outweigh the (cost) benefits. The ESS core scientific team has rejected such a switch for the near future.

## 5.3 Concurrent and Sequential Designs

The issue of the order in which to offer the modes in a mixed-mode design has been the topic of many studies. There are two main bodies of literature. The first one studies whether modes should be offered at the same time (concurrent design) or whether it would be better to use a sequential design, where the subsequent mode is withheld for a time. This is the topic of Section 5.3.1. The second body of literature studies in which sequence modes should be offered in a sequential design. This is the topic of Section 5.3.2.

### 5.3.1 Concurrent or Sequential Order?

In the early days of mixed-mode studies, offering respondents a choice between modes was seen as a respondent-friendly way to create goodwill and boost response rates (De Leeuw, 2018; Olson, Smyth, and Wood, 2012). Empirical evidence shows that offering a choice, at least between paper-and-pencil and web, may actually reduce response rates. Medway and Fulton (2012) showed in a meta-analysis of 19 studies comparing mail surveys with concurrent web-mail alternatives, that the mixed-mode version in all but two cases resulted in (slightly) lower response rates than the single-mode mail survey, on average 3.8 percentage points. Study characteristics like target population, sponsorship, voluntary or mandatory, incentives, and salience of topic for sample members had little influence on the magnitude of the effect. De Leeuw (2018) hypothesizes that offering a choice complicates the respondents' decision. Instead of one decision (respond or not) two decisions need to be taken (respond and in what mode), causing respondents to postpone their decision. Tourangeau (2017) concludes that preventing respondents to procrastinate is an important precursor to high response rates in mixed-mode designs.

In addition to the response effects, the potential for cost savings is limited in web-mail concurrent designs, because a large majority of respondents will choose the mail option (Medway and Fulton, 2012). The effect can be counteracted by making the web option more prominent (Tancreto, Zelenak, Davis, Ruiter, and Matthews, 2011), or by giving an (extra) incentive for choosing the web option (Biemer, Murphy, Zimmer, Berry, Deng, and Lewis, 2018). Biemer et al. (2018) crossed four-mode protocols (web alone, both web and paper, offered concurrently or sequentially) with two incentive conditions (a promise of either $10 or $20 on top of a $5 prepaid incentive). They introduced a protocol called choice+, in which respondents in the concurrent web-mail condition were offered the promised incentive for choosing the web option. The purpose was to find the protocol that would provide the best combination of response rates, the proportion of respondents using the web option, low costs, and good sample representativeness. It was found that the choice+ option attained the highest response rate overall and for both incentive treatments. The proportion of respondents responding by web for the choice protocol was about 27.5%. This was increased to more than 64% for the choice+ protocol. This rate of web returns was about equal to the web-then-paper protocol for the $10 incentive and about 5 points lower than web-then-paper for the $20 incentive. In spite of the additional costs of the $10 promised bonus incentive for choosing web, the choice+ protocol cost 5% less than the choice protocol. Biemer et al. (2018) conclude that for their purposes the choice+ protocol was best suited to address the range of requirements.

In studies comparing sequential with concurrent mixed-mode designs, the results vary. Several studies have found the response rates to be lower for the sequential design than the concurrent design, while others found the sequential approach to yield higher response rates and still others have found similar response rates between sequential web-to-mail designs and concurrent designs. There may also be interactions between respondent characteristics that determine the success of either design. For example, Tancreto, Zelenak, Davis, Ruiter, and Matthews (2011) found that a sequential design starting with web was more successful with respondents with higher internet skills, while the concurrent design was better for people with lower skills. In their extensive overview Patrick et al. (2018) conclude that the evidence in studies comparing sequential with concurrent mixed-mode designs is inconclusive. Like the findings on mode combinations that were discussed in the previous section, similar differences in design and population may play a role in the varying findings that can only be resolved by meta-analyses and additional study.

In spite of the potential drawbacks of concurrent designs, slightly more than half of the mixed-mode designs in the European NSIs use a form of concurrent design where the different modes are in the field at the same time (Murgia et al., 2019; Gravem et al., 2018a). In about half of those cases the respondent has a free choice of modes, in the other half the survey organization determines which mode a person or household is assigned to. For example, CATI-CAPI designs but also web-CATI designs are driven by the availability of a phone number. It is striking that in a significant part (39%) of the truly concurrent designs a web component is part of the mix: web-CATI, web-CAPI, or web-CATI-CAPI designs exist in European official statistics. The web-paper concurrent design that is the subject of much literature is rare: 9% of the concurrent designs described in 2018 are of this form. The DCSS inventory (Blanke and Luiten, 2014) showed that web-paper questionnaire designs are more often used for the census, both within and outside Europe.

Other features can also determine which mode is offered: in Greece, educational level and the age of the reference person determines whether an internet or paper version of the questionnaire is offered. A number of NSIs mention that persons can request a mode change, e.g., by providing their telephone number or indicating preferences for the next wave of a panel survey. Cost considerations are often not the driving principle in these designs. For example, Hungary sends interviewers who offer the respondent to do the questionnaire via web.

Sometimes modes are available at the same time, but the use of the more expensive mode is discouraged, for example, by having the respondent request a written questionnaire, or having the respondent download and print it from a website. Also one of the modes can be offered later, e.g., at the first reminder to the sample unit. This amounts to a delayed concurrent design, see Section 2.3.

Apart from the risk of lower response rates in mixed-mode designs where respondents can choose the design, the European NSIs mention another potential drawback of concurrent designs, namely more complicated logistics and potential planning problems. Especially when fieldwork is outsourced, but also in in-house data collection, a firm idea about the number of respondents in each mode is a prerequisite for the planning of fieldwork. A third point of attention when considering this kind of design is the need for an IT system where respondents can be easily shifted from one mode to another.

In 18 European surveys a sequential mixed-mode design is used. And here we see that web surveys are far more common as first mode: in more than 60% of these designs. The web first round is followed by another, more expensive mode, mostly CATI, but also frequently a paper questionnaire. There are two exceptions where the most expensive mode is used first: the Polish SILC and AES. The sequential designs have a far lower organizational complexity: all sample units are assigned to the first mode and only nonrespondents are switched to the subsequent one.

In an additional 27 cases, the design starts off sequentially, but when the second mode is offered the respondent can still choose to use the first one. Here too, web is often the first mode. The remainder of the European mixed-mode designs are a hybrid form, where web is used as the first mode, and nonrespondents are subsequently divided over other modes, depending on, for example, the presence of a known telephone number.

In summary, many different designs are used within and across European NISs. Web surveys have been introduced in a large number of surveys in most countries. They occur in four kinds of designs:

1. As the first mode in a sequential design. Nonrespondents are followed up in other modes. This design offers opportunities for large cost savings, but, as was described in the previous section, runs the risk of lowering response rates.

2. As a concurrent mode along other options, as a choice for respondents. This design is most often used when the other option is a paper questionnaire. The design will limit potential cost savings, as respondents tend to favor the paper questionnaire, unless specific action is undertaken to counteract that tendency (Tancreto et al., 2011; Biemer et al., 2018). Web as the concurrent mode was the most commonly used mixed-mode design in 2013 in Europe, and still was in 2018.

3. Offered to nonrespondents of earlier CATI or CAPI rounds. This design is chosen when the explicit aim is to improve response rates over the initial mode. This design is rarely used; Austria uses it for the Household Budget Survey and the EHIS.

4. Offered in subsequent waves, after the first wave in another mode explores respondents' preference and collects email addresses.

What mode or what sequence of modes is offered to which sample unit does not need to be an all or one decision: Adaptive or responsive survey designs adapt to characteristics of the sample unit. A simple version would be to not offer web to the (very) elderly, or not to call households with more than three adults. More advanced possibilities are to use insights into measurement errors per group per mode to fine-tune mode allocation. This is demonstrated by Calinescu and Schouten (2016) for the Dutch LFS, who propose mode switches dependent on a given sample and budget based on response probabilities of different subsamples. Adaptive Survey Design in a mixed-mode context is described in Chapter 11, Adaptive Mixed-Mode Survey Design.

## 5.3.2 Sequential, but Which Sequence?

Especially the combination web-mail has enjoyed a lot of experimental attention in recent years, where the influence of starting with web or starting with mail on response rates and survey costs has been the focal point. But studies have also been done on other modes and other mode combinations. Reminiscent of the other subjects in this chapter, literature does not come to a firm conclusion about sequence with regard to response rates (Wagner, Arrieta, Guyer, and Ofstedal, 2014).

One of the topics in a sequential design is how soon to introduce the next mode. This was also a topic in the DCSS and MIMOD inventories. The inventory showed varying practices, with terms mentioned of four days, five days, seven days, three weeks, or one month. It is unclear as to how far these practices are based on studies as to the optimal timing. Statistics Netherlands' mode switch after one month has to do with the logistics of transferring sample persons to different modes, more than with costs or response considerations. In the context of official statistics, speedy shifts may also be related to reference weeks for specific surveys (like the Labor Force Survey), where fieldwork must by necessity be as short as possible.

Wagner, Schroeder, Piskorowski, Ursano, Murray, Heeringa, and Colpe (2017) summarize literature findings and contend that longer periods of time allow a greater proportion of the interviews being conducted in the less expensive web mode. The longer period also allows more reminders. The more the number of reminders, the higher the web response (Sheehan, 2001). A long period is not a necessary prerequisite for frequent reminders, however. Statistics Norway describes the communication strategy in the web-then-telephone Housing Panel, where sample members receive an invitation by text message on Friday, a text and email reminder on the following Monday, a new reminder on Tuesday, and again on Wednesday. The mode switch occurs at the end of the first week. Panel members can still respond in web though. The fourth reminder to that end is sent on Tuesday in week two. The entire communication campaign includes four text messages and four emails in the span of a bit more than one week (Gravem et al., 2018a).

Wagner et al. (2017) experimented with the timing of the mode switch in the fourth wave of a longitudinal study on the mental health of soldiers in the US Army. The survey has a web-then-telephone design. Fieldwork began with a prenotification message sent via mail, text, and/or email. If soldiers did not respond during the web phase, an attempt was made to administer the survey by telephone. The mode switch was experimentally varied to be after one, two, three, or four weeks. Every seven days, a reminder was sent until the case was switched to telephone. Therefore, the number of reminders and the length of time before switching are confounded with the timing of the switch. Each weekly email reminder increased web response rates, albeit to ever-diminishing returns. It was found that the different timings of the switch did not produce differences in final response rates or key estimates, but longer delays before switching did lead to lower costs. The costs decline with longer periods of time in the web mode before switching to telephone.

These findings concur with those of Patrick et al. (2018), who also find that in web-then-mail designs the longer the delay before mail is introduced, the higher the proportion of responses that come via the web.

## 5.4 Costs

Reducing survey costs is one of the main reasons to implement a mixed-mode survey design. From the survey cost perspective, modes vary, perhaps, even more than for survey errors (Murgia et al., 2019). The differences in costs per respondent in web and face-to-face interviewing are very large. Depending on a country's circumstances, face-to-face may be as much as 50–100 times more expensive, depending on the implementation. The difference in costs between modes arises mostly from fieldwork, especially travel costs, and not from design or analysis.

NSI's participating in the DCSS and MIMOD questionnaires express very different reasons for implementing mixed-mode data collection. This has partly to do with the initial mode: increasingly, CATI surveys suffer from coverage problems as it becomes harder to track down telephone numbers, and land lines are replaced by smartphones, see also Chapter 4, Mode-Specific Selection Effects. For CAPI surveys other considerations play a role. Table 5.2 shows the reasons that were given in 2013 and 2018 for implementing or increasing mixing modes, and the reasons given in 2018 for implementing web data collection.

In 2013, reduction of costs was the reason that was mentioned most as a driver for the implementation of mixed mode. In 2018 this reason was not mentioned so much for implementing mixed mode but was again by far the most important reason to implement web data collection. Facilitating respondents was mentioned as the second most important

**TABLE 5.2**

Reasons to Implement Mixed Mode and Web Data Collection

|  | 2013 Mixed Mode | 2018 Mixed Mode | 2018 Web |
|---|---|---|---|
| to increase response rates | 14 | 2 | 4 |
| to increase coverage | 3 | 6 | 1 |
| to reduce costs | 30 | 3 | 20 |
| to facilitate respondents | 19 | 3 | 4 |
| to improve timeliness | 6 | – | – |
| other reasons | 6 | 3 | – |

reason in 2013, but has lost its driving appeal in more recent years. Increasing response rates was mentioned as well in both queries but was not the most important driving force. Coverage issues are mentioned as the most important driver for the introduction of mixed mode for those countries that started after 2013. For the introduction of web questionnaires, the picture is clear cut: costs reduction is by far the most important reason, although four NSIs mention that response rates improved as a result of the introduction.

Although fieldwork costs for web data collection are relatively low, introducing web into the design may have cost implications for other aspects of fieldwork that abate the possible savings. Examples are:

- The sample units responding in web are the 'easiest' respondents. The remaining persons may need more effort to trace and persuade. Because the work is harder, interviewers may expect higher remuneration. Or interviewers are disheartened by the more difficult work and quit their job.

- If web is introduced as the first mode, less work is available for the CAPI field interviewers, and interviewers have to travel longer distances to reach their addresses. This may have a substantial negative influence on the costs of fieldwork. There are a number of alternatives to alleviate that scenario: CAPI interviewers can be trained to do CATI work as well. Alternatively, sample units can be clustered, with the risk of diminished precision, or (regionally) subsampled, with the risk of introducing bias.

Lesser, Newton, Yang, and Sifneos (2016) illustrated that the cost effectiveness of using the web depends on the sample size, overall response rate, percent completing the questionnaire by web, data entry costs, postage/printing costs, and programming cost. All these features should be considered, along with the population under study, in determining the optimal mode and number of contacts when collecting survey data.

Cost saving is also a function of the relative costs of fieldwork. Villar and Fitzgerald (2017) show that there are large differences between (European) countries in the costs of fieldwork in each mode that have to do with the availability of sampling frames of individuals, and the costs of interviewers, but also with the size of the country, the population density, the mode coverage, available software, and interviewer experience.

Fieldwork costs are but one of the components of the costs of running a survey. The introduction of mixed-mode data collection can have cost implications for all other stages, like sampling, data processing, case management, and adjustment. Especially if IT systems need to be bought or developed, that could mean a major investment. Also, the greater complexity of the other stages may potentially reduce the cost savings that can be attained. For example, multiple modes may mean multiple questionnaires, more testing, and more methodological support. Most NSIs that were queried on this subject in DCSS (Blanke and Luiten, 2014) foresaw that the balance of costs and profits of mixing modes would, however, become more profitable after a number of years of increased use of web surveys.

## 5.5 Sample Composition and Data Quality

A large majority of studies on mixed-mode data collection focus on response rates and costs. Some of the studies in addition give attention to the effect of mixing modes on sample composition, data quality, and substantive findings. Sakshaug, Cernat, and Raghhunatan (2019) found in a panel study among young drivers that a second mode can reduce both nonresponse and measurement bias, but the sequence makes a difference: the mail-telephone sequence minimized bias to a greater extent than the telephone-mail sequence. The CATI starting mode especially minimized nonresponse bias, while the mail starting mode did a better job of minimizing measurement error bias. An interesting observation in this study was that a large nonresponse bias in a mixed-mode design can sometimes be offset by an opposite measurement effect, with the result that the total bias becomes negligible.

Bianchi, Biffignandi, and Lynn (2017) studied the effect of introducing a sequential web-face-to-face mixed-mode design over three waves of the longitudinal UK Household Study, in which members were previously interviewed face-to-face. No differences were found between the mixed-mode design and face-to-face design in terms of cumulative response rates and only minimal differences in terms of sample composition. They conclude that the study paints a rather positive picture of the potential for mixed-mode data collection in panel surveys.

Mauz et al. (2018b) compared a concurrent web-paper design of the German Health Update survey with a sequential web-then-paper design. Both conditions were also offered a telephone option – in the concurrent design as a third option, in the sequential design as a second option along the paper questionnaire. The two designs were very similar in terms of response rates, the socio-demographic characteristics of the achieved samples, and the prevalence rates of the health indicators studied.

Suzer-Gurtekin, Elkasabi, Lepkowski, Liu, and Curtin (2019) performed a series of experiments in the University of Michigan's Surveys of Consumers. Groups either received a request to complete the survey by mail or web or a request to complete the survey by web before offering a mail alternative. These approaches were compared in terms of response rates, process measures, sample composition, and key substantive measures. The studies showed no clear differences between response rates or substantive findings on economic attitudes. It was hoped that the web survey would attract different subgroups to participate who would not have participated in the mail survey, but this was not the case. Sample composition was quite similar in the two designs.

Cornesse and Bosjnak (2018) concluded on the basis of a meta-analysis of over 100 studies, that mixed-mode surveys attain better sample representativeness compared to single-mode surveys. They also found that web surveys are, on the other hand, less representative than other modes. The meta-analysis also showed that higher levels of response are associated with better sample representativeness.

In spite of their optimism related to response, sample composition, and costs of the introduction of a web-face-to-face design in the UK Household Study, Biachi, Biffignandi, and Lynn (2017) point out that, at that moment, it was unclear whether measurement could be considered equivalent between the modes. Survey mode can indeed have an effect on measurement error (e.g., Klausch, Hox, and Schouten, 2015). For example, interviewer-administered modes have been shown to lead to higher levels of social desirability bias compared with self-administered modes (Tourangeau, Rips, and Rasinski, 2000; Heerwegh and Loosveldt, 2011; Hope, Campanelli, Nicolaas, Lynn, and Jäckle, 2014; Cernat, Couper, and Ofstedal, 2016). De Leeuw (2018) contends that combining multiple modes can have a beneficial effect on measurement, but it can also introduce unwanted measurement differences. When a different mode is used for a specific part of the questionnaire, and provided that all respondents receive the questionnaire in this mode, measurement can be improved. But when different groups of respondents complete a survey in different modes there is a risk of differential measurement errors. The more modes differ, the larger the risk of measurement effects, especially for sensitive questions. With more neutral questions a mode switch does not necessarily threaten comparability. In longitudinal design there is the risk that a switch from an interviewer mode to web threatens the compatibility over waves. However, much depends on how questionnaires are designed and implemented in the various modes. This is the topic of Chapter 6, Mixed-Mode Questionnaire Design.

Another aspect of self-administrative modes that may have a large impact on measurement quality is within household person selection, in cases where sampling frames of individuals are not available. Villar and Fitzgerald (2017) call within-household random selection of respondents one of the biggest obstacles for self-administration modes in the general population. Without an interviewer present, the selection of the individual has to rely on the person in the household who receives the instructions. Olson and Smyth (2014) studied within-household selection in a self-administered survey among panel members of whom individual information was known. In their study, about 82% of the households selected the correct respondent, while about 18% did not. They showed that there are multiple mechanisms underlying inaccurate selections in mail and web surveys. These mechanisms include that household members may become confused about what they are supposed to do or about who should be considered eligible for participation and that some household members may conceal information about their household. That means, people are sometimes unable and at other times unwilling to follow the instructions. Therefore, in many countries, a change in the mode of data collection can affect whether the final sample is actually a random sample of the population. Olson and Smyth (2014) expect that with better communication, confusion and concealment can be combatted, and conveying legitimacy and trustworthiness of the survey may be increased. A communication strategy is not only relevant to making people select the right household member but also crucial for attaining high response rates in self-administration surveys, especially in web surveys where people need to take the perhaps unfamiliar extra step to login. This is the topic of the next section.

## 5.6 Communication Strategies to Increase Web Response

Once the decision is made to embrace web data collection, measures need to be taken to get the respondent to actually use the mode. The more people fill in the questionnaire on the web, the less effort and money needs to be spent in later phases of the fieldwork. How to do this has been the topic of many experiments. The first step is to withhold offering other modes, like is described in the previous sections. But there are various other aspects in the communication strategy around 'pushing respondents to the web' (Dillman, 2017) that influence the success of the web-push: how to reach the respondent (by letter, email, text message), how to get him to understand the task of logging in to the questionnaire, how to remind him, and how many times, what to say to make them trust you, if and how to give incentives.

Lozar Manfreda, Bosnjak, Berzelak, Haas, and Vehovar (2008) summarize why it is more difficult to get high response rates in web than in other modes. Some of these reasons may by now be less pertinent, but others are still as relevant as a decade ago:

- Security and privacy concerns associated with the internet. Some respondents are anxious about their data being transferred via the internet and may consequently be reluctant to participate in web surveys.

- Less possibilities to employ methods and procedures to increase response rates. For example, giving an unconditional incentive is problematic in electronic communication. However, by mixing the mode of invitation as well (e.g., sending a paper letter), this difficulty can be attenuated.

- Limited web literacy is still a problem in some groups, as confirmed by the PIAAC study of the OECD (OECD, 2013). The accessibility and usability of web questionnaires should perhaps accommodate persons with computer skills corresponding to PIAAC's level 1.

- Increased burden as respondents need to take the extra step to start their computer, find the website, and copy passwords and/or login codes. Their meta-analysis showed that web response was 15% lower than other modes when requested by postal mail, against 5% when requested by email. Much of the burden described here is lower when email invitation is possible.

- Technical limitations associated with the web mode, such as software incompatibilities, misrepresentation of visual elements, and long or irregular loading times.

Millar and Dillman (2011) formulate additional reasons: It's easier to forget the invitation, especially when the invitation is sent through email. The hard copy paper questionnaire is a constant reminder, laying on the desk. It is also more difficult to convince people of your trustworthiness. Additional mailings, but also the hard copy questionnaire, may install that trust. Illustrative is that web response may increase in a sequential web-other mode design, once the second mode is released.

Asking respondents who participated in later modes why they did not participate in web is a treasure trove of inspiration. Statistics Netherlands has done this several times in a number of surveys. Respondents who participated in CATI or CAPI follow-up modes were asked a number of questions about their experiences with the invitation to the web survey: did they try to login, did they ever intend to participate in web, and did anything go wrong in the process. The inventory gave important insights into the reasons for non-participation in web. In one survey among children and young adults, for example, it appeared that almost half of them had not read the invitation letter; a quarter had not even seen the letter. In the general population this percentage was lower, but still substantial with about 10%. In addition, the amount of computer or internet problems was much larger than expected. Well over 10% mentioned some kind of technical problem. Of about half of all the reasons mentioned Statistics Netherlands felt the situation could be attenuated by better procedures or more careful communication. A large number of people mentioned that they simply forgot to fill in the web questionnaire, even after two reminders. We need to think about circumventing that problem too, for example, by using alternative ways of reminding people. We also need to find ways of getting the letter read, and if read, ways to better sell the survey. See also Suzer-Gurtekin et al. (2019) who mention that failure to respond in web and mail surveys is related to failure to open the mail envelope. Nichols (2012) showed that the two primary reasons for nonresponse involved receipt of the envelope containing ACS materials: many respondents claimed they that they either did not receive the envelope or received the envelope but did not open it.

It is the task of the statistical office to make the process and instruction so clear that doing something wrong is virtually impossible. Usability testing is a large part of that process, see also Chapter 6, Mixed-Mode Questionnaire Design.

Dillman, Smyth, and Christian (2014, Chapter 11, guidelines 9 to 13) formulate a number of communication strategy guidelines to increase web response rates:

1. Use multiple contact modes to increase the likelihood of contacts being received and attended to by sample members.

2. Use contact by a mode different than the response mode to increase trust that the survey is legitimate and useful.

3. Send a token cash incentive with an initial postal mail contact to increase trust in the survey. Consider using a second cash incentive in a later contact to improve response rates.

4. Try to obtain contact information for more than one survey mode.

The remainder of this section illustrates these and other guidelines with findings from literature and the European projects on mixed-mode data collection, DCSS and MIMOD, see Section 5.1.

### 5.6.1 Contacting the Sample Person or Household

In DCSS and MIMOD projects, data collection strategy was part of the inventory among NSI's. The invitation letters and other means of invitation of NSI's with web as the first mode were studied. Invitation letters for web surveys need specific attention. In addition to the regular information on the purpose of the survey and why the respondent should participate, information on how to log on to the web, and what passwords and other security measures are necessary need to be described. This may result in a long and complicated letter. We have seen various examples of advance or invitation letters that use different approaches to handle this complication, like breaking up the information in various communications, or structuring the letter with clear headings and short descriptions. In Section 5.6.3 a number of experiments by Statistics Netherlands on this topic are described.

Millar and Dillman (2011) showed that adding an email invitation with a clickable link, on top of the paper invitation, increased response substantially in both single web and concurrent web and mail surveys. Israel (2012), using mail and available email addresses in a mixed-mode design, secured response rates that were equivalent to a mail-only survey. Cernat and Lynn (2014) and Cernat (2015) found that an additional email contact in a web-then-face-to-face panel survey did not so much increase participation above the mailed contact, but did increase the web participation. Patrick et al. (2018) showed that a web-push design plus additional email in a population of 19–20-year-olds had a similar response rate as a mail-only control group, while a web-push design without email had a significantly lower response rate. There are indications that the efficacy of an email invitation is higher in some groups than in others (Dykema et al., 2013; Kaplowitz, Lupi, Couper, and Thorp, 2012), but as long as invitations in other modes are also provided, this should not harm the ultimate sample composition.

The problem is, of course, that most statistical institutes do not have email addresses for their sample units. In longitudinal surveys, additional effort can be made to request email addresses. Bandilla, Couper, and Kaczmirek (2014) asked respondents to the 2012 German General Social Survey (ALLBUS) who reported having internet access at home for their email address for a follow-up web survey. Less than half was willing to provide it, though. Asking did not appear to harm subsequent response rates however. Those who are asked for an email address but declined to provide one did not respond at a lower rate than those not asked for an email address.

The MIMOD inventory showed that almost all European NSIs send a paper invitation, or have one dropped off if an address frame is not available, also if the invitation is for a web questionnaire. Sometimes the letter is accompanied by a flyer explaining how to log on to the internet. Marked exceptions are Denmark, which contacts respondents electronically via a government portal, and Norway, which uses digital contact communication *only* for all surveys. Estland describes that they send an email to those sample units for which the email address is known, and a letter otherwise. Most NSIs do not have access to registries of email addresses or smartphone telephone numbers. The few NSIs that do have that kind of information available let the written invitation be accompanied by an email invitation (Greece, Hungary) and sometimes even an additional text message (Denmark). Statistics Norway and Statistics Denmark have access to quality register information on email addresses and mobile phone numbers for the majority of the population. Statistics Finland expects to have access within shortly. The Norwegian Digital Contact Information Register has made digital communication much easier and contributed to improving response rates in recent years (Gravem et al., 2018b).

### 5.6.2 Reminders

One of the four recommendations Tourangeau (2017) gives for spurring response to web surveys is to make multiple contact attempts, as they are as crucial for mixed-mode surveys as for surveys using a single mode. Lesser et al. (2016) increased the number of mailings from 4 to 5 and witnessed an increase in response between 6 and 9 percentage points in a web-then-mail survey. In a replication two years later, this was 4 to 5 percentage points.

In the DCSS inventory (Blanke and Luiten, 2014), most participants mentioned sending two reminders, even when the period in which web data collection is foreseen is very short. For example, Statistics Canada sent two reminders in four days. The US Census Bureau likewise sent the first reminder after two days. Each reminder results in a clear increase in response.

The most commonly sent reminder was a letter, although the timing of the letter varied widely; it can be sent either after one, two, or three weeks for the first reminder and one, two, or three weeks after the first reminder. Australia embraced modern communication means for the reminders by sending emails and text messages. The Netherlands mentioned that CBS research has shown that additional reminders above two would still bring in additional web response. However, sending more than two reminders had a bad influence on subsequent CATI and CAPI response because sample units start feeling harassed. In occasional web-only surveys, three reminders may, however, be used. When a consecutive design is used, the number of reminders must be handled with care, in order not to unduly harass the sample units who will still need to be approached with other modes.

Five years later, according to the MIMOD inventory (Murgia et al., 2019), emails and text message are used to a larger extent. Thirty-five percent of NSIs that use web data collection send an email as reminder, and an additional 11% send a text message. Most NSIs send two reminders, although one reminder is also often mentioned. Three reminders happen as well, although one NSI indicates that they only do that if there is no follow-up in another mode. Paper reminders are most often another letter, although reminder cards are used as well.

The timing of the reminders is reported to be mostly a function of the length of the fieldwork period. As a result, there is no consensus between European NSIs on when to send a reminder. Out of ten NSIs who described their reminder strategy in MIMOD, eight different timings were mentioned: one day, two days, twice a week, one week, two weeks, and three weeks.

### 5.6.3 Invitation Letters, Flyers, and Envelopes: Statistics Netherlands' Experiments

All Statistics Netherlands' person and household surveys have a consecutive design, starting with web. The data collection strategy for the first web stage comprises the following steps:

1. Statistics Netherlands does not have (access to) a registry of email addresses. Thus, all sample units receive an invitation letter, containing the internet address of the web questionnaire and a personal login, consisting of a login code and a password. For household samples the letter is addressed to 'the inhabitants of [address]'. Person samples are addressed by name, derived from the Municipal Registry. The person or household selected is requested to complete the questionnaire via the internet. Household members in household surveys need to use the same login to gain access to the questionnaire: i.e., household members do not receive an individual login.

2. Two reminders are sent to nonrespondents two weeks and three to four weeks after the invitation letter. The format is the same, although the contents of the letters are slightly different, to express an increasing sense of urgency (Dillman, Smyth, and Christian (2014).

3. One week after the second reminder the access to the web questionnaire is closed.

4. Follow-up in the other modes is mentioned in the invitation letter. Mentioning of the follow-up mode was experimentally shown to increase web response by 2 percentage points. For some respondents, the 'threat' with an interviewer is extra leverage to make them access the web questionnaire. For other respondents, the 'promise' of an interviewer makes them wait patiently for the call or visit. Lynn (2020) likewise finds that mentioning up front that a CAPI interviewer will visit the household if the internet questionnaire is not forthcoming, leads to a significant increase in the number of web responses, without damaging the overall response rates.

The additional content of the URL, the login code, and password, a short instruction on how to log on and where to put the URL, and additional remarks on internet security made the letter long (one and a halve page long) and complicated. An initial attempt to simplify and shorten the letter leads to a disastrously low response rate, however. As the reason for the lower response rate was not understood, it was decided to systematically vary aspects of the letter in a series of experiments. The experiments were performed on the invitation letter and reminder letters of the LFS. The research team consisted of people from the data collection department, data collection methodologists, and communication experts. Ten percent of the sample for the Labor Force (11.500 sample households per month) was used. As the LFS was based on an address sample at the time (the Dutch LFS uses a person sample as of 2020), the letters were addressed to 'the inhabitants of *address*'. Using the LFS for these experiments is sub-optimal, because of the household aspect. This makes it unclear who read the letter and who subsequently made the decision to participate. Promising experimental results were replicated in the LFS for additional detail, and also in person sample surveys. Of all sample addresses, the age, gender, household composition, and ethnic background of the persons living there is known. Subgroup analyses followed the strata that were also used in the LFS sampling procedure: (1) Households with at least one person of 65 years of age or older. The persons can be of Western or non-Western background. (2) Households where at least one person is unemployed and registered at the unemployment office. (3) Households with children from 14 to 26 years old. (4) Households where at least one person is of non-Western ethnic background. (5) All other households. This section will focus on the main effect results, but important subgroup differences will be mentioned.

### 5.6.3.1 Experimental Manipulations

In this section an overview of the experimental manipulations and their underlying reasons are given. In section 5.6.3.2 the results are described.

1. Visually shortened letter, to make all information visible on the front page

   Literature (e.g., Dillman, Smyth, and Christian, 2014) stresses that a short invitation letter with a maximum length of one A4 is crucial. The LFS invitation letter was 1.5 pages long, however. In qualitative research with invitation letters it was noted repeatedly that respondents did not read the back of the letter, thereby missing relevant information. For this experiment the information of the original LFS letter was confined to one A4 by putting all the information pertaining to logging on to the website (address, passwords, and instruction) in a highlighted cadre in the letter, in a space that was otherwise blank.

2. More and strengthened persuasion arguments

   In the standard LFS letter hardly any arguments are mentioned why people should participate in the LFS. The arguments that are mentioned may not be equally attractive to all people ('for the quality of the statistics it is very important that you participate', 'You represent many other people'). In this experimental letter we tried to reinforce the arguments why people should participate, both altruistic arguments ('we need your help ...', 'you help us ...') and more egoistic arguments 'what's in it for you'. In addition, in line with recommendations by Cialdini (2007) a phrase was added that the LFS is one of the most important CBS surveys and that the respondent is one of a small number of households chosen to participate.

3. Simplified linguistic level

   Not only in the Netherlands, but in the European context, it is recommended that the language level of government communication with the general public should not exceed level B1. This is the level that is understood by 80% of the population. In order to attain this kind of clear language only high frequency words should be used, and sentences should be no longer than 15 words. The standard invitation letter for the LFS had a higher complexity (B2), with a large number of words that are too difficult. The letter was redesigned according to these criteria, while maintaining the original contents.

4. Strengthened 'what's in it for me' message

   The second experiment with the increased altruistic persuasion arguments led to positive results for elderly households, but not for other households, see below. In this experiment, we tried the opposite tract: strengthening the 'what's in it for me' message. The experimental letter started with a number of stimulating questions about topics that the design team thought would appeal to the public. The first section read as follows: 'Does everyone get paid the same amount for the same work? How many job vacancies exist and where? How do men and women distribute care and work? With which education do I get the best opportunity to find work? At what age are we going to retire? These are questions about the labor market. Questions that CBS can answer, but only with your help!'. In addition, a footnote indicated three different websites where interesting CBS animations could be found.

5. More detailed login information

From prior research into the reasons why respondents in CATI or CAPI did not answer the questionnaire in web, see section 5.5, it appeared that technical problems and unfamiliarity with the computer or internet were an important reason not to participate in web. Especially the practice to type in the web address in a search engine instead of the address bar led to many a telephone call to the technical helpdesk. In this experiment, the procedure for logging in to the website and entering the login code is described in detail in the letter of invitation, in a clear step-by-step plan with visual support.

6. A small 'reminder' note block in the letter

One of the most common reasons nonrespondents on web gave for not completing the questionnaire in web was that they had forgotten it. In an attempt to help them remember to fill in the questionnaire, a small notepad with adhesive sticky tapes was added to the envelope. The notepad's cover was embossed with a 'don't forget' logo. Adding the notepad had two additional advantages: it could make the envelope more noticeable in the mail, and the notepad served as incentive. This experiment was replicated in a person sample (experiment 7 in Table 5.3 below).

7. Informed consent text

Each invitation letter (and also the reminders) contains a clause in which is explained that the respondent's data are linked to registry information, and which registries these are. What is explained in this clause is determined in collaboration with the Statistics Netherlands' legal department. The clause reads:

Statistics Netherlands not only collects data but also receives data from other institutions. For example, the data from the population administrations, the centers for work and income, the social services, and the payroll administrations of many companies. We automatically combine the information you give in this study with information we receive from other institutions. With this combined information Statistics Netherlands compiles statistics on Dutch society and we work as economically as possible.

From qualitative research we knew that this clause was not fully understood by many people. In addition, people wonder whether Statistics Netherlands will inform the institutions mentioned above of the answers respondents give, instead of receiving information. Some of the institutions are responsible for giving out unemployment benefits, for example. People who fear that those institutions may learn of unofficial side activities through their response may be reluctant to participate in the survey. In close collaboration with the legal department, a new version was drawn up, where the specific examples of the collaborating institutions are removed. The text now reads:

Statistics Netherlands not only collects data but also receives many files from other institutions. With this combined information Statistics Netherlands compiles statistics on Dutch society and we work as efficiently as possible.

**TABLE 5.3**

Percentage of Respondents That Start, Break-Off, and Respond in Control and Experimental Conditions

| | Started | | | Break-off | | | Response | | |
|---|---|---|---|---|---|---|---|---|---|
| | Control | Experiment | p | Control | Experiment | p | Control | Experiment | p |
| 1. shorter letter | 29.9 | 28.4 | ns | 17.1 | 18.2 | ns | 22.7 | 21.3 | ns |
| 2. persuasion arguments | 28.9 | 28.1 | ns | 17.7 | 21.9 | * | 21.8 | 19.9 | ns |
| 3. lower linguistic complexity | 29.7 | 32.7 | *** | 17.3 | 18.1 | ns | 22.7 | 24.4 | * |
| 4. what's in it for me message | 28.7 | 23.3 | *** | 18.0 | 22.1 | * | 21.8 | 17.0 | *** |
| 5. detailed log-in information | 30.0 | 31.9 | ns | 18.1 | 20.9 | ns | 22.8 | 23.1 | ns |
| 6. 'reminder' notepad | 30.2 | 32.1 | ns | 17.1 | 19.5 | ns | 23.1 | 23.3 | ns |
| 7. id in person sample | 39.5 | 44.0 | *** | – | – | | – | – | |
| 8. less explicit informed consent | 30.0 | 31.2 | ns | 18.2 | 15.4 | ns | 22.5 | 24.8 | * |
| 9. adapted envelopes | 29.0 | 29.0 | ns | 18.6 | 21.7 | ns | 21.8 | 21.4 | ns |
| 10. no flyer | 29.2 | 26.5 | * | 18.2 | 16.5 | ns | 21.9 | 20.2 | ns |
| 11. no flyer person sample | 44.0 | 38.0 | *** | – | – | | – | – | |
| 12. different photo on flyer | 28.3 | 29.0 | ns | 19.5 | 17.7 | ns | 21.2 | 22.1 | ns |
| 13. id in person sample | 44.0 | 41.1 | ns | – | – | | – | – | |
| 14. different contents flyer | 27.1 | 27.3 | ns | 18.0 | 19.0 | ns | 20.4 | 20.7 | ns |
| 15. id in person sample | 44.0 | 42.7 | ns | – | – | | – | – | |

* $p < .10$, ** $p < .05$

*** $p < .01$, ns not significant

8. Experiments with envelopes

   The study into the reasons that respondents did not participate in web, but did participate in other modes showed that a large number of them had simply not seen the (three!) letters. In two experiments this issue was addressed by using envelopes that stood out more, one in the LFS and a replication experiment in a person sample. An (A5) envelope was developed in the style that is also used in publications. It showed a colored banner and a statistic. The invitation letter envelop and the two reminder letters each had a different color and a different statistic. In the replication experiment the envelopes were slightly changed: no statistical facts were depicted as they were deemed to make the envelopes look 'too commercial'. Instead, the text in the banner read: 'your answer counts!'. The reminder envelopes had again different colors and different texts. The text in the banner of the first reminder was, 'Won't you forget us? Your answer counts'. The third reminder had the text, 'Your answer counts. You can still join!'. In addition, the urgency was depicted by the color red of the banner and a running figurine.

9. Experiments with flyers

   Three experiments were performed: sending no flyer at all, using another front, and using another content. Standardly, an invitation letter is sent with a flyer included in the envelop. In the experimental condition, the flyer was left out. We had three reasons for proposing this experiment: If leaving out the flyer is feasible, that would save a lot of trees and money. The survey methodology literature is not clear on the effect of adding a flyer. In addition, in an experiment with another survey we found that adding a flyer had no effect on response rates for most social groups. The situation in the two surveys was not entirely comparable, as the two flyers are very different. Hence the new experiment. The standard flyer is a folded A4 sheet of paper. The text on the front page reads, 'why does CBS ask you?' and shows a photo of people cycling and walking.

10. Different picture on the flyer

    The research group felt that the photo on the flyer was not really attractive, with two grumpy cycling men. And in addition, we felt that the picture did not do justice to the image that Statistics Netherlands wants to convey. In this experiment, the photo was replaced with the picture of an interviewer and a respondent at the door. A secondary goal was to create a more ethnically neutral flyer.

11. Different contents of the flyer

    In this experiment the flyer with the new photo was compared with a flyer with different content. This flyer was survey dedicated, with content relevant to the LFS. The rationale for this experiment was that more and more survey dedicated flyers were being developed. This costs a lot of time and effort, and we wondered if that was worth our while. The contents were in addition presented in a more figurative and less linguistic form.

12–14. Experiments 9–11 were replicated in a person sample, the Safety Monitor. The standard approach in the control group was not to use flyers. For this experiment, three conditions were compared to this control group: sending the standard Statistics Netherlands flyer ('Why does Statistics Netherlands ask you?'), sending a graphic folder (analogous to the one in experiment 11, but with a content dedicated to the Safety Monitor), and again a flyer with a third different picture from the standard and the experimental picture from experiment 10. This time, the picture was chosen by a panel of field interviewers who chose the picture that they thought would most appeal to respondents.

In all experiments, the two reminder letters were sent in the same experimental format.

### 5.6.3.2 Results

Three dependent variables are examined: the number of households starting the questionnaire, the number of households breaking off filling in the questionnaire, and the resulting response rate. The first measure is a more direct measure of the effect of the letter. The second measure may show if the letter rises expectations that are disappointed in the questionnaire. Households are considered to have responded if all household members have responded.

Subgroup analyses are made on the strata used in sampling, as described above, in addition to a number of background variables: age and gender. These variables consist of the mean value of these variables for the household core of one or two persons. Hence, the variable Gender consists of the values men (either single men of households consisting of two men), women, and mixed-gender households. In addition, for each household, urbanicity, income, and ethnic composition in terms of the percentage of non-Western migrants of the neighborhood are determined on postcode 6 grids. These very fine-grained grids of (parts) of streets are standardly used in nonresponse analyses and have proven to be highly predictive of response rates. Linking sample households with background variables was not always unequivocally possible. In a number of cases it could not be determined of which persons a household consisted or it looked like more than one household lived on a certain address. In those cases, the household information was recoded as missing and analyzed as a separate value.

The experimental group was about 1100 sample households per experiment, and the control group consisted of about 11,000 households. Experimental strata subgroups were about 65, 250, 420, 210, and 450 households, respectively, with small fluctuations per experiment.

Analyses were performed with logistic regressions and analyses of variance. For overview and simplicity, only the level of significance is mentioned in Table 5.3. Because in some cases the number of households in subgroups is relatively low, significance levels of $< .10$ are reported as well.

1. It is clear that just visually shortening the letter did not lead to increased response rates, but had in contrast borderline negative overall results, and negative results for some subgroups. Putting the login information in the cadre meant that less information on the correct procedure to log on to the web was given. As a result, experiment 5 was designed, in which extra care was taken to describe the login process.

2. Strengthening the persuasion arguments did not have an overall effect on the percentage of households starting with the web questionnaire. It did, however, have an undesired effect in terms of a larger number of households breaking off the questionnaire. This phenomenon was observed in all subgroups, hence almost reaching significance in the overall results. Probably, the appeal for help did not agree with the subsequent contents of the questionnaire. There was,

however, an increase of more than 8 percentage points of elderly households starting with the questionnaire. They also exhibited less partial response and agreed more often to participate in round 2 of the LFS, resulting in a marginally significant higher percentage of elderly households entering the second wave.

3. This experiment had very positive results, to the extent that in all strata the number of households starting the questionnaire was higher, significant in several strata, and significant overall. Because of the positive results, this experiment was replicated with the same results. Table 5.3 shows the results of the combined experiments. Although there was a significantly higher number of households that started filling in the questionnaire with the simpler letter, there was regretfully also a higher number of households that broke off, almost reaching significance in the group of unemployed people. As a result, the overall increase in response was no longer significant overall, although still significant in the subgroup of households with children in the age group 14–26. The higher number of break-offs could be the effect of the LFS questionnaire, which also contains complicated language. It could also be the effect of having to fill in the questionnaire on the web. People with low literacy also have low computer skills (e.g., Buisman and Houtkoop, 2014). Nevertheless, these results induced us to make it a policy to write all invitation letters, flyers, and other communication with respondents with this lower linguistic complexity. In addition, a program was started to diminish the linguistic complexity of survey questions.

4. Strengthening the 'what's in it for me' message did not lead to higher response rates, but on the contrary to significantly *lower* response rates. The percentage of households starting the questionnaire was significantly lower, and the percentage break-off was marginally higher.

5. The more detailed log-in instruction led to a significantly higher number of households with children from 14 to 26 (i.e., with middle-aged respondents) starting the questionnaire. On the other hand, it had an almost significant negative effect on elderly households. There were no differential effects of the instruction on break-off, so as a result the response results mirror those of the start results. This result was somewhat unexpected, as we had actually targeted the elderly with this instruction. Perhaps the instruction had the unwanted effect of overwhelming these households. Lesser et al. (2016) find similar results in an experiment where adding an information card with additional guidance to access the online questionnaire led to a five percentage point decrease in response rates. They hypothesize that the additional materials in the mailing package may have led to information overload or increased perceptions about task difficulty.

The analyses of the other background variables showed that especially the households living in neighborhoods with a high percentage of non-Western migrants were susceptible to this measure. This is a highly welcome result, as this is a very difficult group to get web response from. No other interactions were found.

For some groups in society, a clear instruction on how to log on to the web is necessary. In addition, the login procedure has to be as simple as possible. Following the practice of Statistics Denmark, Statistics Netherlands has also adopted the practice to direct respondents in all web surveys to one and the same URL ('your answer'). Passwords and login codes redirect to the correct questionnaire. The result is that the (now frequently used) internet address is also found in search engines, and can be accessed from there as well.

6. The 'remember' block note increased the start rate in the unemployed group. However, the same group also had a marginally significantly higher percentage break-off, so that the net effect for this group was nil. The group with children 14–26 also showed a somewhat higher break-off. The block note obviously did not succeed in its intended role of getting people to remember filling in the questionnaire. However, there was another noteworthy result, i.e., that the experimental condition showed a clearly higher response in the beginning of the fieldwork. After the first reminder the difference becomes less and after the second reminder the difference disappears. That means that the notepad was instrumental in inducing respondents to respond right away.

7. A replication experiment was performed in a person survey: the Safety Monitor. Two thousand sample persons of 16 years and older were sent the memento block note with the invitation letter. In this experiment the response rates were significantly higher with the block note. Subsequent analyses showed that the incentive did not have an effect in the age group of 65 years and older, had a marginally significant effect in the age groups from 30 to 65 (p < .10), but had especially effect in the youngest age group from 16 to 29 (p < .001). The increase in response in the latter group was 6 percentage points. In addition, the incentive had no effect with persons with a non-western migration background, a marginal effect with persons with a native background (p = .05), and a significant effect with persons with a western migration background (p < .05). The incentive was not successful in all income groups but was particularly successful in the middle-income group (p < .001).

8. The adapted informed consent phase led to (marginally) significant differences between experimental and control conditions for some groups, both in percentage of starting, break-off, and response. In all cases these effects indicated that the adapted informed consent phrase led to more starting, less break-off, and higher response. Additional analyses showed a marginally significant interaction with Age for both starting and response (p < .10), indicating that the groups of 45–65-year-old people and the group with missing background information were susceptible to the adaptation. This adaptation was subsequently implemented in all CBS surveys.

9. In the stratum of households with 14–26-year-old children the new envelopes led to a marginally higher percentage of starters, and a significantly higher response. Additional analyses also showed that for (single) men and households in the highest income groups the new envelopes led to a lower number of starters. In the replication the envelopes resulted in a slightly increased response overall and in most subgroups. The difference failed to reach significance, however. In spite of these results, CBS decided to continue using these envelopes standardly, for persons, household samples and business surveys. The new envelops fit in with the corporate identity in other communication products, like website and reports. In slight alteration compared to the experiment, the color of the envelope is 'CBS blue'. The text on the banner is 'your answer counts'. See below for an example of the new envelopes, that are used for invitation letters and reminder letters alike.

10. Not sending a flyer in the LFS led to a marginally significant lower login percentage, although this did not translate to a lower response percentage. Additional analyses showed that people living in a neighborhood with a high number of non-Western inhabitants reacted negatively to the absence of the flyer.

11. The new picture on the flyer led to a significantly higher response in the age group of 65 and older but did not have any effect on the other groups. The hoped-for positive effect on the group with a non-Western background was not observed. No further interactions with other background variables were observed.

12. The new content of the flyer did not have any effect at all.

13. In the replication study in a person sample both the standard flyer and the flyer with a different picture had a significantly higher response than the condition with no flyer. The difference between no flyer and a graphic flyer was not significant.

The results of these flyer experiments show that, just as with interventions to the letter, interventions in the flyer do not lead to spectacular results, but can significantly stimulate or deter specific groups. Some conclusions: Whether or not enclosing a flyer leads to increased response rates depends on the survey, the contents, and on the picture on the front of the flyer. In addition, not all subgroups in the population are equally susceptible to the flyer. Careful testing should determine when and to whom to send a flyer. These experiments made us aware of the potential effect of the image used for the flyer. Field tests are not always feasible for new flyers, but at the very least a short pretest of both image and contents is recommended. Whether 'dedicated' flyers lead to a higher response than a general one could not be concluded unequivocally with these experiments. Replication in other studies is recommended.

The results of these experiments show that there are but few interventions that had a generally positive effect on all groups in society. The two exceptions were lower linguistic complexity and adapting the informed consent section. In all other interventions, positive effects for one group were counterbalanced by negative effects for other groups. These findings suggest possibilities for differential approaches where each group gets the letter that most appeals to them. For example, elderly persons would get a letter that appeals to their altruistic tendencies, in a standard envelop. Middle-aged households would get a letter with a clear login instruction in an envelope that stands out in the mail. Some groups get a flyer, and other groups do not. This, however, implies a large increase in the amount of work it takes to design and keep letters updated. An automated system to get the right letter to the right respondent is almost a prerequisite.

Another finding of these experiments was that small differences may have large effects. One example is the phrasing of the informed consent section, in small letters outside the main letter text. Another striking finding is the effect of something perhaps as trivial as the photo that is chosen for a flyer. The takeaway message from all these experiments is that small differences in wording and look and feel can have large effects. If at all possible, new or alternative versions of letters and other materials should be carefully tested.

All these experiments have had repercussions for the LFS letter. It is now considerably different from the LFS letter with which we started and has contained elements of many experiments: a new introduction, more appeals to altruism, headings for relevant subsections (on how to participate, the safety of data, and where to go with questions), another name for the survey for respondents (not 'Labor Force Survey', but 'work'), the offer of an incentive (although not the tested block note), and it is substantially shorter.

Although the form and contents of letters, flyers, and envelops can and do play an important role in securing (web) response, their influence is by far surpassed by the influence of a well-chosen incentive. The following sections describe this topic.

### 5.6.4 Incentives in Web and Mixed-Mode Surveys

Especially when cost considerations are driving the use of web surveys in a mixed-mode design, we want as many respondents in the web mode as possible. Giving or promising an incentive is one possible way to make this happen. The specific mode mix may determine which incentive is the most effective in increasing web response rates against acceptable costs. That may not always be the incentive that raises (web)response the most.

In the DCSS inventory (Blanke and Luiten, 2014), the use of incentives was very uncommon in the NSIs. In the years since, with the onset of web data collection, the situation has changed dramatically. The MIMOD inventory (Murgia et al., 2019) shows that nearly half of the European NSIs offer incentives. Mostly, these incentives are given conditionally, after responding.

Hundreds of experimental studies have shown that incentives are an effective tool for increasing response rates. Several meta-analyses have shown that incentives work in all modes: mail, web, and interviewer modes. Incentives are classified on two dimensions: the timing of giving them and whether they are monetary or non-monetary. Prepaid or unconditional incentives are offered to all sample persons or households, before the start of the survey. They can be a small gift, a gift certificate, or a small amount of money, enclosed in the invitation letter. Postpaid, conditional, or promised incentives are offered to those who respond to the survey. A distinction is further made between conditional incentives to all respondents or lottery incentives where a limited amount of incentives is raffled among respondents. Monetary incentives include vouchers, cash, or loyalty points that can be exchanged for money. A wide range of non-monetary incentives are described in the literature: bikes, tablets, trips, and mousepads, but also donations to charities or study results.

Several other considerations are relevant as well: the value of the incentive (from $1 to several hundreds of dollars or euros in cash or in gifts), and the mode of the survey (mail, web, telephone of face-to-face). In the following section a short summary is given of the literature on the effect of incentives on response rates, sample composition bias, and data quality in web and mixed-mode surveys. Subsequently, an overview is given of the incentive practice in European NSIs, and, finally, the results of a number of experiments by Statistics Netherlands with conditional, unconditional, and lottery incentives in mixed-mode cross-sectional and panel surveys are discussed.

Meta-analyses by Fox et al. (1988), Hopkins and Gullickson (1992), Church (1993), Edwards et al. (2002), and Jobber et al. (2004), and reviews by Singer, Van Hoewyk, Gebler, Raghunathan, and McGonagle (1999), and Singer, Van Hoewyk, and Maher (2000) have shown that in traditional survey modes, monetary incentives are 2–5 times as effective as non-monetary incentives, and that monetary prepaid incentives are 2–3 times as effective as monetary postpaid incentives. The effects of incentives are larger in self-administered (mail) surveys than in interviewer surveys. The analysis by Pforr et al. (2015) for ten German face-to-face studies showed that these findings were replicated many years later, and in different cultural circumstances. The 15–19 percentage points increase in response rates for unconditional monetary incentives in mail surveys are the same rates that were found in research with unconditional incentives in web surveys by Gajic, Cameron, and Hurley (2010) and several experiments by Statistics Netherlands that are described in the following sections.

Incentives play a role in increasing response rates but may also help pushing respondents to the cheapest mode in a mixed-mode design. Biemer et al. (2018) studied the effect of incentives in a 4 × 2 factorial design, where the mode initially offered (either web alone or both web and paper, sequentially and concurrent) was crossed with an incentive variation: either $5 prepaid and $10 promised or $5prepaid and $20 promised. The fourth design offered the web and paper options concurrently but offered an additional $10 if the respondents would answer in web, the choice+ design. This latter condition had the highest response rate in both the low and high incentive condition, it increased the response rate overall by 4.4 percentage points, and in addition successfully pushed respondents to web: wherein the concurrent protocol 27% of respondents answered in web, in the choice+ condition this was more than 64%. Interesting to note is that the costs per response in the web-only condition *decreased* as a result of offering $20 instead of $10. In addition, the costs in the choice+ group were not greater than the concurrent choice group, despite the additional $10 incentive.

While the results for unconditional incentives are unequivocal in showing that even incentives as small as $0.25 have a significant positive influence on response rates, results for conditional incentives are more diverse. Especially the literature on lottery incentives shows mixed findings. Singer and Kulka (2002) give an overview of the results of nine lottery experiments in mail surveys. Four of them had positive effect, but five more did not have an effect. However, the incentives in these experiments had very little value. Singer and Ye (2013) updated the review by including new experiments that tested lotteries in web surveys. Among the five new studies only two showed positive effects of lotteries. Bošnjak and Tuten (2003) compared $2 promised with $2 prepaid at first contact with a lottery of 2 × $50 and 4 × $25 and found that the lotteries were more successful than the other conditions. Göritz (2006, 2015) performed two meta-analyses, based to a large extent on web surveys in access panels that offer small prizes. She concluded that lotteries are usually mildly effective.

Gajic, Cameron, and Hurley (2010) compared a no-incentive condition with $2 included in the invitation letter, a low lottery cash draw of 10 × $25 and a high lottery cash draw of 2 × $250 in a web survey among the general public. They found that the prepaid incentive led to the highest response rate and the lowest dropout rate (+14.4 percentage points compared to no incentive, + 13.4 percentage points compared to low lottery + 8.3 percentage points compared to the high lottery). The highest dropout rate was found in the low lottery condition. Gajic et al. calculated the costs per complete record in the four conditions and compared those to the cost effectiveness of each, that is how much extra an incentive costs per additional completed survey. By this criterion it could be determined which incentive should be used to obtain the most completed surveys for a given budget. In this case that was not the incentive with the highest response rate, but the high lottery incentive. Gajic et al. conclude that prepaid incentives should be the incentive of choice when a high response rate is desired and costs are not a tight constraint. On the other hand, the high lottery is better suited to situations in which as many responses as possible should be obtained given a fixed budget.

Sauermann and Roach (2013) likewise experimented with the probability of winning and size of prize in a web survey among graduate students and postdoctoral researchers. A no-incentive condition was compared to five conditions that each had a total payoff of $500, but differed in the chance of winning and in the size of the prize (100 × $5, 50 × $10, 20 × $25, 10 × $50, and 5 × $100). Subjects were told neither about the size of the subject pool nor the expected number of respondents and so had no objective idea about the chance of winning. The response rate was highest for the condition with the largest prize and the lowest chance of winning: the response went from 25% for the no-incentive condition to 31% in the condition with the highest incentive. The no-incentive and the 100 × $5 conditions had the lowest response rates, but the 10 × $50 lottery did not significantly differ from those conditions. Sauermann and Roach conclude that a fixed budget for lottery prizes is more effective if used for a small number of large prizes than for a large number of small prizes. Göritz and Luthe (2013) experimented in three lotteries with cash prizes that were either paid in one lump sum or split into multiple smaller prizes. Response was higher with a lottery than with the control group (OR = 1.18), when raffling the payout in a lump sum (OR = 1.30) and with higher single prize sizes (OR = 1.02 per €10).

In longitudinal surveys, the effect of incentives may be different from those in cross-sectional surveys: the incentive may have a carry-over effect on other waves. Also, once an incentive is introduced, it may be difficult to withdraw it in later waves. Laurie and Lynn (2008) summarize the use of incentives in longitudinal surveys.

### 5.6.4.1 Effect on Sample Composition and Data Quality

Whether or not incentives improve sample representation, by attracting traditionally underrepresented groups, remains unclear. Several experiments have shown that incentives may improve representation (Dillman, 1996; Storms and Loosveldt, 2004; Gajic, Cameron, and Hurley, 2010; Berlin et al., 1992; Mack, Huggins, Keathley, and Sunsukchi, 1998; Singer, van Hoewyk, and Maher, 2000, Petrolia and Bhatacharjee, 2009; Suzer-Gurtekin et al., 2019). Other experiments have shown that incentives in mail and interviewer modes do not have much influence on sample composition (Brick et al., 2005; Cantor et al., 2008; Furse and Stewart, 1982; Goetz et al., 1984; James and Bolstein, 1990; Singer et al., 1999 in five studies, Lynn, 2020). The experiments that are described in Section 5.6.4.3 show that in some circumstances the use of incentives may actually deteriorate sample composition.

A number of studies have looked into the issue of data quality as a result of giving or promising an incentive. Theoretically, the incentive could reduce measurement errors if they create a sense of obligation to the researcher, that causes respondents to put in more effort. On the other hand, there is a risk of increased measurement error if the incentive convinces respondents who are only motivated by the incentive to participate. Indicators of data quality are skipping items, 'don't know' answers or refusals, straight-lining sets of items (giving the same answer; e.g., the middle answer in a grid of questions), speeding through the survey, early break off, and divergent scale scores. Singer, Van Hoewyk,

Gebler, Raghunathan, and McGonagle (1999) concluded in a meta-analysis of incentives in telephone and face-to-face surveys that incentives do not appear to affect the quality of responses, measured by item nonresponse or the number of words in response to open-ended questions. Another study by Singer, Van Hoewyk, and Maher (2000), however, showed that both promised and prepaid incentives reduced item nonresponse. Medway (2012) studied 12 indicators of respondent effort in 3 studies with unconditional incentives. She concluded that prepaid incentives had minimal impact on measurement error. Cole, Sarraf, and Wang (2015) studied data from surveys among a higher education population, in the National Survey of Student Engagement, who received a survey through their schools. Half of these schools offered incentives, mostly in the form of lotteries. The results show that both first-year and senior students had significantly less missing data, showed less straight-lining, took more time filling in the questionnaire, and showed better scale quality when an incentive was offered. Lemcke, Schmich, and Albrecht (2018) experimented with conditional incentives and a lottery in a mixed-mode (web and mail) survey among the general population. The incentives were a lottery of $100 \times €50$, stamps, and a voucher of €10. The €10 voucher diminished the number of missing values. The other incentives did not have any effect on either missing values, straight-lining, rounding, social desirable answers, and measurement invariance in scale questions.

The overall impression from this short summary is that incentives, be they unconditional, conditional or lottery incentives, either have no effect on data quality, or a small positive effect: respondents sometimes, but not always, take more time, have less missing data, report more fully in open questions, and show less straight-lining. The goal of attaining better data quality will not be the primary reason to include incentives in a study, but the fact that they either have no influence on data quality or may even increase data quality is reassuring for survey managers who want to use incentives to increase response rates.

### 5.6.4.2 *Incentives in European NSIs*

The MIMOD query (Murgia et al., 2019) showed that most countries differentiate the kind of incentive per survey. For example, 13 NSIs give an incentive for the burdensome Household Budget Survey, while the Labor Force Survey is rarely incentivized. Most of these incentives are conditional upon responding. A few countries offer unconditional incentives in a number of cases: the UK for the HBS; Sweden, also for the HBS; Switzerland for the first wave of SILC; and the Netherlands for the second wave of SILC. The kind of conditional incentives vary wildly from small gifts like a pen or a packet of coffee, to amounts like €10, €15, €30, and €50, to raffles of tablets and small amounts (€50) and large amounts (€750) of money. These different forms and amounts are undoubtedly related to the various cultural circumstances and prosperity levels in the countries. Regretfully, it is unclear if these incentive practices are based on experiments to find the optimal form and amount. Finland mentions that they have experimented with lotteries of gift cards and online coupons for meals, but that the results were unsatisfactory.

In the next sections a number of experiments are described with unconditional incentives, conditional incentives, and lottery incentives that were performed by Statistics Netherlands in the years since the introduction of web surveys, both in cross-sectional surveys and in the Labor Force Survey that has a panel wave structure. Attention is given to the response implications for the web survey and the follow-up telephone and/or face-to-face rounds, to sample distribution, response quality, and fieldwork costs.

### 5.6.4.3 Incentive Experiments by Statistics Netherlands

Soon after the introduction of mixed-mode experiments including web, Statistics Netherlands started experimenting with incentives. The general design with web as the first mode, followed by interviewer modes, made that substantial cost savings can potentially be attained by pushing as many people as possible to web. Experiments have been performed with three different kinds of incentives: unconditional incentives, raffled conditional incentives, and conditional incentives for all participants. In these experiments the impact of the incentive on response rates in the web part of the survey and the total response rate were studied, but also the impact on sample representativeness, data quality and fieldwork costs.

#### 5.6.4.3.1 Unconditional Incentives: €5 Gift Certificates

Three experiments were performed in three different surveys of the general population: the survey of Housing and Living Conditions, The Travel Survey, and the Survey of Social Cohesion. See Appendix 1 for a short description of these surveys. The three surveys had a sequential mixed-mode strategy in which persons are first invited in an invitation letter and two written reminders to participate in a web survey. Nonrespondents were followed up by telephone mode one month after the initial invitation letter if a telephone number could be found, or else by face-to-face mode two months later. Nonrespondents in the telephone mode were not transferred to face-to-face. Sample persons were drawn from the Municipal Registries and letters were addressed by name.

One of the first experiments, in the 2011 survey of Housing and Living conditions, was not strictly experimental: an entire sample month of the six-month fieldwork received an unconditional €5 gift voucher. Comparing with the previous five months gave an impression of the impact of the incentive, see Table 5.4. The incentive led to a substantial increase of about 12 percentage points in the web response, and also a small increase in the CATI response, one month later. No effect was seen in the CAPI response, two months later.

Striking in this experiment was the reaction of some subgroups to the incentive, where the incentive led to a substantial *lowering* of response rates in first-generation immigrants of Turkish and Moroccan backgrounds (see Table 5.5).

An experiment in one month of the continuous Travel Survey in 2014 showed comparable response results, see Table 5.6. The incentive led to an increase of almost 17 percentage points in web, but also to higher response rates in the subsequent CATI and CAPI rounds.

In a subsequent experiment in 2016 in the Travel Survey, a no-incentive condition was compared with a €5 unconditional incentive and a raffle of iPads. See the next section for a description of the raffle condition. See also Table 5.7.

Again, a comparable response gain of almost 19 percentage points in web was found for the unconditional incentive. An additional response gain in the subsequent CATI round, but a slight decrease in CAPI, led to a total response increase of 12 percentage points with

**TABLE 5.4**

Effect of a €5 Unconditional Incentive on the Web, Telephone, and Face-to-Face Response of the Housing and Living Conditions Survey

| | Response 5 Previous Months | n | Response Incentive Month | n |
|---|---|---|---|---|
| web | 25.2 | 61,199 | 37.5 | 12,261 |
| CATI | 47.7 | 17,532 | 49.6 | 4038 |
| CAPI | 41.5 | 16,948 | 40.3 | 2566 |

**TABLE 5.5**

Effect of a €5 Unconditional Incentive on the Total Response of the Housing and Living Conditions Survey, by Ethnic Background

|  | Response Previous 5 Months | n | Response Incentive Month | n |
|---|---|---|---|---|
| rest population | 58.1 | 59,624 | 67.5 | 11,935 |
| 1e generation Turkish | 49.1 | 1,079 | 31.3 | 207 |
| 1e generation Moroccan | 41.6 | 864 | 37.9 | 174 |

**TABLE 5.6**

Effect of a €5 Unconditional Incentive on the Web, CATI and CAPI Response of the Travel Survey 2014

|  | Incentive | | No Incentive | |
|---|---|---|---|---|
|  | Response | n | Response | n |
| web | 37.4 | 551 | 20.7 | 551 |
| CATI | 49.0 | 247 | 45.7 | 194 |
| CAPI | 57.0 | 128 | 49.2 | 177 |

**TABLE 5.7**

Effect of a €5 Unconditional Incentive and a Raffle of iPads on the Web, CATI, and CAPI Response of the Travel Survey 2016

|  | No Incentive | | €5 Unconditional | | iPad Raffle | |
|---|---|---|---|---|---|---|
|  | Response | n | Response | n | Response | n |
| web | 18.0 | 2530 | 36.7 | 1500 | 23.1 | 1500 |
| CATI | 47.6 | 1128 | 54.3 | 499 | 50.6 | 625 |
| CAPI | 47.8 | 812 | 45.8 | 393 | 50.9 | 479 |
| Total | 54.6 | 2530 | 66.8 | 1500 | 60.5 | 1500 |

the unconditional incentive. The lottery incentive increased web response by 5 percentage points and the total response with 6 percentage points.

Analyses of sample composition again showed that the unconditional incentive has a significantly lower effect with people from non-western migration background and a significantly higher effect with people in higher-income groups. The conditional incentive interacted less strongly with background variables. The incentives did not have any influence either on measures of data quality or on substantial variables.

### 5.6.4.3.2 *Conditional Incentives: Raffles*

The unconditional incentive has a large positive influence on the response rates but is obviously also very expensive. A lottery incentive on the other hand would not increase response to the same extent but would be far less costly. Even though the literature on lottery incentives shows mixed results, the literature also suggests the conditions under which a lottery incentive could be a success. The raffle design should meet three preconditions:

- The raffled prize should be large (Stevenson, Dykema, Cyffka, Klein, and Goldrick-Rab, 2012; Laguilles, Williams, and Saunders, 2011; Laguilles, Williams, and Saunders, 2011; Sauermann and Roach, 2013; Gajic, Cameron, and Hurley, 2010).

As a thank you for your help, you will have a chance to win **one of the iPads or gift cards worth EUR 400** that we are providing. If you win, you can choose which prize you want to receive.

*How do you participate?*
We ask you to fill in where you have been on one specific day. For you this is a **Friday (...etc...)**

**FIGURE 5.3**
Depiction of the choice-incentive in the invitation letter.

- The prize should be salient in the invitation letter (Zang, Lonn, and Teasley, 2016).
- Respondents should learn of their winnings right after filling in the questionnaire. The latter precondition was based on the findings of Tuten, Galešić, and Bošnjak (2004), who found that immediate versus delayed notification (one month later) led to significantly higher response rates for the immediate condition.

These conditions were used to design the lottery experiments. In the experiments described below, the 'large prize' was operationalized as a recent iPad or the equivalent sum in gift vouchers, and 'salience' in the invitation letter is accomplished by depicting a number of iPads in the margin of the invitation letter (see Figure 5.3 for an example). The third condition, letting the respondent know right away if he or she won the iPad, was accomplished by pre-determining in the sample who will win if they respond. One iPad is raffled per 2,000 sample persons, with a minimum of 2 iPads. This to be able to speak of a lottery of iPads in the invitation letter. In larger surveys, where a substantial number is raffled, the number is mentioned in the invitation letter. Qualitative research has shown that people gauge their chance of winning to be higher in that circumstance. The chance of winning is not made explicit in the invitation letter but can be found on the survey website.

Experiments were performed in various surveys and populations. It was found that a raffle of iPad mini's among a general population sample of 10–20-year-olds led to an increase in web response rates of 12 percentage points: from 21% in the control group to 33% in the experimental group. Experiments with raffled iPads in other populations and the entire age range consistently led to increases of about 7 percentage points. An experiment with raffled vouchers worth €250 increased response to a lesser amount, i.e., 5 percentage points. Although theoretically the monetary incentive should have led to a higher increase than the iPad lottery, in this experiment the second precondition of salience was not adhered to. Instead of the picture of iPads that is prominent in the invitation letter, the raffle of vouchers was only mentioned in the text. In an experiment in 2020 this was remedied: in three conditions the iPad was compared with a voucher of the same amount, and with a choice between those. See Figure 5.3 for an impression of how this was depicted and made salient in the letter.

The experiment showed that the voucher condition had a slightly higher response rate than the iPad condition, but that the highest web response rate was attained in the choice condition: 25.9%, 26.8%, and 27.9% web response, respectively (de Regt, 2020).

*5.6.4.3.3 Conditional Incentives: Gift Certificates for Respondents*
In certain circumstances a gift upon completion may be preferable, especially in surveys where relatively much is asked from the respondent. In the Survey of Income and Living

Conditions (SILC), with a web-CATI design, an experiment in 2016 with promised €10 gift certificates increased web response by 9 percentage points and overall response by 11 percentage points (see Table 5.8). The incentive increased the CATI response somewhat but not significantly.

In addition, the incentive significantly increased the number of people that were recruited for subsequent waves, both in web and in CATI, see Table 5.9.

### 5.6.4.3.4 Sample Composition

In the experiments with unconditional and conditional promised incentives it was found that some subgroups react more strongly to the incentive than other subgroups and some groups may actually show decreasing response rates. This resulted in increased differences in response propensity between subgroups. Differential reaction to incentives may be an expedient result, if the incentive brings in respondents of difficult groups. Mostly this was found not to be the case, however. In earlier experiments in CAPI (Wetzels, Schmeets, van den Brakel, and Feskens, 2008), it was found that an unconditional incentive of postal stamps could increase overall response by almost 8 percentage points, but the incentive did not have any effect on persons of non-Western ethnic background. We saw the same phenomenon in the incentive experiments in both mixed-mode and unimode web surveys. In general it can be said that we find a high positive correlation between the response rates of subgroups in the control group and the response gain in the incentive conditions, for both the unconditional and the conditional promised incentives: the higher the response without incentive, the stronger the reaction to the incentive. This phenomenon is especially strong in relation to income. In a mixed-mode setting with interviewer follow-up, a differential effect of incentives can be counterbalanced, see also Chapter 4, Mode-Specific Selection Effects. But in web-only surveys, or web-mail surveys, this may be an unwanted result.

The raffled incentives seem to suffer less from differential influence. For example, in the Travel Survey, the increase in web response with the unconditional incentive was 21 percentage points for native Dutch persons, while the increase for people from the non-Western

**TABLE 5.8**

EU-SILC Response by Unconditional €10 Incentive by Mode

| Response Rates | No Incentive | n | Incentive | n | p |
|---|---|---|---|---|---|
| web | 17 | 8135 | 26 | 8133 | *** |
| CATI | 34 | 2699 | 36 | 3280 | ns |
| total | 27 | | 35 | | *** |

* p < .05, ** p < .01, *** p < .001, ns not significant

**TABLE 5.9**

Panel Recruitment by Incentive and Mode

| Panel Recruitment | No Incentive | n | Incentive | n | p |
|---|---|---|---|---|---|
| web | 67 | 8135 | 79 | 8133 | *** |
| CATI | 80 | 2699 | 84 | 3280 | * |
| total | 72 | | 81 | | *** |

* p < .05, ** p < .01, *** p < .001, ns not significant.

background was 6 percentage points. The increase as a result of the iPad raffle in the same survey was more homogeneous: 6 percentage points increase for native Dutch persons against 3 percentage points for persons with the non-Western background. Likewise, in the survey amongst children and young adults, there was no interaction between ethnic background and raffled incentive condition.

### 5.6.4.3.5 Data quality

In all experiments data quality of web responses with and without incentive were compared. The rate of missing items, the time respondents took to fill in the questionnaire, and in some cases the amount of straight-lining was studied. In none of the experiments did we find an effect of the incentive on these measures of data quality, not even in the experiments amongst children and young adults who would learn right after filling in the questionnaire if they had won an iPad. If anything, there were more indications that data quality was higher in the incentive conditions.

### 5.6.4.3.6 Fieldwork Costs

Giving all sample persons a gift certificate of €5 is obviously very costly. However, in a mixed-mode data collection where face-to-face is one of the modes, the increase in web response is such that more than that amount could be saved in fieldwork costs. In addition, the overall increase in response allows for a smaller sample, resulting in substantially lower fieldwork costs. The promised incentive, conditional upon the response, is cheaper than the unconditional incentive. However, the gain in response was not high enough to overcome the costs of the incentive in the web-CATI design of the EU-SILC experiment. On the contrary, the incentive experiment turned out to be 16% more expensive than the arm without incentive. Finally, the raffled incentive of either iPads or large gift certificates is very cost effective. Although the increase in (web) response is not so high as with the unconditional incentive, the costs are a mere fraction. Statistics Netherlands has chosen this kind of incentive as the 'default' incentive in new surveys.

### 5.6.4.4 An Incentive Experiment in the Longitudinal Labor Force Survey[2]

The incentive of raffled iPads that was described in Section 5.6.4.6 was also tested in the Labor Force Survey. In contrast to the person samples in the cross-sectional survey experiments, the LFS is a household survey with a wave structure. Households are interviewed in five consecutive quarters before rotating out of the sample. The survey is voluntary. Sample addresses are drawn from the Municipal Personal Records Database which contains personal details of everyone who lives in the Netherlands. The target population consists of people residing in the Netherlands, aged 15+ and living in private households. Only when all household members (15+) complete the questionnaire, the household is considered to have responded. Proxy answering is allowed.

The LFS uses a sequential mixed-mode strategy in which households are invited in an invitation letter and two written reminders to participate by web. Nonrespondents are followed up by either telephone or face-to-face interviews, depending on the availability of a telephone number and the size of the household: larger households are interviewed face-to-face. Nonrespondents in the telephone mode are not transferred to face-to-face.

Not all nonresponding households of the web phase are followed up in the subsequent mode, however. In order to make optimal use of the cheaper web mode, and also to be able to keep the number of sample units that go to CAPI and CATI stable each month, a relatively large sample is drawn for the first web phase, of which a subsample is drawn for the follow-up phases. The subsample is stratified by interviewer region and known telephone, but is otherwise random. About half of the nonrespondents in web are not followed up in other modes.

Table 5.10 shows the wave 1 response rates for three months for the three modes and the total response after three waves. The total response in this table is low, as it is the response from the initial sample, not compensating for the subsampling of CATI and CAPI addresses, or adjusting for ineligibility in CATI and CAPI. Would we compensate for subsampling and ineligibility, total response rate would amount to 50–55%. Weighting for inclusion probability, therefore compensating for the oversampling of groups in the population with low response propensities and the undersampling of people of 65 years and older would increase the response rate further. The calculation shown here was deemed to be the simplest one to compare the two conditions.

As can be seen, the incentive leads to a significant increase in the response of almost three percentage points in web but leads to an equally large decrease of response in CATI. In CAPI there is also a decrease in response, but this does not reach significance. Because the web sample is far larger than that of the other two modes, the incentive still leads to a significantly higher overall response. The increase in web response is smaller than in person surveys. It is probable that the household component is responsible for this finding. The decrease in response in the other modes is remarkable.

The interpretation of this finding is that some people who would otherwise have answered in CATI or CAPI were drawn to web by the incentive. Even if the incentive would not have had an effect on the overall response rate, this outcome would still have been desirable, to the extent that it pushes people to the cheapest mode.

Another important aspect of the success of an incentive design in a panel survey is the number of households that are recruited for subsequent waves. Table 5.11 shows the

**TABLE 5.10**

Wave 1 Response in Web, CATI, CAPI, and Overall, by Incentive

|  | No Incentive | | Incentive | | |
|---|---|---|---|---|---|
|  | Sample n | Response % | Sample n | Response % | p |
| CAWI | 16.335 | 21.4 | 19.965 | 24.3 | *** |
| CATI | 3.159 | 35.7 | 1.212 | 32.7 | * |
| CAPI | 2.932 | 39.0 | 3.527 | 37.7 | ns |
| Total[1] | 16.335 | 35.3 | 19.965 | 37.0 | *** |

$* p < .05, ** p < .01, *** p < .001$, ns not significant

**TABLE 5.11**

Panel Recruitment by Mode and Incentive Condition

|  | No Incentive | | Incentive | | |
|---|---|---|---|---|---|
|  | n responses | Recruited | n responses | Recruited | p |
| CAW I | 3485 | 67,5 | 4841 | 74,3 | ***: |
| CATI | 1125 | 91,5 | 1212 | 90,6 | ns |
| CAPI | 1130 | 90,5 | 1304 | 91,1 | ns |
| total | 5740 | 76,7 | 7357 | 80,0 | *** |

$* p < .05, **p < .01, ***p < .001$, ns not significant

**TABLE 5.12**

Response Rates by Incentive in Wave 2 and Wave 3

|         | No Incentive | | Incentive | | |
| --- | --- | --- | --- | --- | --- |
|         | Sample n | Response % | Sample n | Response % | P |
| wave 2 | 3.944 | 76.6 | 3.877 | 76.4 | ns |
| wave 3 | 3.198 | 83.1 | 3.166 | 83.4 | ns |

* p < .05, ** p < .01, *** p < .001, ns not significant

percentage of households in the three modes that express a willingness to be contacted again and provide their telephone number. The incentive leads to a substantial increase in the number of households who are recruited for wave 2.

Table 5.12 shows results for the wave 2 and wave 3 responses. The mode in these waves is CATI. The table shows that there is no effect at all of the incentive in the later waves.

It is to be expected that an incentive that you did not win in the previous wave does not influence the decision to take part in wave 2 or 3. The lack of effect of the incentive has an important implication: it means that the higher number of households that were recruited do not drop out in the second wave, even though they mostly did not win the incentive.

Like for the cross-sectional experiments, the effect of the incentive on sample distribution, and on data quality was studied. It was found that the incentive increases representativeness for several important variables, i.e., age and income. For some other variables no difference in representativeness was found. Over all variables, the design with incentives showed less variation in subgroup response propensities than the design without incentives.

The incentive leads to higher data quality: especially the number of 'don't know' answers diminishes significantly. The effect is strongest in web.

Although the literature shows mixed findings on the effect of lottery incentives, Statistics Netherlands has found a way to consistently make these incentives work, both in cross-sectional person surveys and longitudinal household surveys.

## 5.7 Summary

The main takeaway messages from this chapter are:

- Circumstances like tradition, complexity of the subject matter, survey length, the kind of sample, or the presence of a panel structure influence the optimal (mixed mode) design for a survey.
- Starting with a cheaper mode and subsequently using more expensive modes for nonresponse follow-up will result in greater costs saving than other sequences in cross-sectional surveys. In panel surveys it may make sense to start with more expensive modes of data collection for the first wave and use less expensive modes in later waves.

- Starting with less expensive modes may lead to lower response rates. This tendency can be counteracted by measures like additional reminders or reminders in a different mode, and careful survey design.
- The evidence in studies comparing sequential with concurrent mixed-mode designs is inconclusive as to the effects on response and bias. Differences in design and population may play a role in the varying findings.
- Concurrent designs have more complicated logistical and planning challenges. In addition, an IT system is needed where respondents can be easily shifted from one mode to another.
- Using web surveys puts extra strain on the communication strategy, as people need instruction on how to access the survey. The search for the optimal strategy is still ongoing. It is clear that the survey organization needs to make all steps of the process as simple as possible.
- Communication content may have a differential impact on various groups. If possible, study group differences in susceptibility, and use an adaptive strategy to target specific groups with specific communication means.
- Both conditional and unconditional incentives play an important role in increasing web response, but a risk of increased bias exists. Lottery incentives may be used effectively if applied under the right conditions.
- Incentive strategies that differentiate kind, timing, and level of the incentive for different groups may help reduce bias but may have ethical or legal restrictions.

## Notes

1. The European Social Survey (ESS) is an academically driven cross-national survey that has been conducted across Europe since its establishment in 2001. Every two years, face-to-face interviews are conducted with newly selected, cross-sectional samples. The survey measures the attitudes, beliefs, and behavior patterns of diverse populations in more than thirty nations.
2. This work was supported by Eurostat Grant 07131.2017.003-2017.596 'Quality improvements for the Labour Force Survey'. See Luiten and Goffen (2018) for a complete report.

# 6

## Mixed-Mode Questionnaire Design

### 6.1 Introduction

This chapter focuses on designing questionnaires for mixed-mode surveys. Questionnaire design is one of the four stages of survey design in the *plan-do-check-act* as discussed in Chapter 2, Designing Mixed-Mode Surveys (see Figure 6.1). Mixed-mode questionnaire design in this book is defined as designing questions that are intended to measure the same concepts and variables but are administered in different modes within one survey. As discussed in Chapters 2, Designing Mixed-Mode Surveys, and 3, Mode-Specific Measurement Effects, the way respondents understand and answer questions can be affected by mode-specific characteristics such as the presence of an interviewer, whether they hear or read questions and answer options, or how questions are visually presented on a screen or on paper. The mode of administration may result in mode-specific measurement error and ultimately in mode effects. Mixed-mode questionnaire design is an important means to prevent or reduce mode-specific measurement errors.

Section 6.2 provides a brief overview of the general goals and challenges of questionnaire design. Section 6.3 continues part of the discussion of Chapter 3, Mode-Specific Measurement Effects, and focuses on how questionnaire design can affect mode-specific measurement errors. Section 6.4 discusses general considerations for mixed-mode questionnaire design. This includes a list of criteria to assess the smartphone fitness of questionnaires. The topic of Section 6.5 is the pre-testing of mixed-mode questionnaires. Next, in Section 6.6, a set of guidelines for mixed-mode questionnaire design is provided. Finally, Section 6.7 summarizes the main points of this chapter.

### 6.2 General Goals and Challenges in Questionnaire Design

#### 6.2.1 Why Questionnaire Matters

As Krosnick and Presser (2010) put it, the questionnaire is the heart of the survey, scripting the conversation between researchers and respondents. As such, it is not surprising that questionnaire design is an important topic for both survey methodologists and survey practitioners. Numerous handbooks have been written about questionnaire design and there is a large and growing body of research focusing on how various characteristics of questions and questionnaires affect respondents' behaviors and data quality (e.g. Fowler, 1995; Fowler and Cosenza, 2008; Tourangeau, Rips, and Rasinski, 2000; Krosnick and

DOI: 10.1201/9780429461156-6

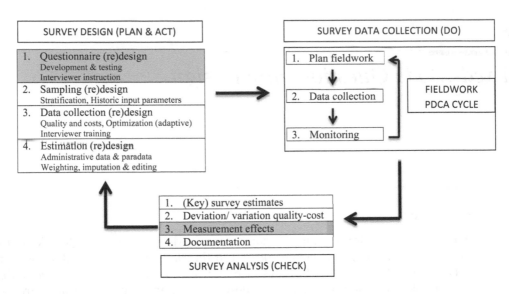

**FIGURE 6.1**
The plan-do-check-act cycle for surveys.

Presser, 2010; Dillman, Smyth, and Christian, 2014; Smyth, 2016; Schaeffer and Dykema, 2020). In survey practice, designing, programming, pre-testing, and evaluating questionnaires is an important and time-consuming part of the research process.

As discussed in Chapter 2, Designing Mixed-Mode Surveys, questionnaire design can affect unit-nonresponse, item nonresponse, and measurement error. The main goal when designing a questionnaire is designing questions that respondents are willing and able to answer, that are understood and answered in the same way by all respondents, and in the way as was intended by the researcher. Or, stated differently, to produce a valid and reliable measurement of the concepts of interest with minimal measurement error. However, it is good to bear in mind that besides data quality, there are at least two other important reasons to invest in questionnaire design that are less often discussed in the literature: costs and the reputation of the survey organization. First, a good questionnaire design takes time and other resources, but also saves costs further along the process. An unappealing or unclear questionnaire will require more reminder letters or a follow-up in a more expensive mode. Difficult or annoying information can cause a peak in helpdesk calls and complaints that need to be processed. Also, problematic questions may lead to more effort needed for data processing and cleaning. Second, data collection communication is part of the survey organization's corporate communication. The professionalism and user-friendliness of the letters and questionnaires that are sent to thousands of members of the public can impact the reputation of a survey organization.

## 6.2.2 Striking a Balance between Conflicting Goals and Stakeholders

Designing questionnaires usually involves striking a difficult balance between multiple, sometimes conflicting, interests. For instance, it would make sense to measure the financial position of households through several questions that remind the respondent of all possible forms of income for each household member. However, a greater number of questions also brings additional costs (e.g., for more design time and interview time) and sometimes a greater respondent burden. An excessive respondent burden can lead to careless answers, item-nonresponse, or even breaking off the questionnaire. When developing questionnaires, several types of questionnaire users and stakeholders must be taken into consideration. Some stakeholders, like the respondent and the client, will be obvious. Others might be easily overlooked. Below, a brief summary of the most common stakeholders and their needs is given.

### Clients (Partly Overlap with Data Users)

Clients want to have data on a specific subject, preferably as quickly as possible, as cheap as possible, comparable with a previous year or other similar research. Typically, clients also like to be flexible in the design process and to be able to make last-minute changes.

### Respondents

Respondents need a questionnaire with relevant questions that are non-threatening, easy to understand, and easy to answer. Respondents should feel respected and important: their answers matter and the questionnaire is tailored as much as possible to their individual situation. Completing the survey questions should take as little effort as possible. The questionnaire should preferably be brief, easy to navigate (if self-administered), and interesting to answer.

### Interviewers

Conducting a standardized interview can be a very demanding task. Interviewers should conduct the interview exactly as scripted, but also need to keep respondents as engaged and motivated as possible. The questionnaire should facilitate the work of interviewers as much as possible. To a large extent, interviewers' needs are similar to those of the respondents. However, an interviewer will have some additional needs. In general, it is important that the questionnaire allows for an interview that will have an as natural and logical conversational flow as possible. Questions should be easy to be read aloud and communicate the expected type of response. Interviewer instructions and possibly additional training may be needed to provide interviewers with the necessary information and skills to work with a specific questionnaire. Also, if interviewers work with different questionnaires, some standardization over questionnaires will make their task easier and help them to focus as much as possible on the interview.

### Programmers

Programmers need a survey that can be programmed with the tooling that is available and within the given time frame. The design of the survey should be clear, well thought out, and error-free. A survey that is thoroughly designed in advance will help to reduce the workload of programmers and prevent re-work.

### Data Users (Partly Overlap with Clients)

There are various types of data users: the data analysts who prepare the raw data for analysis, the researchers analyzing the data, and users of the research results. They all need a questionnaire that yields reliable and valid data that answer the research questions. For ongoing or repeated surveys, a primary concern for the data users is the comparability of the measurement over time. Users actually working with the raw data need a complete dataset with the relevant metadata that complies as much as possible with existing procedures, software, and syntaxes. Changes in the questionnaire over time should be clearly visible, amongst other things, in order to correctly adjust existing syntax. Using several questions and filters in the questionnaire to calculate one single variable may facilitate the response process, but complicates analysis. Eliminating mode differences in the data facilitates the analysis.

### *Project Managers*

The project manager responsible for the survey needs a questionnaire that can be developed within a certain time frame, fitting within the overall planning of other projects and within a certain budget. Mostly, this means finding an optimal balance between quality, costs, and timeliness. The persons responsible for the data collection need a survey that delivers metadata which allows for monitoring of the progress of the project. This includes auxiliary information like identification keys for sample units or contact information in case an evaluation or follow-up study is needed.

### *Questionnaire Designers*

The questionnaire designer should consider all stakeholders and their needs. These needs can be conflicting. For example, even simply altering the wording of a question in order to make an obvious improvement might cause a break in time series or interfere with the analysts' data processing software. Also, a questionnaire that is perfectly tailored to the respondent will take too much time to design and program. If the survey is dull or too difficult, respondents will not participate or give unreliable responses, which negatively affects the quality of the data. However, if the questions do not comply with the researcher's need, there will be no need to conduct the survey at all.

### 6.2.3 Questionnaire Design as an Iterative Process

Questionnaire design is part of the *plan-do-check-act cycle for surveys* as discussed in Chapter 2, Designing Mixed-Mode Surveys. Figure 6.2 zooms in at questionnaire design as an iterative process (see also Brancato et al., 2006; Giesen et al., 2012). The questionnaire design process starts with specifying the objectives and concepts of the research domain and choosing an overall research design. Only if it is known what should be measured for which units, can one start designing a questionnaire. It is considered good survey practice

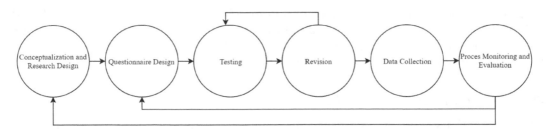

**FIGURE 6.2**
Questionnaire design as a process. *Source*: adapted from Brancato et al. (2006).

to test questionnaires prior to the data collection in various ways (e.g., European Statistical System Committee, 2018; U.S Office of Management and Budget, 2006). Usually, after testing, one or more revisions are necessary before the data collection can start. During and after the data collection the quality of the data collection should be monitored and evaluated. Especially for official statistics, where questionnaires often are used for a long period of time, regular evaluation of how the questionnaire is performed in the field is important. Typically, this evaluation focuses on the effectiveness of the data collection (how well does the questionnaire (still) measure the intended concepts) and the efficiency (the costs of the data collection). Collecting paradata during fieldwork, such as data on the mode and device of data collection, is paramount for gaining insight into whether and for which groups of respondents question items did not perform well. Obviously, this type of paradata is also crucial for assessing mode-specific measurement errors.

## 6.3 Questionnaire Design and Mode-Specific Measurement Errors

### 6.3.1 The Cognitive Response Process Model

Chapter 3, Mode-Specific Measurement Effects, discussed how mode-specific measurement effects result from the interplay of mode features, characteristics of the survey and its items, and the characteristics of the respondent. To further elaborate how questionnaire design can affect measurement errors in general, and mode-specific ones in particular, let us return to the cognitive response process model (e.g. Tourangeau, Rips, and Rasinski, 2000). This model (see also Figure 6.3) distinguishes four steps in the cognitive response process that respondents may go through when they answer a survey question:

1) *Comprehension*: Interpreting a question, identifying its meaning;

2) *Retrieval*: Collecting the information relevant for answering the questions, either from memory or from other sources (e.g. files, household members);

3) *Judgment*: Assessing and integrating the retrieved information into an 'internal answer'; in this step the respondent may decide to use a specific heuristic to arrive at an answer; and

4) *Reporting*: Mapping the 'internal answer' onto the response format, e.g., choosing between predefined answer categories or reporting a number in the requested quantity; in this step the internal answer may be edited, for example, for reasons of social desirability.

Few respondents will always run through all steps consciously. When asked, 'How old are you?' most people will have an answer ready quickly. It may take more effort to understand and answer a more complex question like, 'Did you contact or interact with public authorities or public services over the internet for private purposes in the last 12 months for the following activities: a) Obtaining information from websites or apps b) Downloading/printing official forms c) Submitting completed forms online?' For more complicated questions respondents may need to reiterate some steps in the response process.

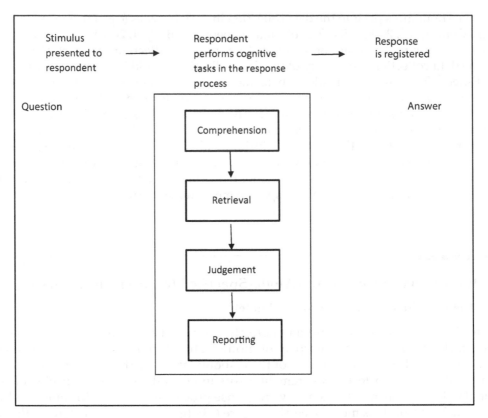

**FIGURE 6.3**
Cognitive response process Tourangeau et al. (2000).

Measurement errors arise when respondents do not perform all steps of the cognitive response process in the way intended by the researcher. As discussed extensively in Chapter 3, Mode-Specific Measurement Effects, modes have different features that can affect the response process. These features are related to the presence or absence of an interviewer who can help and motivate the respondent, but may also invoke social norms about desirable answers, the speed and pace of the response process (with telephone interviews typically having the highest pace), the presentation of the questionnaire (aural or visual), and the timing (the freedom to complete the questionnaire anytime).

### 6.3.2 Main Sources of Mode-Specific Measurement Error

Two well-known sources of measurement error are affected by mode: socially desirable answering behavior and satisficing.

Socially desirable answering occurs when respondents adapt their true answers to a question to present themselves in a favorable light (Tourangeau et al., 2000). This tendency may lead to either underreporting of behavior (e.g. alcohol use) or overreporting (e.g. segregation of waste). The literature shows rather consistently that interviewer-administered modes are more prone to socially desirable answering than self-administered modes (see e.g. Tourangeau and Yan, 2007; Kreuter, Presser, and Tourangeau, 2008). Related, but conceptually different, sources of measurement error are positivity bias (the tendency to give a positive rating; Tourangeau et al., 2000) and the 'Mum about Undesirable Messages' (MUM) effect (reluctance to communicate information that is likely to be undesirable for the receiver; Ye, Fulton, and Tourangeau, 2011). Ye et al. (ibid) performed a meta-analysis of 18 experiments comparing telephone interviews and other data collection modes. One of the goals of their study was to assess if answering behaviors found in telephone surveys should be attributed to extreme response style (tendency to select the most extreme answer options; Roberts, 2016) or to positivity bias. Their results show that respondents in telephone interviews are more likely to choose extreme positive answers than respondents in other modes, but that this effect is not found for more negative answer options.

Respondents do not always go through all steps of the response process as thoroughly as required, a phenomenon known as satisficing (Krosnick, 1991). This suboptimal engagement with the questionnaire may range from weak satisficing, when steps of the response process are only performed superficially, to strong satisficing, when one or more steps are completely skipped. The result has a broad range of manifestations, such as primacy effects (tendency to select the first acceptable response option), recency effects (tendency to select one of the last response options), acquiescence (tendency to agree), and selecting non-substantive answers (e.g., don't know). The tendency to satisfice is assumed to be affected by task difficulty, respondent ability, and respondent motivation (ibid). Roberts and colleagues (2019) systematically analyzed 141 methodological empirical studies that studied response quality in relation to predictors of satisficing. A majority of the studies that investigated task difficulty, motivation, and/or respondent ability in relation to satisficing behaviors found significant effects (respectively 74%, 68%, and 61% of the studies). Further, the review shows most empirical support for satisficing manifesting as primacy/recency and selecting don't know and least support for acquiescence and choosing middle alternatives in rating scales. As discussed in Chapter 3, Mode-Specific Measurement Effects, satisficing may be affected by mode in various ways, e.g., the presence of an interviewer can motivate and help the respondent, or the fast pace of CATI may make the response process more difficult.

### 6.3.3 Even Small Design Differences May Matter

Questionnaire designers should be aware that respondents may use all types of information provided when answering survey questions. They may take cues about the meaning of questions or the required level of detail from an example that was used in the advance letter, the location of a question in the questionnaire, the order of the answer options, or the size of an answer field. Several studies indicate that respondents are more inclined to use these types of contextual information when understanding the question or providing the correct answer is more difficult (e.g., Holbrook, Krosnick, Moore, and Tourangeau, 2007; Galesic and Bosjnak, 2009).

Redline and Dillman (2002) distinguish four types of 'language' used in questionnaires: verbal, numeric, symbolic, and graphical (see Figure 6.4). A large body of research has shown that any change in these types of language (or design elements) in a questionnaire may alter survey outcomes (e.g., Redline and Dillman, 2002; Toepoel, Das, and van Soest, 2009; Toepoel and Dillman, 2011) For example, Beatty, Cosenza, and Fowler (2019) show how relatively small changes in wording in telephone surveys can affect both research results and respondent behavior, such as asking for clarification. One of their experiments focuses on a question about the frequency of doctor contacts. One version of the survey question asked how many times 'have you seen or talked with a primary care doctor' and the other version asked 'have you seen or talked on the telephone with a primary care

| Type of language | Examples |
|---|---|
| Verbal | Textual items such as headings, questions and explanations. |
| Numeric | Numbering of blocks, questions, response categories and scale points etc. |
| Symbolic | Check boxes, summation signs, arrows to the next question, a clickable icon for help text etc. |
| Graphic | Font size, brightness, colour, position on the page, question spacing etc. |

**FIGURE 6.4**
Four types of language in questionnaires *Source*: Redline and Dillman (2002).

doctor'. The experiment showed that just adding the words 'on the telephone' significantly increased the percentage of respondents reporting no doctor contacts in the last 12 months from 9.2% in the first version to 24.7% in the second version of the question. This experiment confirmed findings from cognitive testing that several respondents had interpreted the question as asking only about telephone contacts. Obviously, the type of 'language' used in questionnaires varies over modes, especially between self-administered and interviewer-administered modes.

To understand how visual design affects the response process, Tourangeau, Couper, and Conrad (2004, 2007, 2013) provide five useful heuristics:

1. Middle means typical: respondents see the middle option of an answer scale as either representing the conceptual midpoint or the most typical answer and may use this as a reference point when deciding where to place themselves on an answer scale;

2. Left and top mean first: respondents expect some logical order, progressing from (depending on the scale orientation) from the left/top to the right/bottom;

3. Near means related: respondents interpret design elements (e.g., questionnaire items and instruction) to belong together based on their physical proximity;

4. Up means good: respondents expect a more positive evaluation/value for items that are higher in an options list; and

5. Like means close: respondents assume that visually similar options are also conceptually close.

See Dillman et al. (2014) for an extensive discussion of the impact of various visual design choices in questionnaire design.

### 6.3.4 Device-Specific Measurement Errors

Traditionally, web survey questionnaires were designed for use on a PC (desktop or laptop). But with the emergence of mobile devices, most web surveys have become mixed-device surveys. As shown in Figure 6.5, in 2020 around 20–25% of the first logins in Statistics Netherlands' continuous surveys are from smartphones and around 10–8% from

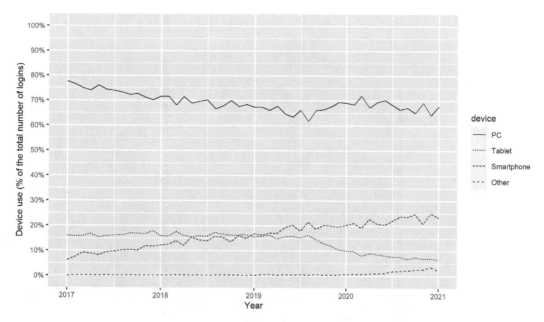

**FIGURE 6.5**
Device use at first login – continuous CAWI household/person surveys Statistics Netherlands 2012–2020 (Labor Force Survey, Health Survey, Consumer Confidence Survey (since 2017), Social Cohesion Survey, Life Style Survey (since 2015).

tablets (other devices may be game consoles or smartwatches, but are mostly new makes of smartphones that are not recognized; see Roberts and Bakker, 2018 for a more detailed description and analysis of device-specific logins and break-offs Statistics Netherlands surveys).

Couper, Antoun, and Mavletova (2017) distinguish four sets of device-related characteristics that may affect measurement error: (1) size of the display screen, affecting, e.g., how much text is visible and the ease of reading text; (2) technology features, e.g., connectivity (speed and reliability) and the data entry interface (e.g., touchscreen or keyboard); (3) user characteristics and behavior that affect the willingness and ability to complete a mobile web survey, e.g., familiarity with the device; and (4) context of use, e.g., physical mobility, presence of others, and distractions. Research into the effects of the device on response behavior is still relatively new. Also, findings may be hard to generalize over studies and over time, as technology, the way people use mobile devices, and the design of questionnaires vary. Based on the current evidence, the impression is that well-designed questionnaires can prevent device-specific measurement errors (e.g., Couper, Antoun, and Mavletova, 2017; Tourangeau et al., 2017).

Figure 6.6 illustrates this point. The figure shows that for questionnaires that have not yet migrated to the mobile-friendly Blaise5 style sheets (shown in the left panel) the break-off for smartphones is much higher than for tablets and PCs. For questionnaires using the latest versions of Blaise5 stylesheets, the break-off on smartphones is still a bit higher, but the difference is much smaller. Figure 6.6 also shows that in 2017, the year Statistics Netherlands introduced the Blaise5 stylesheet, the design was still being fine-tuned. After about 3–4 months the design for the tablet and PC was finished and the smartphone design

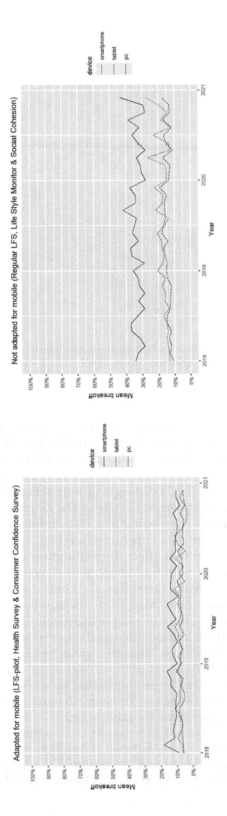

**FIGURE 6.6**

Break-off by device for continuous household surveys Statistics Netherlands 2017–2020 – Comparison of surveys for which the stylesheets are not adapted (left) and are adapted (right) for Mobile Devices (other devices' excluded).

followed roughly 6–8 months later. Even when the design for the smartphone was not finished, the break-off is in early 2017 already lower than for the unadapted version. Also, the break-off for the PC and tablet is lower in the new design.

Antoun and colleagues (2017) performed a systematic review of empirical studies and developed the following five heuristics for the effective design of surveys for smartphones: (1) readability (text should be large enough to promote easy reading), (2) ease of selection (touch targets should be large enough to tap accurately), (3) visibility across the page (content should fit the width of the screen, horizontal scrolling should be prevented), (4) simplicity of design elements (simple from a technical point of view to prevent device-specific errors and simple for respondents to use, e.g., not requiring complex touch gestures), and (5) predictability across devices (questionnaires should function in a predictable way across different devices).

## 6.4 Questionnaire Design for Mixed-Mode Surveys

### 6.4.1 Minimizing Differences or Minimizing Error?

De Leeuw (2018, p. 81) strikingly summarizes the many challenges of mixed-mode questionnaire design as designing equivalent questionnaires with the purpose to 'maximize data quality in a specific mode and minimize differences in data across modes'. Given mode-intrinsic characteristics, such as the presence of an interviewer or the aural or visual mode of presentation, design elements always need to be more or less adjusted to a specific mode. When a specific design choice yields better data in one mode but is not possible or not desirable in another mode, questionnaire designers theoretically can choose between two approaches (e.g., Tourangeau, 2017): the unified mode approach or the best practices approach. The *unified mode – or Dillmans unimode – approach* focuses on *minimizing measurement differences between modes* by presenting a unified stimulus across modes (e.g., Dillman et al., 2014; Dillman and Edwards, 2016). Alternatively, the *best practices approach* focuses on minimizing *error* within each mode, even if that means using somewhat different stimuli in different modes. Tourangeau, Conrad, and Couper (2013) and Tourangeau (2017) argue that the best practice approach is preferable if an overall point estimate is the main goal of a survey. The reasoning behind this is that combining the best estimates for each mode provides the best overall estimate. They advise the unified mode approach if comparison between groups is important. As mode composition in the response can change over time, comparability over time is an additional reason to prefer a unified mode design.

An example may illustrate this theoretical discussion. The Dutch Health survey has a sequential mixed-mode design; a random sample of the general population is first invited to complete a web survey and after one month the nonrespondents are assigned to a face-to-face interview. Among many other variables, the Health survey measures respondents' self-reported height and length in order to assess obesity. Some respondents may routinely 'flatter' their weight if they complete a web survey, but may be less inclined to do so in face-to-face interviews, possibly because the interviewer can see them and can more or less check the plausibility of their answer. If respondents report more obesity in face-to-face interviews than in web interviews and the face-to-face answers would be a better measurement, what would be the best approach to deal with this in a mixed-mode questionnaire design? In a best practices approach the mode differences would be accepted, as they are the best possible measurements for each mode individually and would jointly lead to the best possible estimate of obesity in the population. However, in a unified mode approach, researchers would be concerned about the comparability of the measurements. Suppose researchers want to study differences in obesity levels for different levels of educational attainment. If education is also associated with the likelihood of responding via the web, this may result in different measurement errors for obesity for different educational groups. This could affect the estimation of the correlation between education and obesity. For this reason, in a unified mode approach combining CAPI and CAWI/PAPI, one could consider asking the questions on length and weight in a self-administered part of the face-to-face survey.

In practice, estimate comparisons between groups and over time are paramount for most surveys. Therefore, mixed-mode questionnaire design generally focuses on minimizing measurement differences between modes, i.e., the unified mode approach. Also from a practical and cost-efficiency point of view it is preferable to design questionnaires for various modes as identical as possible. It is more expensive and error-prone to develop and maintain mode-specific versions of the same question.

It must be noted that a pure identical design for all modes is often simply not possible. For example, some instructions must be mode-specific (e.g., 'press enter to continue'). Also, certain mode-specific conventions are highly desirable to facilitate the response process (e.g., using automated routing in electronic questionnaires, or not repeating all answer options in CATI for long lists of items with the same answer options). Choosing a unified mode approach thus is not a cure-for-all mode-specific design decisions. The general recommendation is to only use mode-specific adaptations where this cannot be avoided or where it will improve the response process in a specific mode *without affecting the way respondents interpret and answer the question.*

### 6.4.2 Mixed-Mode Requirements in the Questionnaire Design Process

Mixed-mode requirements should be considered in each stage of the questionnaire design process.

1. In the stage of conceptualization and research design, the risks for mode-specific (and device-specific) measurement errors should be carefully considered given the topic and other requirements of the specific research project. Mixes of modes that are likely to cause large mode-specific measurement errors should be prevented in the first place. An important example is that for highly sensitive topics, such as drug use or domestic violence, mixing interviewer and self-administered modes should be avoided (see also Chapter 2, Designing Mixed-Mode Surveys).

2. In the stage of questionnaire design, preferably all relevant modes should be considered from the first draft of the questionnaire. It is practical to start drafting questions with the mode and device that has the strongest limitations. Typically, PAPI (no automated routing), CAWI (no interviewers), and smartphones (small screen) are most challenging to design for. But aural modes also pose limitations on for example the number of answer options respondents can process. Following the unified mode approach, the aim is to provide respondents in each mode as much as possible with the same stimulus. This means that when developing texts, question format, visual design, etc., their appropriateness for each mode should be considered in order to keep the design in each mode as identical as possible.

3. In the stage of testing all modes should be tested (see also Section 6.6).

4. In the data collection stage, relevant paradata information should be collected to allow evaluation of the questionnaire over the various modes and devices.

5. After data collection, analyses should check for mode- and device-specific measurement error (see also Chapter 9, Analysis of Mixed-Mode Data Analysis) and questionnaire designers should reflect on how questionnaire design may have impacted any effects found; insights from the analysis should feed into the design of a new version of the same or similar questionnaires. As technology, respondents' expectations and preferences, and even the meaning of questions can change over time, this evaluation and reflection should be repeated regularly (for example, annually) for ongoing surveys.

Good documentation can greatly facilitate the design process. It is therefore recommended to document all questionnaire texts for all modes in one document or system. This will prevent unintended mode differences. See Textbox 6.1 for the so-called 'omni mode' questionnaire development procedure used by Statistics Netherlands (Cremers, 2016). In this procedure the questionnaire design for all modes is documented in one document and one questionnaire is programmed, while still allowing for mode-specific adaptations (e.g., showing interviewers which part of the text they should read out).

Also, the exact design of tested and fielded questionnaires should be documented. To understand mode or device differences, it is often necessary to be able to see and compare how a question was exactly presented to respondents and in which context. Documenting the presentations of questions in web questionnaires with many routings and imputations and on various devices is in practice very challenging. However, for a few crucial routings (e.g., for employed and unemployed, victims and non-victims), for all modes, and for the most commonly used devices, it should be possible to document screenshots. Interviewer instruction and training materials should also be documented and be available for methodological research and data users. These types of meta information can be crucial to judging the quality and usefulness of data. Additionally, this documentation can facilitate future applications and revisions of the questionnaire.

---

**TEXTBOX 6.1   A PRACTICAL EXAMPLE: THE 'OMNI MODE' QUESTIONNAIRE DESIGN PROCEDURE AT STATISTICS NETHERLANDS**

In 2008 Statistics Netherlands started to use CAWI alongside the more traditional CATI and CAPI modes. Initially, the design documents for the 'old' modes were kept and an additional design document was made for CAWI. The questionnaires were changed minimally following a unified mode approach, and the designs were recorded in two separate documents. Working with two design documents proved to be very inefficient:

• It more or less doubled the work. For each questionnaire two documents were needed for the design and two questionnaires had to be programmed and tested.

• Many iterations were necessary. The design usually started with CAWI. Once done, this design had to be adjusted to CATI and/or CAPI. Often, in this process it was discovered that some questions were not suitable for CATI or CAPI and, following the unified mode principle, the CAWI design had to be changed as well.

- The method was error prone. When last-minute adjustments were made to one design this could be forgotten for the other design. These errors also meant more work for the programmers and other departments, such as data processing, analysts, etc.

In order to optimize the design process, a new way of designing was necessary. This is called the omni mode procedure at Statistics Netherlands. These main changes made were:

- Same text for answer categories for all modes: Originally, special punctuation marks were used in CAPI and CATI to indicate which text interviewers should read out. The omni mode procedure omits the special punctuation marks for interviewers and instead uses color to show which text should be read out. This is indicated in the design by the statement {/Read} (meaning 'Read aloud'). The programmer uses a special layout code for these statements so that in CAWI these texts have a normal text color. Also, the texts for question and answer options were formulated identical for all modes. Initially, there was concern that interviewers might not like this solution because the answer categories do not mimic a natural conversation anymore. However, an evaluation with interviewers indicated that this was not a problem at all. On the contrary: it was much clearer now for respondents that these were options they could choose from. As a result interviewers had to repeat the answer categories less often.

- Nonresponse attributes are the same in all modes: if, e.g., 'don't know' is not visible in CAWI, this option can also not be chosen in CATI/CAPI.

- Questions instead of checks: CATI/CAPI no longer uses pop-up fields for checks and controls, but questions are shown.

- Mode-specific imputations: Of course, it will not always be possible to make the exact same design for all modes. For example, in CAWI you might want to use an introduction like 'Below you will find several examples of ...' while in CATI/CAPI you may want to use the text 'I will now give you several examples of ...'. These differences are indicated by mode-specific imputations: $P for CAPI, $T for CATI, and $W for CAWI.

### 6.4.3 Mixed-Device Questionnaire Design

With the increase in respondents using mobile devices, research agencies need to consider to what extent their web questionnaires are suitable for mobile devices. Based on this assessment, decisions can be made about redesign priorities and data collection strategies. If it is impossible or very burdensome to complete a questionnaire on mobile devices, the use of these devices should be discouraged (e.g., explicitly advising respondents to complete the questionnaire on a PC, not providing a QR code for login) or even technically blocked. Both strategies, however, are far from ideal. Research has shown that many respondents ignore instructions about the preferred device (e.g., de Bruijne and Wijnant,

2013). Moreover, blocking mobile devices may – increasingly – cause coverage errors and selective nonresponse. See Peterson, Griffin, LaFrance, and Li (2017) for an extensive discussion of how to deal with smartphone participation in web surveys.

The European MIMOD – Mixed Mode Designs for Social Surveys – project (Schouten et al., 2018) developed a list of dimensions and indicators to assess the smartphone fitness of questionnaires. This list provides a practical starting point to review existing or draft questionnaires for device-related measurement risks. Adapting questionnaires for mixed-device data collection is still a relatively new field. This list will probably be further developed and adjusted as more research into device effects and experience in assessing smartphone fitness of questionnaires become available.

---

**TEXTBOX 6.2   DIMENSIONS AND INDICATORS TO ASSESS SMARTPHONE FITNESS OF QUESTIONNAIRES (ADAPTED FROM SCHOUTEN ET AL., 2018)**

1. Screen size: This dimension evaluates the amount of information presented on a screen and thus the overall visibility of the items and the need to scroll. Partial invisibility of survey items may lead to confusion, underreporting of particular answer categories, and respondent fatigue. Indicators:

   a. Use of introductions and/or instruction;

   b. Question length;

   c. Number of answer categories;

   d. Answer category length;

   e. Matrix questions (on smartphones matrix questions require scrolling or navigation is changed through accordion, carrousel, or other modifications of the navigation); and

   f. Horizontal answer scales.

2. Touch navigation: This dimension evaluates the conflict between visibility on the screen and the simultaneous need to use the screen for navigation. Such navigation may lead to typing errors and respondent fatigue. Indicators:

   a. The number of open questions; and

   b. Items with many answer categories.

3. Burden: Response burden in combination with timing and location affect respondent motivation, concentration, and the likeliness to break-off. Indicators:

   a. Number of survey items;

   b. Person or household survey: In a household survey, multiple persons need to answer questions, making the survey longer and/or requiring the need to switch between persons or allow for proxy reporting;

   c. Interaction with a classification database: Survey items with many and diverse answer categories, e.g., occupation, educational level, or type of economic activity, often employ interaction with a classification database positioned at the server of the survey institute. Such interaction requires internet traffic and especially for smartphones may slow down the interview;

d. Task complexity: Survey items that require extra effort from the respondent, e.g., calculations or searching for information in personal administration, take more time and may not be compatible with the time and location in which the respondent completes the questionnaire; and

e. Perceived enjoyment/relevance/burden: Surveys differ in how they are perceived by respondents, generally assessed in scores on enjoyment, relevance, and/or burden. Surveys that score weak on enjoyment/relevance/burden may especially be problematic on smartphones because survey completion can be more demanding on smartphones and smartphone users may be more easily distracted.

## 6.5 Testing and Evaluating Mixed-Mode Questionnaires

### 6.5.1 Testing and Evaluation as Part of the Questionnaire Design Process

Questionnaire testing and evaluation can serve a variety of objectives, can take place in all stages of the design process, and may involve various methods. Typically, both qualitative and quantitative methods are used. For example, in the step of conceptualization it can be very useful to conduct qualitative interviews or focus groups to explore how members of the target population think and talk about the concept of interest or the feasibility of a new data collection methodology. Once draft questions are available, cognitive interviews are a well-established method to gather qualitative information about how respondents understand and answer questions (see also e.g. Willis, 2005; Collins, 2015; U.S. Office of Management and Budget, 2016a). When a computer-assisted questionnaire has been programmed, it must be technically tested for correct construction. Small pilots test the questionnaire in field conditions and collect data on, for example, completion duration and data quality. Interviewer and respondent debriefings after pilots can provide valuable information about the quality of the questionnaire. Experiments may be needed to decide between alternative questionnaire designs or to assess the impact on time series when the design of an ongoing survey is changed (see Chapter 7, Field Tests and Implementation of Mixed-Mode Surveys).

When a questionnaire goes into production, the need for evaluation of the questionnaire does not stop. Procedures during field work should allow early detection of any unexpected problems that need quick fixes (e.g., a new version of the questionnaire or additional instruction for interviewers). During field work various types of paradata should be collected that can be used to evaluate the questionnaire. For example logs of break-offs (where, who, which device) or a log of which warnings were triggered can be indicators of question problems (see Chapter 8, Re-interview Design to Disentangle and Adjust for Mode Effects). Many survey institutes routinely add a few evaluation questions at the end of a questionnaire or re-contact a small sample of respondents and nonrespondents for an evaluation of the questionnaire and/or field work procedures. Sometimes, qualitative methods are used during or after field work, for example, to help understand some implausible patterns in the data. See Textbox 6.3 for an example of how qualitative interviews and interviewer debriefing helped understand a mode-specific breach in time series in bicycle ownership reported in the Dutch Mobility Survey.

Of course, many parts of questionnaires and field work procedures are standardized and can be re-used. For example, survey agencies typically use stylesheets for their questionnaires that have been developed to work on a certain range of devices and operating systems. Also, the explanation and procedures on how to find a web questionnaire, login, save and submit, etc., can to a large extent be standardized and need not be tested for each new questionnaire. However, when developing and updating these standards it is important to test if users are able to work with them and how user friendly they are.

Which mix of testing and evaluation methods to choose for a specific project depends on the needs of the specific project and on the time and resources available. Our recommendation is to always include at least some testing with respondents from the target population in circumstances that are as similar to the field work conditions as possible. For a comprehensive overview and discussion of questionnaire testing and evaluation methods, see e.g. Tourangeau et al., 2020; d'Ardenne and Collins, 2020, chapter 6 in Brancato et al., 2006 and U.S. Office of Management and Budget, 2016b.

---

### TEXTBOX 6.3   UNDERSTANDING HOW QUESTION FORMAT AND MODE AFFECTS THE NUMBER OF BICYCLES IN A HOUSEHOLD VIS-VISSCHERS AND MEERTENS, 2016

The Mobility Survey includes questions about different sorts of vehicles. Through the years the survey has seen many changes, among other things, the changes in design from single-mode to mixed-mode data collection. In 2013 additional questions about electronic bicycles were added. Before, the questions were about bicycles in general, without specifying the kind of bicycle (see Table 6.1).

The survey has produced rather steady timeline data. And also a stable division of the responses between the different modes. See Table 6.2 for data for the variables *'Number of bicycles in the household'* and *'Does Respondent own a bicycle'* in the period 2011–2014. In 2013 and 2014 a rather big break of the time series occurred for the variable 'Does Respondent own a bicycle'. Table 6.2 shows that the break is mostly due to a breach in the data for the Web-mode.

#### QUESTIONNAIRES EVALUATION METHODS USED

- Review of questionnaire design and survey methodological issues using checklists and best practices. Emphasis on the formulation of the questions, question order, and the exact changes between the old and new questions.

#### TABLE 6.1

Survey Questions on Ownership of Bicycles before and after 2013

| Survey Questions up until 2012 | Survey Questions 2013–2014 |
|---|---|
| *Now we will ask several questions about vehicles.* | *Now we will ask several questions about vehicles.* |
| *How many bicycles are in your household?* [0..97] | *How many electronic bicycles are in your household?* [0..97] |
| *Do you yourself own a bicycle?* [Yes/No] | *Do you yourself own an electronic bicycle?* [Yes/No] |
| | *How many other bicycles are in your household? Do not include electronic bicycles.* [0..97] |
| | *Do you yourself own another\* bicycle?* [Yes/No] |

\*'other' in Dutch can also mean 'different'.

**TABLE 6.2**

Time Series Data of the Mobility Survey

| Number of Bicycles in the Household | | 2011 (%) | 2012 (%) | 2013 (%) | 2014 (%) |
|---|---|---|---|---|---|
| | Value | | | | |
| Total all modes | 1 or more | 94.3 | 94.6 | 93.8 | 93.7 |
| Web | 1 or more | 96.7 | 96.8 | 96.5 | 96.0 |
| CATI | 1 or more | 93.5 | 93.3 | 92.9 | 93.1 |
| CAPI | 1 or more | 91.3 | 92.6 | 91.4 | 91.4 |
| **Does the respondent own a bicycle?** | | 2011 (%) | 2012 (%) | 2013 (%) | 2014 (%) |
| | Value | | | | |
| Total all modes | Yes | 95.1 | 95.3 | **85.3** | **86.2** |
| Web | Yes | 96.6 | 96.8 | **71.1** | **71.6** |
| CATI | Yes | 95.4 | 95.5 | 94.7 | 95.3 |
| CAPI | Yes | 91.0 | 91.4 | 89.8 | 90.7 |

- Interviewer debriefing in which the Mobility Survey was evaluated. What do the interviewers think of the mode effect? Do they recognize it? How can the questions be improved?
- Interviews re-approaching actual respondents of the Mobility Survey by telephone interviewers with standardized survey questions followed by a short open interview.
  - Test questionnaire with alternative survey questions, and instructions for open interview.
  - Compare answers to original Mobility Survey questions to answers to alternative questions and to answers in an open interview.

## CONCLUSIONS

- Context: The context of the original question has been disturbed.
- Order: Respondents think they have already answered the question.
- Comprehension: Misinterpretation of the word 'other'. From interviewer debriefing: *'They don't mean a different or special bicycle, but a normal one'.*
- Mode: In web there is no interviewer to correct the misinterpretations.

New questions were suggested and tested:

**Alternative questions:**

Now we will ask several questions about vehicles.

How many bicycles are in your household? Include all kinds of bicycles.
[0..97]

How many of those are electronic bicycles? Electronic bicycles are all bicycles with electronic aid.
[0..97]

Do you yourself own a bicycle?

- Yes, a non-electronic bicycle
- Yes, an electronic bicycle
- Yes, both
- No

Conclusion after implementation of the new questions:

- The alternative questions generate the expected amount of bicycles, time series returned to the level of 2012.
- Interviewers reported that the new question required less guidance from the interviewer to answer correctly.

## 6.5.2 Mixed-Mode Questionnaire Testing and Evaluation

Compared to single-mode questionnaires, the pre-testing and evaluation of mixed-mode questionnaires faces two additional challenges: to evaluate the questionnaire in each mode and to test for mode effects. Also, self-administered surveys and especially web surveys make pre-testing more complicated. These surveys require the evaluation of not only the content but also of the usability of the questionnaire (see Geisen, Romano and Bergstrom, 2017; Geisen and Murphy, 2020; Nichols et al., 2020 for more information on the usability testing of web surveys). Ideally, questionnaires should be tested for all relevant modes and devices. Even with well-established stylesheets, unforeseen interactions may occur with mode, device, and questionnaire content that can affect the response process.

In practice resources are scarce and testing on all modes and devices may not be feasible for all projects. To guide decision making about test plans for mixed-mode questionnaires the following steps are recommended:

### Step 1 Make an Explicit Mode Risk Assessment

Are there any reasons to expect mode or device-specific measurement errors? This assessment should consider the specific characteristics of the survey (e.g., topic, question formats, visual presentation) as well as indicators of mode effects that may be known from previous waves of the same or a similar survey.

### Step 2 Decide if the Test Should Compare Modes

If this is the case, the test design should allow for comparisons over modes. One way to do this is to assign comparable groups to different modes and test in a similar way by randomizing and matching (Gray, 2015). Alternatively, the same respondents can be asked the same or similar questions in different modes (see, for example, Campanelli, Blake, Mackie, and Hope, 2015; Gravem et al., 2018a; Gravem et al., 2019).

### *Step 3 Decide Which Modes to Test, and When in the Development Process*

Possible choices may be to:

a) Focus test capacity on the highest risk modes, e.g., CATI (high pace) and smart-phone (small screen);

b) Test iteratively, starting with testing wording and content with easily available modes for draft questionnaires that are not programmed yet (paper and inter-viewer modes), and use these results for the next version of the questionnaire that will be programmed and tested in CAWI; and

c) Test highest risk modes only, and do only a desk review of other modes.

### *Step 4 Test Relevant Modes as Realistically as Practically Possible*

As stated before, the choice of methods for questionnaire evaluation depends on the test goals and resources. However, in our experience, for all new or redesigned questionnaire it is highly recommended to include at least some testing in which respondents from the target population complete the questionnaire in a way that mimics the field situa-tion as much as practically possible. Some questionnaire issues can only be detected if respondents are actually interviewed or complete the questionnaire as they would in field conditions. Observation of their behavior during this task combined with qualitative inter-viewing usually yields valuable insights in if and why measurement errors occur (see also Maitland and Presser 2020; d'Ardenne and Collins, 2020, for a discussion of the importance of testing methods that respectively closely observe the response process and mimic the mode of survey administration).

Specific issues to consider to test the questionnaire as realistically as possible:

a) Assign test respondents to modes and devices they are familiar with and are likely to use if the survey is in production. Preferably let them use their own devices for the test. If this is not possible, make sure they get familiar with the test device before testing the questionnaire. This can be accomplished by first having them perform another short task on the device that requires similar input and naviga-tion as the questionnaire.

b) At some point in the questionnaire design process, test the usability and content of self-administered questionnaires simultaneously. Both usability and question-naire-content issues may not be detected if content and usability are tested only separately.

c) Carefully consider to what extent the test design interrupts the response process and how this may affect the findings. Commonly used techniques like thinking aloud and concurrent probing (asking about the response process directly after a respondent has answered a survey question) provide valuable data that may not be available anymore after the questionnaire is completed. However, these tech-niques may also interfere with the way respondents would normally go through a questionnaire. To have the best of both worlds, it can be a good solution to use think aloud and concurrent probing for some test respondents and to only observe with retrospective probing for other test respondents.

d) Conduct CATI test interviews over the telephone and in a way that the respondent and interviewer cannot see each other.

e) If it is likely that interviewers or respondents will use the questionnaire in various conditions (e.g. interviewers standing at the door, respondents sitting in a car with bad light conditions), consider testing the questionnaire in these conditions. Especially for diary studies on smartphones it is likely that respondents will enter data under various conditions during the day.

f) For questionnaires like diary studies that require multiple uses over a longer period of time, test both the initial interaction with the instrument as well as how respondents use the instrument after they have become more experienced with it. This will give insight into the learning curve and into the user needs of both new and more experienced users.

### Step 5 In Analyses and Reporting, Distinguish between Usability and Content Findings, and Reflect on Mode and Device Specificity Regarding Findings

In the test data and analyses, it is important to register in which mode and on which device and operating system the questionnaire was tested. For the analysis of any issues found in the test clearly distinguish – as far it is possible – to what extent usability and question content play a role. Fixing usability problems usually takes a different expertise than fixing question content issues – so it is practical to make these findings easily distinguishable in a report. For each issue found, reflect on whether there is any likelihood that mode or device will affect this issue. For example, if it is found that wording is too long or complex, it is likely that these problems will be stronger in aural modes and on a small screen. Include in the reports test materials, screen shots, etc., that facilitate decisions on the development of the questionnaire.

## 6.6 Guidelines for Mixed-Mode Questionnaire Design

As with any questionnaire, there is no simple cookbook that will guarantee a high-quality mixed-mode questionnaire. Different goals may compete and design decisions must be made within the real world of deadlines, software requirements, and limited resources. The set of guidelines presented here is intended to help researchers make these decisions in the mixed-mode questionnaire design process. There may be good reasons to deviate from these guidelines. However, it is recommended to check the design against these guidelines and explicitly consider costs and benefits of applying each guideline in a mixed-mode survey.

### 6.6.1 Keep the Stimulus and Response Task as Similar as Possible in All Modes

This implies, among other things, that:

- All wording of questions, answer options, instructions, etc., should be as identical as possible.
- If text formatting is used to communicates different meanings for different text (e.g., italics for instruction) these different meanings should be communicated in all modes.

- Use as much as possible mode-neutral wording for instructions (e.g. use 'choose' instead of 'click on'), transition texts, etc.
- The order of questions and answer options should be as identical as possible.
- If edit checks are used, they should be used identically in each mode (to the extent possible).
- When interviewer-administered modes are mixed with self-completion modes, make sure that all response options presented visually in one mode are also presented aurally or on a show card in the other mode(s).
  - If answer options like 'don't know' or 'rather not say' are explicitly presented in self-administered modes, they should also be explicitly presented in interviewer modes. If these options are not explicitly offered, but accepted by an interviewer after probing, the self-administered mode should simulate such a probe (see for example de Leeuw, Hox, and Boeve 2016).
  - Do not mix open questions that are coded by interviewers in interviewer modes (e.g., asking, 'What is your marital status?' without reading out all answer options) with closed questions (explicitly presenting the answer options) in self-administered modes.
  - All labels in response scales should preferably be offered explicitly both visually and aurally.
  - Do not mix matrix questions on paper or PC and an item-by-item presentation in interviewer modes or on smartphones. Consider the use of an accordion or carrousel presentation of items in matrix questions to present one item at a time in CAWI modes (see de Leeuw and Berzelak 2016 for an example).
  - Do not mix a 'check-all that apply' format in self-administered mode with a list of yes/no questions in an interviewer mode. If interview administered modes are combined with self-administered modes, use a yes/no format in all modes for these types of questions.
  - If an instruction is always visible in self-administered modes, it should also be offered at least once in interviewer modes.
  - Avoid the use of text that can be offered at the discretion of the interviewer (e.g., help text that interviewers can read out if they think is needed). If this cannot be avoided, consider making this information also available in self-administered modes in a less prominent way (e.g., behind a help-button, in a footnote, etc.).

### 6.6.2 Organize the Questionnaire Design Process to Prevent and Detect Mode-Specific Measurement Errors

- Consider mode-specific measurement errors in each step of the questionnaire design process (design, development, pre-testing, monitoring, and evaluation).
- Test the questionnaire with respondents in all relevant modes and on all relevant devices.
- Preferably design questions and questionnaires from the start on with the required modes and devices in mind; if new modes or devices are added to an existing questionnaire, consider if a complete redesign is necessary.

- Document all texts for all modes in one document or system to work more efficiently and to prevent unintended differences.
- Document all main variations of the way the questionnaire is presented to respondents (e.g., save screenshots of some of the main routings of a questionnaire at various devices).

### 6.6.3 Prevent Social Desirability Bias

If a questionnaire contains questions that are likely susceptible to socially desirable responses, it is better not to mix interviewer and self-administered modes. If this cannot be prevented, consider adding a self-administered module to a face-to-face interview for the sensitive questions. In theory, a telephone interview could be combined with a self-administered module, for example, by asking respondents to complete a part of the survey on the web or via a paper questionnaire, or by switching to an interactive voice response mode without an interviewer (Kreuter, Presser, and Tourangeau, 2008). In practice, this is seldom done because of the risk of nonresponse related to switching to a different mode.

### 6.6.4 Prevent Satisficing

Satisficing can be prevented by lowering the cognitive burden of the questions and by increasing the motivation of the respondent (see e.g. Krosnick and Presser, 2010; Roberts, 2016 for an extensive discussion of factors affecting satisficing).

- Encourage respondents to respond carefully (e.g., with instructions, feedback when speeding).
- Keep questionnaires short.
- Use short questions in plain language.
- Start with interesting and easy questions that motivate the respondent.
- Divide cognitive burdensome tasks into less difficult smaller tasks.
- Use item specific measurements instead of indirect measurement (e.g., ask to rate one's health on a scale ranging from very bad to very good, instead of asking to rate a statement like 'my health is excellent' on a strongly agree-strongly disagree scale).
- Try to minimize the response options for closed questions.
- Use show cards in face-to-face interviews for long and complex answer options.
- Avoid:
  - Hypothetical questions;
  - Questions that need difficult calculations or a lot of memory retrieval;
  - Proxy questions;
  - Matrix questions; and
  - Extensive lists of rating scales.
- Do not put the most important or cognitive burdensome questions at the end of the interview/questionnaire, because the respondent's motivation will be exhausted by then.
- Self-completion forms should be easy to fill in, as not to add to the burden of the task.

## 6.7 Summary

The main takeaway messages from this chapter are:

- Designing for mixed-mode adds considerable challenges to each stage of the questionnaire design process.
- Even small differences in how a questionnaire is presented to a respondent can cause differences in measurements. Therefore, questionnaire design is an important means to preventing mode-specific measurement errors and thus mode effects.
- Mixed-mode questionnaire design should aim to maximize data quality in each mode, while minimizing differences between modes.
- To prevent mode-specific measurement errors, mode-specific adaptations to questionnaires should only be made where this cannot be avoided or where it will improve the response process in a specific mode *without affecting the way respondents interpret and answer the question.*
- If at all possible, do not just add a new mode to an existing survey, but do a total redesign of the questionnaire if a mode is added.
- To prevent mode-specific measurement errors in mixed-mode surveys, it is even more important than in single-mode surveys to apply general guidelines of good questionnaire design. The main aim should be to design the questionnaires in a way that respondents are willing and able to understand questions as intended, retrieve the information needed, formulate the correct answer and report this in the required format.

# Part IV

# Analysis

# 7

# Field Tests and Implementation
# of Mixed-Mode Surveys

## 7.1 Introduction

This chapter considers experimental methods aimed at quantifying non-sampling errors associated with different data collection modes. These are typically relative mode effects, since in an experiment, alternative modes (treatments) are compared with a standard mode. Different data collection modes typically require different questionnaire designs. As a result, not only is the data collection phase in Figure 7.1 involved, but also the questionnaire (re)design phase. Randomized experiments are particularly useful in quantifying interactions between multiple changes in survey design, e.g., between different questionnaires and different data collection modes. Large-scale field experiments where experimental units are selected with a probability sample can be applied to test the effects of one or more adjustments in a survey process on response rates or parameter estimates of an ongoing survey. Conducting this type of field experiments therefore requires careful planning of the field work. To obtain unbiased estimates of treatment effects (or relative mode effects) data collection under different modes must strictly obey the definitions of the different treatments. As a result the three steps under survey data collection mentioned under "do" in Figure 7.1 are also involved. After data collection, approximate design-unbiased estimates for the key survey estimates under the different treatments are obtained. Finally, differences between these survey estimates can be analyzed to draw inferences on relative mode effects. Therefore, steps 1 and 3 in the survey analysis shown in Figure 7.1 are also involved.

In the case of repeated surveys, time series methods can be considered a cost-effective alternative for experiments to quantify the impact of implementing alternative data collection methods. In this case steps 1 and 3 of survey design and step 3 of survey analysis are involved.

In Chapters 3, Mode-Specific Selection Effects, and 4, Mode-Specific Measurement Effects, the impacts of data collection modes on measurement errors and selection errors are discussed. The present chapter describes the design and analysis of field experiments to test the effect of different data collection modes and questionnaire designs. These kinds of field tests are useful in developing mixed-mode data collection strategies and mixed-mode questionnaire designs, as discussed in Chapters 5, Mixed-Mode Data Collection Design, and 6, Mixed-Mode Questionnaire Design, and complement the cognitive and usability testing strategies treated in Chapter 6. The field experiments discussed in this chapter are appropriate for observing differences that are the net result of mode-dependent measurement errors and selection bias. Experimental designs that attempt to disentangle

DOI: 10.1201/9780429461156-7

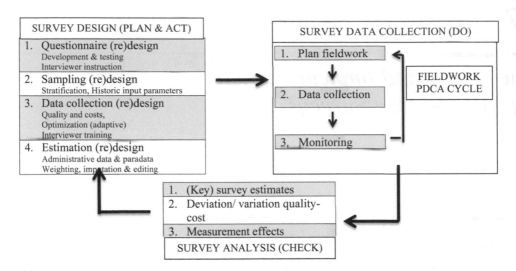

**FIGURE 7.1**
The plan-do-act cycle for surveys.

mode-dependent measurement errors and selection bias require repeated measurements and are described in Chapter 8, Re-interview Design to Disentangle and Adjust for Mode Effects.

This chapter requires a more advance background in statistics, in particular, knowledge of design-based and model-assisted survey sampling (Särndal et al., 1992), and basic knowledge of structural time series analysis (Commandeur and Koopman, 2007), and small area estimation (Pfeffermann, 2002, 2013).

In survey methodology literature many references to experimental studies on improving the quality or efficiency of survey processes can be found – for example, studies to compare the effect of different questionnaire designs, data collection modes, or approach strategies on the main outcomes of a sample survey, with the purpose of reducing measurement errors (Chapter 3) and nonresponse bias or improving response rates (Chapter 4). Many experiments conducted in this context are small scale or conducted with specific groups. Empirical research into survey methods is valuable as conclusions can be generalized to populations larger than the sample that is included in the experiment. This can be achieved by selecting experimental units randomly from a larger target population, which naturally leads to randomized experiments embedded in probability samples. This enables the generalization of conclusions observed in an experiment to larger target populations and is particularly important if experiments are conducted to improve survey methods or to obtain quantitative insights into the different sources of non-sampling errors in survey research. From this point of view, they complement the insights obtained with the cognitive and usability testing strategies discussed in Chapter 6.

The experimental methods outlined in this chapter are important for improving empirical research aimed at survey field work methods. This is of interest to empirical survey researchers who work in the area of questionnaire design and field work strategies as well as survey agencies that conduct single-occasion surveys. On top of that there is a particular application in the context of repeated sample surveys. At national statistical offices such experiments are particularly useful for quantifying systematic effects in a series of repeated surveys due to adjustments in the survey process. Surveys conducted by national statistical institutes are generally conducted continuously or repeatedly in time with the purpose of producing a consistent series. One important aspect of such repeated surveys is the comparability of the outcomes over time. In order to preserve the comparability of estimates, the underlying process of repeated surveys is often kept unchanged as long as possible. It is inevitable, however, that adjustment or redesign of this process is needed from time to time, since the existing procedures become gradually outdated or more cost-effective methods are required. A typical example is the transition from single-mode to mixed-mode data collection to increase response rates and reduce administration costs.

Besides sampling errors, survey samples contain different sources of non-sampling errors that have a systematic effect on the outcomes of a survey (Chapter 3 and 4). As long as the survey process is kept constant, this bias component is not directly visible. If, however, one or more components of the survey process are modified, the biases induced by these non-sampling errors are changed. Modifications in the survey process therefore generally have systematic effects on the survey estimates, disturbing comparability with figures published in the past. To avoid confounding real period-to-period change in the parameters of interest with changing measurement bias due to alteration in the survey process, it is important to quantify the effect of changing the field work strategies and data collection modes. Systematic differences in survey outcomes due to differences in measurement procedures are further referred to as discontinuities.

The method used to measure discontinuities depends on the type of change in the survey process. In the case of changing field work strategies or data collection methods, it can be expected that the microdata are not consistent under the old and new approaches. In such situations, a straightforward and reliable approach to quantifying discontinuities is to collect data under the old and new approaches alongside each other at the same time. This is referred to as parallel data collection. Redesign of long-standing surveys like, e.g., the US Current Population Survey and the US National Crime Victimization Survey are accompanied by a parallel run (Dippo et al., 1994; Kindermann and Lynch, 1997).

Sections 7.2 and 7.3 are focused on the design and analysis of experiments embedded in probability surveys with the purpose of testing the effect of alternative data collection modes, eventually in combination with different questionnaire designs or other field work strategies. The advantage of embedded experiments is that they increase the validity of the results since they allow generalizing conclusions to the target population where the sample is drawn from, as explained above. Embedding experiments in ongoing sample surveys is efficient since a regular survey serves as the control group in the experiment and is simultaneously used for regular publication purposes of the survey. Another strong point of parallel data collection is the low-risk level for regular publications during the change-over to the new design. This approach can avoid the risk of a period without data for regular publication should the new approach turn out to be a failure. Through a well-designed experiment the risk of failing to detect a discontinuity is minimized since the design of the experiment gives full control over the minimum detectable difference at a pre-specified significance and power level. A further major advantage of parallel data collection is that it facilitates the production of timely estimates for impact measurements. Estimates for the discontinuity can be made directly after finalizing the field work. If the sample size meets the pre-specified precision requirements for estimating the discontinuities, then there is no need for revision of the estimated discontinuities. Updates of imprecise estimates are possible with time series methods when the results of the subsequent editions of the new survey become available, as explained below.

A disadvantage of a parallel run is that it is costly, since additional data collection is required. Obtaining sufficiently precise estimates for the discontinuities often requires sample sizes for the new approach that come close to the regular sample size. In most cases the available budget does not allow to conduct a sufficiently large parallel run. In the most extreme case, there is no parallel data collection at all. In that case time series models can be used to separate real period-to-period change from differences in bias due to a redesign of the process, for example, with state-space intervention models. In many situations there is a budget for a small parallel run. In that case, model-based inference procedures, known from the realm of small area estimation, can be used to obtain estimates for the discontinuities that are more precise than simple direct estimates. This information can also be combined with state-space intervention models. In this case the information of the entirely observed series before and after the parallel run is used to further improve the estimates for the discontinuities. Section 7.4 introduces the time series modeling approach for estimating discontinuities. In Section 7.5 small area estimation methods for estimating domain discontinuities with small parallel runs are reviewed. The methods introduced in this chapter are applied to the analysis of a small parallel run to estimate discontinuities due to the introduction of a new data collection approach in the Dutch Crime Victimization Survey in Section 7.6. The chapter concludes with a summary in Section 7.7.

## 7.2 Design of Field Experiments

In an embedded experiment, a probability sample is drawn from a finite target population. This sample is randomly divided into two or more subsamples according to a randomized experiment. In survey literature, such experiments are also referred to as split-ballot designs or interpenetrating subsampling, and date back to Mahalanobis (1946), but see also Cochran (1977, section 13.15), Hartley and Rao (1978), and Fienberg and Tanur (1987, 1988, 1989).

Randomized experiments are undertaken under a clearly specified protocol, which sets out in advance what is to be tested, decision rules for the test outcomes, the procedures to be followed, and the analysis to be undertaken. The following topics need to be set out when an experiment (whether or not part of a sample survey) is set up:

- Clear definitions of and a decision about the number of treatment factors and treatment levels;
- Clear specification of the hypotheses about the main effects and interactions between the different treatment factors that need to be analyzed;
- Dependent variables (parameters for which hypotheses about treatment effects are tested);
- Differences between the parameter estimates, i.e., the main effects and their interactions, that at least should result in a rejection of the null hypothesis of no treatment effects;
- Power and significance levels to test these hypotheses;
- Experimental design (randomization of sampling units over the treatments and level of randomization);
- Decisions concerning the use of the field staff in the data collection of the experiment;

- Minimum required sample size; and
- Method of analysis, including a decision whether a design-based or model-based approach is applied.

This results in the specification of the hypotheses to be tested. A typical approach in the design and analysis of experiments is to pre-specify and quantify the objective of the experiment to avoid unnecessary post-hoc analyses. A general framework and practical guidelines for this process of planning and conducting experiments is given by Robinson (2000).

Before a large-scale field experiment is planned to test hypotheses about discontinuities, the survey process of the redesign must be definite. This implies that pilots to test a new approach strategy or questionnaire must precede field experiments that are aimed at testing differences in the target parameters due to the survey redesign. These kinds of experiments come in the final stage to assess the mixed-mode data collection strategy and questionnaire design, as discussed in Chapters 5 and 6. It is perilous to combine both purposes in the same experiment. The results of the experiment might indicate that the new survey process must be adjusted. In this stage of a survey redesign, however, there is often no time or budget to conduct a new large-scale field experiment to investigate discontinuities of the revised survey process.

The most straightforward approach is to split the sample into subsamples by means of a completely randomized design (CRD). Generally, this is not the most efficient design available. The power of an experiment might be improved by using sampling structures such as strata, clusters, or primary sampling units as block variables in a randomized block design (RBD) (Fienberg and Tanur, 1987, 1988). An RBD improves the precision of experiments, since blocking removes the between-block variation from the standard errors of the treatment effects. This is comparable to stratified sampling to remove between-stratum variation from the population estimates. Unrestricted randomization by means of a CRD might also result in practical complications, like long travelling distances for interviewers. This can be avoided by using small geographical regions as block variables.

The field staff requires special attention in the planning and design stage of an experiment. To draw conclusions that can be generalized to a situation where the new approach is implemented as a standard, it is advisable to use the entire field staff or a representative sample thereof. Newly recruited staff, on the other hand, might be precluded for this reason. It is also advisable to provide sufficient training to ensure that the field staff has sufficient experience with data collection under the new approach. One might also anticipate that the data collected under the new approach in the first period of the experiment cannot be used in the analysis, since the interviewers must adapt to or gain sufficient experience with the new methods.

From a statistical point of view it is attractive to use interviewers as a block variable in an RBD, since this removes the interviewer variance component from the analysis of the experiment. A major drawback is that this implies that each interviewer has to collect data under both the regular and the new methodology, which might give rise to confusion. In the case of testing mixed data collection modes where some of the modes are not interviewer driven, interviewers naturally cannot be considered as block variables. If it is decided that interviewers are assigned to one treatment only, then this must be done randomly to avoid one of the treatments being systematically favored with experienced interviewers or handicapped with newly recruited staff. See Van den Brakel and Renssen (1998) and Van den Brakel (2008) for more details about issues concerning the field staff in embedded experiments.

In each application the right trade-off between the number of treatments in one experiment and the accompanying practical problems must be established carefully. Users generally expect that the effect of each separate factor that has varied in the survey process can be quantified. This generally requires a factorial design, which is difficult to apply in the fieldwork of a survey process, since the number of treatment combinations grows rapidly. In practice it is usually necessary to combine the factors that changed into one treatment and test the total effect against the standard alternative in a two-treatment experiment. This implies that the effects of all factors in the experiment are confounded and cannot be estimated separately.

Another consideration is the minimum required sample size. An indication is required about the size of the treatment effects that should at least result in a rejection of the null hypothesis at pre-specified levels of significance and power. Based on these, the minimum subsample sizes can be determined by an appropriate power calculation, see for example Montgomery (2001).

The standard literature on design and analysis of experiments applies model-based inference procedures for the analysis of experiments (Montgomery, 2001; Hinkelmann and Kempthorne, 1994, 2005). In this case estimates for the discontinuities are obtained from the estimated treatment effects of a linear model underlying an appropriate ANOVA for the applied experimental design. A drawback of this approach is that the sample design is ignored, which might result in biased estimates for the discontinuities if the sample design is not self-weighting, as well as incorrect variance estimates if, for example, stratification or clustering is ignored. A design-based inference procedure that accounts for the sample design, as well as the superimposition of the applied experimental design on the sampling design, might therefore be a more appropriate approach for the analysis of embedded experiments.

## 7.3 Design-Based Inference for Field Experiments

### 7.3.1 Measurement Error Model and Hypotheses Testing

Consider a finite population of $N$ units. Let $u_i$ denote the true but not directly observable target variable of the $i$-th population unit ($i = 1, \ldots , N$). An experiment is conducted to test the effect of one factor on $K \geq 2$ treatment levels at the true population mean $\bar{U} = 1/N \sum_{i=1}^{N} u_i$. If, for example, the purpose is to test the differences between four different data collection strategies, there is one factor 'data collection mode' on four different levels, so $K = 4$. Estimating and testing systematic differences between finite population parameters observed under different data collection modes (or other differences in the survey process) implies the existence of measurement errors. Consequently, a measurement error model is required to explain systematic differences between a finite population parameter observed under different data collection modes.

Let $y_{ik}$ denote the observation obtained from respondent $i$ assigned to treatment $k$. One approach is to assume that responses obtained in an experiment can be modeled as

$$y_{ik} = u_i + \gamma_k + \varepsilon_{ik}, \tag{7.1}$$

with $u_i$ the true intrinsic value of the variable of interest of respondent $i$, $\gamma_k$ a systematic treatment effect or measurement bias related to the $k$-th treatment or data collection mode, and $\varepsilon_{ik}$ a random measurement error for respondent $i$ observed under treatment $k$. This model can be easily extended to allow for interviewer effects (Van den Brakel and Renssen, 2005). Without loss of generality, interviewer effects are ignored here since in many mixed data collection mode, interviewers participate only in some modes. In such situations interviewer effects cannot be controlled for, e.g., through blocking. In such cases systematic interviewer effects become part of the treatment effects, i.e., the systematic differences between the different data collection strategies, and are absorbed in $\gamma_k$.

Let $E_m$ and $Cov_m$ denote the expectation and (co)variance with respect to the measurement error model. It is assumed that $E_m(\varepsilon_{ik}) = 0$, $Var_m(\varepsilon_{ik}) = \sigma_{ik}^2$, $Cov_m(\varepsilon_{ik}\,\varepsilon_{i'k}) = 0$, and $Cov_m(\varepsilon_{ik}\,\varepsilon_{ik'}) = \sigma_{ikk'}$. Note that variances and covariances of measurement errors can vary between respondents, see Van den Brakel and Renssen (2005) for details. From the assumptions it follows that measurement errors of different population units are independent. Let $\bar{Y}_k$ denote the population mean of $\bar{U}$ observed under the $k$-th treatment. From the measurement error model it follows that $\bar{Y}_k = \bar{U} + \gamma_k + \bar{\varepsilon}_k$, with $\bar{\varepsilon}_k$ the means of the random measurement errors. Then $\bar{Y} = (\bar{Y}_1, \ldots, \bar{Y}_K)^t$ denotes the $K$ dimensional vector with population means observed under the different treatments of the experiment.

The purpose of the experiment is to test the hypothesis that the population means observed under the different treatments (data collection strategies) are equal against the alternative that at least one pair is significantly different. Only systematic differences between the data collection modes in $\gamma_k$ should lead to a rejection of the null hypothesis. Random deviations due to measurement errors should not lead to significant differences in the analysis. Therefore, hypotheses are formulated about $\bar{Y}$ in expectation over the measurement error model, i.e.,

$$H_0 : CE_m\bar{Y} = 0$$
$$H_1 : CE_m\bar{Y} \neq 0$$

(7.2)

where $C = (j\,|-I)$ denotes a $(K-1) \times K$ matrix with the $(K-1)$ contrasts or treatment effects, $0$ and $j$ vectors of order $(K-1)$ with each element equal to zero and one respectively and $I$ an identity matrix of order $(K-1)$. Since $E_m\bar{Y}_k = \bar{U} + \gamma_k$ it follows that $CE_m\bar{Y} = (\gamma_1 - \gamma_2, \ldots, \gamma_1 - \gamma_K)^t$ exactly corresponds to the treatment effects, in this application the relative differences between measurement bias of the different data collection modes. Let $\hat{\bar{Y}}$ denote a design-unbiased estimator for $E_m\bar{Y}$ and $V\left(C\hat{\bar{Y}}\right)$ the covariance matrix of the contrasts between $\hat{\bar{Y}}$. Now hypothesis (7.2) can be tested with the Wald statistic

$$W = \hat{\bar{Y}}^t C^t V\left(C\hat{\bar{Y}}\right)^{-1} C\hat{\bar{Y}}$$

(7.3)

In the case of a two-treatment experiment, hypothesis (7.2) can be formulated as

$$H_0 : E_m\bar{Y}_1 = E_m\bar{Y}_2$$
$$H_1 : E_m\bar{Y}_1 \neq E_m\bar{Y}_2 \text{ or } E_m\bar{Y}_1 < E_m\bar{Y}_2 \text{ or } E_m\bar{Y}_1 > E_m\bar{Y}_2$$

(7.4)

which can be tested with a t-type test statistic

$$t = \frac{\hat{\bar{Y}}_1 - \hat{\bar{Y}}_2}{\sqrt{Var\left(\hat{\bar{Y}}_1 - \hat{\bar{Y}}_2\right)}}$$

(7.5)

### 7.3.2 Parameter and Variance Estimation

To test hypotheses (7.2) and (7.4) using test statistics (7.3) and (7.5), design-based estimators for the population means in $\bar{Y}$ and its covariance matrix of the contrasts are required. A general approach is to use the general regression (GREG) estimator proposed by Särndal, Swensson, and Wretman (1992). As a first step, the first-order inclusion probabilities that a sampling unit is drawn in the sample and assigned to one of the $K$ treatments are computed. Consider a sample $s$ of size $n$ drawn by a generally complex sample design that can be described with first- and second-order inclusion probabilities $\pi_i$ and $\pi_{ii'}$ of the $i$-th and $i,i'$-th sampling unit(s), respectively. In the case of a CRD, $s$ is randomly divided into $K$ subsamples $s_k$ of size $n_k$. From this it follows that the inclusion probability that the $i$-th sampling unit is included in $s_k$ equals $\pi_i^* = \pi_i(n_k/n)$. In the case of an RBD, $s$ is deterministically divided in $B$ blocks of size $n_b$ ($b = 1,\ldots,B$). Subsequently, within each block $n_{bk}$ sampling units are randomly assigned to subsample $s_k$. As a result the inclusion probability that the $i$-th sampling unit is included in $s_k$ equals $\pi_i^* = \pi_i(n_{bk}/n_b)$. See Van den Brakel and Renssen (2005) for a more detailed motivation of these inclusion probabilities.

Let $x$ denote a $H$-vector with auxiliary information and $\bar{X}$ the corresponding $H$-vector containing the finite population means known from an administrative data source or census. Furthermore, $\beta$ denotes a $H$-vector with the regression coefficients of the linear regression model $u_i = \beta^t x_i + e_i$. The $n_k$ observations collected in subsample $k$ can be used to estimate the population mean $\bar{Y}_k$ with the GREG estimator, which is defined as

$$\hat{\bar{Y}}_{k;r} = \hat{\bar{Y}}_k + \hat{\beta}_k^t\left(\bar{X} - \hat{\bar{X}}_k\right), k = 1,\ldots,K, \tag{7.6}$$

with

$$\hat{\bar{Y}}_k = \frac{1}{N}\sum_{i=1}^{n_k}\frac{y_{ik}}{\pi_i^*}, \text{ and } \hat{\bar{X}}_k = \frac{1}{N}\sum_{i=1}^{n_k}\frac{x_i}{\pi_i^*}$$

the Horvitz-Thompson estimators for $\bar{Y}_k$ and $\bar{X}$, and

$$\hat{\beta}_k = \left(\sum_{=1}^{n_k}\frac{x_i x_i^t}{\pi_i^*}\right)^{-1}\sum_{=1}^{n_k}\frac{x_i y_{ik}}{\pi_i^*} \tag{7.7}$$

a Horvitz-Thompson type estimator for the regression coefficients $\beta$ of the underlying linear model that explains the relation between $u_i$ and $x_i$. Now $\hat{\bar{Y}}_r = \left(\hat{\bar{Y}}_{1;r},\ldots,\hat{\bar{Y}}_{K;r}\right)^t$ is an approximately design-unbiased estimator for $\bar{Y}$ and also for $E_m\bar{Y}$ by definition. An estimator for the covariance matrix of the contrasts between the elements of $\hat{\bar{Y}}_r$, where the covariance is taken over the sampling design, the experimental design, and the measurement error model, is given by

$$\widehat{Cov}\left(C\hat{\bar{Y}}_r\right) = C\hat{D}C^t \tag{7.8}$$

In the case of a CRD, $\hat{D}$ is a $K \times K$ diagonal matrix with elements

$$\hat{d}_k = \frac{1}{n_k(n_k-1)}\sum_{i=1}^{n_k}\left(\frac{n\hat{e}_{ik}}{N\pi_i} - \frac{1}{n_k}\sum_{i'=1}^{n_k}\frac{n\hat{e}_{i'k}}{N\pi_{i'}}\right)^2 \equiv \frac{\hat{S}_{E_k}^2}{n_k} \tag{7.9}$$

with $\hat{e}_{ik} = y_{ik} - \hat{\beta}_k^t x_i$. In the case of an RBD, $\hat{D}$ is a $K \times K$ diagonal matrix with elements

$$\hat{d}_k = \sum_{b=1}^{B} \frac{1}{n_{bk}(n_{bk}-1)} \sum_{i=1}^{n_{bk}} \left( \frac{n_b \hat{e}_{ik}}{N\pi_i} - \frac{1}{n_{bk}} \sum_{i'=1}^{n_{bk}} \frac{n_b \hat{e}_{i'k}}{N\pi_{i'}} \right)^2 \equiv \sum_{b=1}^{B} \frac{\hat{S}_{Ebk}^2}{n_{bk}} \tag{7.10}$$

A full proof of this result is given by Van den Brakel and Renssen (2005) under the assumption of a weighting model for the GREG estimator for which it holds that a constant $H$-vector $a$ exists such that $a^t x_i = 1$ for all elements in the population. This is a relatively weak condition, since it assumes that the size of the finite target population is known and is used as auxiliary information in the GREG estimator. As an alternative the residuals $\hat{e}_{ik}$ can be multiplied with correction weights of the GREG estimator, which are also called $g$-weights (Särndal, Swensson, and Wretman, 1992, result 6.6.1). Note that the variance estimator has an attractive simple structure as if the $K$ subsamples are drawn independently from each other, where sampling units are selected with unequal selection probabilities with replacement. No joint inclusion probabilities or design covariances between the different subsamples are required, which simplifies the analysis considerably. See Van den Brakel and Renssen (2005) for a more detailed discussion and interpretation of this result.

Comparing the covariance structure (7.9) and (7.10) illustrates the importance of blocking on sampling structures like strata, clusters, or PSUs. Note for example that the variance reduction of stratified sampling is only preserved in the variance of the treatment effects of an embedded experiment if the strata of the sample design are used as block variables. In the case of unrestricted randomization of a CRD, the between-stratum variance will be reintroduced in the variance of the treatment effects.

The condition that a constant $H$-vector $a$ exists such that $a^t x_i = 1$ for all elements in the population precludes the ratio estimator and the Horvitz-Thompson estimator for the proposed design-based inference procedure. As an alternative for the HT estimator, a GREG estimator with weighting scheme $x_i = (1)$ for all elements in the population can be used. This weighting scheme is known as the common mean model and only uses the size of the finite population as a priori knowledge (Särndal, Swensson, and Wretman, 1992, Section 7.4). Under this weighting scheme it follows that

$$\hat{\bar{Y}}_{k;r} = \left( \sum_{i=1}^{n_k} \frac{1}{\pi_i^*} \right)^{-1} \left( \sum_{i=1}^{n_k} \frac{y_{ik}}{\pi_i^*} \right) \equiv \tilde{y}_k \tag{7.11}$$

which can be recognized as the ratio estimator for a population mean, originally proposed by Hájek (1971). In this case the covariance matrix can be estimated by (7.9) or (7.10) with $\hat{\beta}_k^t x_i = \tilde{y}_k$.

In the case of small sample sizes, it might be efficient to pool the population variances $\hat{S}_{Ek}^2$ and $\hat{S}_{Ebk}^2$ to obtain more stable estimates. For a CRD, $\hat{S}_{Ek}^2$ can be pooled over the treatments:

$$\hat{S}_{Ep}^2 = \frac{1}{(n-K)} \sum_{k=1}^{K} \sum_{i=1}^{n_k} \left( \frac{n\hat{e}_{ik}}{N\pi_i} - \frac{1}{n_k} \sum_{i'=1}^{n_k} \frac{n\hat{e}_{i'k}}{N\pi_{i'}} \right)^2 \tag{7.12}$$

As an alternative for $\hat{S}_{Ek}^2$ in (7.9). For an RBD, the population variances can be pooled within each block

$$\hat{S}^2_{Ep} = \frac{1}{(n_b - K)} \sum_{k=1}^{K} \sum_{i=1}^{n_{bk}} \left( \frac{n_b \hat{e}_{ik}}{N \pi_i} - \frac{1}{n_{bk}} \sum_{i'=1}^{n_{bk}} \frac{n_b \hat{e}_{i'k}}{N \pi_{i'}} \right)^2 \tag{7.13}$$

as an alternative for $\hat{S}^2_{Ebk}$ in (7.10).

### 7.3.3 Wald Test and Related Tests

The design-unbiased estimators for the subsample means and the covariance matrix give rise to the following Wald statistic:

$$W = \hat{\bar{Y}}^t_r C^t \left( C \hat{D} C^t \right)^{-1} C \hat{\bar{Y}}_r \tag{7.14}$$

In Van den Brakel and Renssen (2005) it is proved that this expression can be simplified to

$$W = \sum_{k=1}^{K} \frac{\hat{\bar{Y}}^2_{k,r}}{\hat{d}_k} - \left( \sum_{k=1}^{K} \frac{1}{\hat{d}_k} \right)^{-1} \left( \sum_{k=1}^{K} \frac{\hat{\bar{Y}}_{k,r}}{\hat{d}_k} \right)^2 \tag{7.15}$$

Under general complex sampling designs there is no theorem that states that $\hat{\bar{Y}}_r$ has a multivariate normal limit distribution. Therefore, it can be conjectured that $\hat{\bar{Y}}_r$ is asymptotically multivariate normally distributed and consequently $W$ is asymptotically chi-squared distributed with $(K - 1)$ degrees of freedom.

For the analysis of two-sample experiments and to test hypotheses defined by (7.4), the following design-based t-type statistic is obtained:

$$t = \frac{\hat{\bar{Y}}_{1;r} - \hat{\bar{Y}}_{2;r}}{\sqrt{\hat{d}_1 + \hat{d}_2}} \tag{7.16}$$

For general complex sample designs it is conjectured that $t$ is asymptotically standard normally distributed.

In the case of a CRD embedded in a self-weighted sample design with (i) the use of the ratio estimator for a population mean defined by (7.11), (ii) the pooled variance estimator defined by (7.12), it can be shown that $W/(K - 1)$ is equal to the F-statistic of an ANOVA for a one-way layout, if it is conjectured that $W$ is asymptotically chi-squared distributed with $(K - 1)$ degrees of freedom.

In the case of RBD embedded in a self-weighted sample design where (i) sampling units are allocated proportionally to the treatments over the blocks (i.e., $n_{bk} / n_b = n_{b'k} / n_{b'}$ for all $b, b'$), (ii) the use of the ratio estimator for a population mean defined by (7.11), (iii) the pooled variance estimator defined by (7.13), it can be shown that $W/(K - 1)$ is equal to the F-statistic of an ANOVA for a two-way layout with interaction (if it is conjectured that $W$ is asymptotically chi-squared distributed with $(K - 1)$ degrees of freedom).

In the case of a self-weighted sample design and a two-treatment experiment designed as a CRD with the use of estimator (7.11), it can be shown that (7.16) equals Welch's t-statistic (Miller, 1986). If in addition the pooled variance estimator (7.12) is used, then it follows that (7.13) is equal to the standard t-statistic. In both cases it is conjectured that $t$ is asymptotically standard normally distributed.

## 7.3.4 Extensions

Population parameters are often defined as the ratio of two population means or totals. Testing hypotheses about ratios of two survey estimates requires different point and variance estimators for the Wald statistic. Let $R_k = \bar{Y}_k / \bar{Z}_k$ denote the ratio of two population means observed under treatment $k = 1, \ldots, K$. The $K$ ratios observed under the different treatments can be collected in the $K$ dimensional vector $\mathbf{R} = (R_1, \ldots, R_K)^t$. Hypotheses about differences between the ratios are formulated in a similar way as in (7.2) and (7.4), where the numerator and denominator both denote the population mean in expectation over the measurement error model. These hypotheses are tested with a design-based Wald test (7.15) or t-type test (7.16), where the GREG estimates $\hat{\bar{Y}}_{k,r}$ are replaced by the ratio of the GREG estimates of $\bar{Y}_k$ and $\bar{Z}_k$. The GREG estimator for $\bar{Y}_k$ is defined in (7.6). The GREG estimator for $\bar{Z}_k$ is defined similarly. The variance components in (7.15) and (7.16) are defined as

$$\hat{d}_k^{(R)} = \frac{1}{\hat{\bar{Z}}_{k;r}^2} \frac{1}{n_k(n_k-1)} \sum_{i=1}^{n_k} \left( \frac{n\hat{e}_{ik}}{N\pi_i} - \frac{1}{n_k} \sum_{i'=1}^{n_k} \frac{n\hat{e}_{i'k}}{N\pi_{i'}} \right)^2 \tag{7.17}$$

in the case of a CRD and

$$\hat{d}_k^{(R)} = \frac{1}{\hat{\bar{Z}}_{k;r}^2} \sum_{b=1}^{B} \frac{1}{n_{bk}(n_{bk}-1)} \sum_{i=1}^{n_{bk}} \left( \frac{n_b\hat{e}_{ik}}{N\pi_i} - \frac{1}{n_{bk}} \sum_{i'=1}^{n_{bk}} \frac{n_b\hat{e}_{i'k}}{N\pi_{i'}} \right)^2 \tag{7.18}$$

in the case of an RBD. The residuals in (7.17) and (7.18) are defined as $\hat{e}_{ik} = (y_{ik} - \hat{\beta}_k^{yt} x_i) - \hat{R}_{k;r}(z_{ik} - \hat{\beta}_k^{zt} x_i)$. Here $\hat{\beta}_k^z$ denotes the Horvitz-Thompson type estimator for the regression coefficients of the regression function of $z_{ik}$ on $x_i$, and is defined in a similar way as (7.7).

Expressions for the Hájek estimator are now obtained in a straightforward way by taking $\tilde{R}_k = \tilde{y}_k / \tilde{z}_k$, where $\tilde{y}_k$ is defined by (7.9) and $\tilde{z}_k$ is defined analogously. An approximation of the covariance matrix of the contrasts between the subsample estimates is defined by (7.11) with residuals $\hat{e}_{ik} = (y_{ik} - \tilde{y}_k) - \hat{R}_{k;r}(z_{ik} - \tilde{z}_k)$.

So far, the ultimate sampling units are randomized over the treatments in an experiment. Due to fieldwork restrictions, clusters of sampling units, instead of separate sampling units, might be randomized over the treatments. In the case of testing different data collection modes it might be impractical to assign different household members of a sampled household to different treatments, in particular if one or more data collection modes are interviewer driven. At the cost of the reduced power of the experiment, clusters of sampling units can be randomized over the treatments. Consequently, experimental units do not coincide with the ultimate sampling units of the sampling design. The fact that clusters of sampling units are randomized over the treatments must be accounted for in the GREG estimator for $\bar{Y}_k$ (and $\bar{Z}_k$ in the case of ratio's) and in the variance approximations of these GREG estimators. Details for different situations are given by Van den Brakel (2008).

From standard experimental design theory it is well known that it is efficient to test different treatment factors simultaneously in one factorial design instead of conducting separate single-factor experiments (Hinkelmann and Kempthorne, 1994; Montgomery, 2001). It can be expected that different design parameters in a survey process interact with each other, e.g., when different questionnaire designs and data collection modes are compared empirically. Factorial setups are indeed appropriate if more than one factor in the survey is adjusted and tested in an embedded experiment, since fewer experimental units are required to test the main effects of the treatment factors while interactions between the factors can be analyzed. Another advantage of testing different treatments simultaneously in a factorial design is that the validity of the observed results is extended, since the effects are observed over a wider range of conditions (Hinkelmann and Kempthorne, 1994).

The most simple set-up is a $K \times L$ factorial design, where the effects of two factors are tested simultaneously. The first factor, denoted $A$, contains $K \geq 2$ levels, for example factor 'data collection mode' at four different levels: CAPI, CATI, PAPI and CAWI. The second factor, denoted as $B$, contains $L \geq 2$ levels, for example 'questionnaire', at three different levels, which implies that three different types of questionnaires are compared. The purpose of the experiment is to test the main effects of the two factors and the interactions between both factors on the main parameter estimates of the ongoing survey. In a full factorial design, all $K \times L$ possible treatment combinations are tested. In the case of an embedded experiment, a probability sample is drawn from a finite target population. Subsequently, this sample is randomly divided into $K \times L$ subsamples according to a CRD or an RBD. The design-based inference approach for one-factor designs, described in Section 7.3, can be extended for the analysis of factorial designs. Point and variance estimators proposed in Subsection 7.3.2 are now applied to the $K \times L$ treatment combinations. Hypotheses of main effects and interaction effects can be tested with design-based Wald tests, equivalent to the set-up in Subsection 7.3.3. Only the specification of the contrast matrices that specify the hypotheses on main and interaction effects becomes more complex. See Van den Brakel (2013) for the details of a design-based inference approach for embedded factorial designs, including extensions to higher-order factorial designs, i.e., with more than two factors.

### 7.3.5 Software

The design-based analysis procedures described in this section are implemented in a software package, called X-tool. This package is available as a component of the Blaise survey processing software package, developed by Statistics Netherlands (Statistics Netherlands, 2002). X-tool supports the design-based inference to test hypotheses about differences between population parameters observed under different survey implementations in randomized experiments embedded in generally complex probability samples. X-tool handles experiments designed as CRDs and RBDs. It is possible to analyze experiments where the ultimate sampling units, as well as clusters of sampling units, are randomized over the different treatments. Subsample estimates for means, totals, and ratios are based on the Hájek estimator or the GREG estimator. The integrated method for weighting individuals and households of Lemaître and Dufour (1987) can be applied under the GREG estimator to obtain equal weights for individuals belonging to the same household. Also, a bounding algorithm based on Huang and Fuller (1978) can be applied to avoid negative correction weights. See Van den Brakel (2008) for more details on the functionality of X-tool.

Alternatively, these methods can be implemented in R, where point and variance estimates for the GREG estimator can be obtained with the R-package 'survey', Lumley (2010). Once the point and variance estimates are available, the implementation of the test statistics is relatively straightforward.

## 7.4 Time Series Methods

A strong advantage of quantifying discontinuities with a parallel run is the low level of risk of disturbing the regular survey during the implementation of a new data collection method. An experimental set-up offers full control over the minimum detectable differences via variance and power calculations. In the case of sufficiently large parallel runs, estimates for discontinuities are timely and not subject to revisions. A major drawback, however, is that this method is not cost neutral since additional data are collected. Due to budget and field work capacity constraints, parallel runs are not always possible. In such situations, the time series of observations obtained with the repeated survey can be used to estimate the discontinuity using the autocorrelation in the series. A convenient way to estimate a discontinuity in a time series is through a state-space model fitted with a Kalman filter. The state-space approach can also combine information from an embedded experiment with the information from the time series observed before and after the change-over.

With a structural time series model, a series is decomposed into a trend component, seasonal component, other cyclic components, regression component, and an irregular component. For each component a stochastic model is assumed. The trend, seasonal, and cyclic components are usually time dependent, but this approach also allows the regression coefficients to be time dependent. If necessary, ARMA components can be added to capture the autocorrelation in the series beyond these structural components. See Harvey (1989) or Durbin and Koopman (2012) for details about structural time series modeling and fitting these models with a Kalman filter.

Modeling time series observed with repeated surveys requires a measurement error model. Following the notation of Section 7.3, $\bar{Y}_{k,t}$ denotes the true population mean of $\bar{U}_t$ for period $t$, observed under the $k$-th survey implementation. Furthermore, $\hat{\bar{Y}}_{k,t;r}$ denotes the GREG estimator for $\bar{Y}_{k,t}$, using the sample observed in period $t$. A measurement error model for the observed series is given by $\hat{\bar{Y}}_{k,t;r} = \bar{Y}_{k,t} + e_{k,t}$, where $e_{k,t}$ denotes the sampling error of $\hat{\bar{Y}}_{k,t;r}$. Inserting the measurement error model for $\bar{Y}_{k,t}$ from Subsection 7.3.1 gives $\hat{\bar{Y}}_{k,t;r} = \bar{U}_t + \gamma_k + e_{k,t}$, where it is assumed that $\gamma_k$ is time independent.

To minimize the amount of notation, the methods are explained for a time series observed at an annual frequency. If $\hat{\bar{Y}}_{k,t;r}$ are annual estimates, $\bar{U}_t$ refers to annual population means, which can be modeled with a stochastic trend, say $L_t$, and a white noise, say $I_t$, for the unexplained variation, i.e., $\bar{U}_t = L_t + I_t$. Frequently applied models for $L_t$ in econometric time series modeling are the local level model, the smooth trend model, and the local linear trend model, see Durbin and Koopman (2012, Ch. 3) for expressions. These are stochastic models that can change gradually over time and are therefore appropriate to model the low-frequency variation in the time series. These models have the flexibility to model business cycles, but it is also possible to include a separate component in addition to $L_t$. If $\hat{\bar{Y}}_{k,t;r}$ are monthly or quarterly estimates, a stochastic seasonal component to account for seasonal effects will generally be required. Frequently applied models for seasonal patterns are the so-called dummy seasonal model and the trigonometric seasonal model, see Durbin and Koopman (2012, Ch. 3) for expressions. Both are stochastic models and therefore allow for seasonal patterns that gradually change-over time.

Inserting the structural time series model for $\bar{U}_t$ in the measurement error model gives $\widehat{Y}_{k,t;r} = L_t + I_t + \gamma_k + e_{k,t}$. In the case of cross-sectional surveys, the population white noise and the sampling error are confounded and can be combined into one error term, say $v_{k,t} = I_t + e_{k,t}$. It is assumed that the disturbance terms are normally and independently distributed, $v_{k,t} \cong N\left(0, \sigma_{k,t}^2\right)$. To account for heteroscedasticity due to changing sampling sizes or design over time, the variances of the disturbance terms are time dependent. This can be accomplished by making the variance proportional to the variance of the GREG estimates; i.e., $\sigma_{k,t}^2 = Var\left(\widehat{Y}_{k,t;r}\right)\sigma^2$, where $Var\left(\widehat{Y}_{k,t;r}\right)$ is estimated from microdata and used as a priori information in the time series model.

The subscript $k$ differentiates between observations before and after the change-over to a new data collection method. So $\widehat{Y}_{k,t;r}$ refers to sample estimates observed under the old approach and $\widehat{Y}_{k',t;r}$ to estimates observed under the new data collection approach. To separate period-to-period change from the systematic differences in measurement bias due to the introduction of a new data collection approach, an intervention component is added to model the discontinuity in the outcomes at the moment that the new design is introduced. The most straightforward approach is a level intervention, which means that a dummy indicator, say $\delta_t$, is added to the model that changes from zero to one at the moment of the change-over to the new survey process. The regression coefficient, say $\beta$, of this intervention variable can be interpreted as the discontinuity induced by the redesign of the survey process. Note that the measurement bias $\gamma_k$ induced by a particular treatment or survey implementation cannot be observed using the survey data only. Similar to parallel runs and experiments, this approach only allows estimating the relative difference in measurement bias between the survey process before and after the change-over, i.e., $\beta = \gamma_k - \gamma_{k'}$ (where $k$ and $k'$ refer to the data collection mode before and after the change-over, respectively). The measurement bias $(\gamma_k)$ of the survey approach before the change-over will typically be absorbed in the trend component of the population parameter, i.e., $\tilde{L}_t = L_t + \gamma_k$. This gives rise to the following structural time series model for the observed series: $\widehat{Y}_{k,t} = \tilde{L}_t + \beta\delta_t + v_{k,t}$. Note that $\beta\delta_t$ is a regression component with $\delta_t$ the auxiliary variable and $\beta$ the regression coefficient. Under the assumption that the other time series components correctly model the evolution of the population variable, the regression coefficient can be interpreted as the discontinuity due to the introduction of a new data collection strategy.

This state-space intervention model is proposed by Harvey and Durbin (1986) for analyzing the effects of seatbelt legislation on road casualties in the UK. Van den Brakel and Roels (2010) applied this to series obtained with repeated surveys to estimate discontinuities for situations where there is no parallel data collection.

Estimating discontinuities in a series of related target variables independently from each other according to univariate time series models will give rise to inconsistencies between the estimates. Consider, for example, discontinuities that are quantified for a series at the national level and its breakdown over different subclasses or domains. In this case there might be a requirement that the sum over the domain discontinuities is equal to the discontinuity at the national level if the variable is defined as a total. Another example is a categorical variable measured on an ordinal scale of which the population value of interest is the distribution over the $J$ categories of these variables. In this case the sum over the $J$ discontinuities must equal zero. There are several ways to impose these restrictions.

One approach is to benchmark the initial estimates obtained with a univariate time series model to the required restriction with a Lagrange function. Another approach is to combine the related series in a multivariate time series model and augment this model with an additional restriction on the regression coefficients of the intervention parameters. Both approaches are discussed in detail in Van den Brakel et al. (2008) and Van den Brakel and Roels (2010). Bollineni et al. (2016) proposed a multivariate model for the domains only. Discontinuities for the national level are derived as a linear combination from the estimated domain discontinuities.

The discontinuity in the series is modeled with an intervention variable that describes the moment when the survey process is redesigned. This approach assumes that the other components of the time series model approximate the real development of the population variable reasonably well and that there is no structural change in, e.g., the trend or the seasonal component at the same moment at which the new survey is implemented. If a change in the real development of the population variable exactly coincides with the implementation of the new survey, then the model will wrongly assign this effect to the intervention variable which is intended to describe the redesign effect. Information available from a series of correlated variables outside the survey can be used to evaluate the assumption that there is no structural change in the real evolution of the parameter. There are different ways to incorporate auxiliary information in the model. One possibility is to extend the time series model with a regression component for the auxiliary series, i.e., $\hat{\bar{Y}}_{k,t} = \tilde{L}_t + bx_t + \beta\delta_t + v_{k,t}$, where $x_t$ denotes the auxiliary series and $b$ the regression coefficient. An auxiliary series can also be included as a dependent variable in a multivariate model, where both series are modeled with their own trend, seasonal component, and white noise. The correlation between both series can be modeled, e.g., by specifying a full covariance matrix for the disturbance terms of the trend components. In this way, additional information from the auxiliary series is used to better separate discontinuities from real developments and reduce the risk that a structural change in the evolution of the series of the target parameter is wrongly assigned to the intervention variable. Details of multivariate state-space models with correlated disturbance terms can be found in, e.g., Harvey (1989, Section 8.5) or Koopman et al. (2007, Section 9.1).

A widely applied approach to fit structural time series models is to write them in state-space form and analyze them with the Kalman filter. Expressions for the state-space representation of structural time series models can be found in Durbin and Koopman (2012). The Kalman filter is a recursive procedure that runs from period $t = 1$ to $T$ and gives, for each time period, an optimal estimate for the state variables based on the information available up to and including period $t$. These estimates are referred to as the filtered estimates. The filtered estimates of past state vectors can be updated, if new data after period $t$ become available. This procedure is referred to as smoothing and results in smoothed estimates that are based on the complete time series. To start the filter at $t = 1$, the filter must be initialized with starting values for the state variables and its covariance matrix for $t = 0$. In the absence of additional information, for non-stationary state variables, i.e., the states for the trend, seasonal, and regression coefficients, a so-called diffuse initialization is generally used. This implies that the expectation of the initial states for these state variables are equal to zero and the initial covariance matrix of the states is diagonal with elements diverging to infinity.

So far, the time series method is introduced for situations without parallel data collection. In many cases, there will be some budget for a small parallel run. This will, however, be insufficient to meet pre-specified precision requirements. An advantage of the structural time series modeling approach is that it has the flexibility to combine information in the entire series with the information obtained with parallel data collection. As explained above, in the absence of any a priori information, the Kalman filter is started at $t = 1$, using a diffuse initialization of the non-stationary state variables. One way to incorporate the initial information about the discontinuities available from the parallel run in the model is to use an exact initialization for the regression coefficients $\beta$. This can be done by using the direct estimate for the discontinuity obtained from the parallel run in the initial state vector for $\beta$ and the estimated variance of this direct estimate as an uncertainty measure for $\beta$ in the covariance matrix of the initial state vector. In this way the Kalman filter combines the information from the parallel run with the information available in the entirely observed series before and after the change-over to the new design.

A major advantage of the time series approach is that no additional data collection is required, which makes this approach very cost effective. The time series modeling approach uses all available data under both the old and the new approach, since the entire observed series is used to estimate discontinuities. This information can be combined with the information from a parallel run through an exact initialization of the Kalman filter.

Not conducting parallel data collection and relying only on a time series model to estimate discontinuities has several disadvantages and risks. If after the change-over, the new approach turns out to be a failure and it is decided to fall back on the old approach, then there is a period where no data are available for the production of official statistics. Another factor that contributes to an increased level of risk is that real developments and estimates for the discontinuities are confounded if the real evolution of the population parameter deviates from the assumed time series model and a sudden turning point is incorrectly absorbed in the estimate for the discontinuity. Furthermore, estimates for the discontinuities will change if new observations under the new survey become available. As a consequence, revisions of discontinuity estimates must be accepted. The size of these revisions mainly depends on the dynamics of the trend component. As the trend component becomes more volatile, only local observations before and after the change-over influence the level of the trend. This reduces the size of revisions of the estimated discontinuities. See Van den Brakel and Roels (2010) for more details. As a result of these revisions, final estimates are not timely. Finally, this method does not offer control over the precision and size of the minimum observable differences, contrary to a parallel run designed as an embedded experiment, where power calculations offer full control over the minimum observable differences in the design phase of the experiment.

Several software packages are available to fit structural time series models. Most standard structural time series models can be fitted with STAMP (Koopman et al., 2007). For more advanced models, more advanced software is required. One option is to implement these models in OxMetrics in combination with the subroutines of SsfPack 3.0 (Doornik, 2009; Koopman et al., 1999, 2008). Another possibility is to implement these models in R (R Core Team, 2017) using packages like KFAS (Helske, 202021) or DLM (Petris, 2010).

## 7.5 Small Area Estimation Methods

In Section 7.4 it was already mentioned that in many applications there is a budget for a parallel run with a reduced sample size. In these cases, the regular survey, used for official publication purposes, will be conducted in full scale while the alternative approach is conducted with a limited sample size. The design-based approach, set out in Subsection 7.3, is appropriate to quantifying discontinuities at the national level. The sample size allocated to the alternative approach, however, will often be insufficient to produce design-based estimates of discontinuities of adequate precision for subpopulations or domains.

In the case of small domain sample sizes, model-based estimators can be used. These estimators employ sample information observed in other domains through an explicit statistical model and thus increase the effective sample size in the separate domains. In survey methodology, this type of estimation techniques is known as small area estimation, see Rao and Molina (2015) for a comprehensive overview.

In small area estimation, two types of models are commonly used. The first one is the basic area-level model, also known as the Fay-Herriot model (Fay and Herriot, 1979), where the input data for the model are the direct estimates for the domains. The second one is the nested error regression model of (Battese et al., 1988), which is often referred to as the basic unit-level model. In this model the input data are the observations obtained from the sampling units.

The extent to which model-based small area predictions result in a reduction of the mean squared error strongly depends on the auxiliary information available to define models that explain the variation between the domains of interest. In most applications, models borrow strength from auxiliary information which is available from censuses and registrations. In the case of a parallel run, however, direct estimates for the same variables and the same domains are available from the regular survey. For the planned domains, these direct estimates will be sufficiently precise and are potential auxiliary variables. This type of auxiliary information is available at the domain level. Therefore, the discussion below is restricted to Fay-Herriot models.

Using similar notation as in Sections 7.3 let $\hat{\bar{Y}}_{d,k}$ and $\hat{\bar{Y}}_{d,k'}$ denote the direct estimate under the regular approach $k$ (conducted at the regular large sample size) and alternative approach $k'$ (conducted at a reduced sample size) for domain $d$. To improve the precision of the $\hat{\bar{Y}}_{d,k'}$ these direct estimates are modeled in a Fay-Herriot multilevel model: $\hat{\bar{Y}}_{d,k'} = \bar{Y}_{d,k'} + e_{d,k'} = \alpha' X_d + \upsilon_{d,k'} + e_{d,k'}$, with $e_{d,k'}$ the sampling error, $X_d$ a vector with auxiliary variables to explain the domain variables, and $\alpha$ a vector with regression coefficients. Finally, $\upsilon_{d,k'}$ is the random component that models the unexplained variation between the domains. Obviously, $\hat{\bar{Y}}_{d,k}$ is highly correlated with $\hat{\bar{Y}}_{d,k'}$ since it measures the same population parameter only using a different measurement approach. Using $\hat{\bar{Y}}_{d,k}$, among others, as auxiliary variables in the vector $X_d$ leads to a Fay-Herriot model containing auxiliary information with error: $\hat{\bar{Y}}_{d,k'} = \alpha' \hat{X}_d + \upsilon_{d,k'} + e_{d,k}$. See Ybarra and Lohr (2008) for details of the empirical best linear unbiased estimators, say $\tilde{\bar{Y}}_{d,k'}$, for the Fay-Herriot model using auxiliary variables observed with sampling error.

Domain discontinuities are now estimated as $\hat{\Delta}_d = \hat{\bar{Y}}_{d,k} - \tilde{\bar{Y}}_{d,k'}$. The variance of the domain discontinuities is given by $Var(\hat{\Delta}_d) = Var(\hat{\bar{Y}}_{d,k}) + MSE(\tilde{\bar{Y}}_{d,k'}) - 2cov(\hat{\bar{Y}}_{d,k}, \tilde{\bar{Y}}_{d,k'})$. Since $\tilde{\bar{Y}}_{d,k'}$ uses $\hat{\bar{Y}}_{d,k}$ or related estimates from the regular survey as auxiliary variables in the model to construct a small area prediction, there is a strong positive correlation between $\tilde{\bar{Y}}_{d,k'}$ and $\hat{\bar{Y}}_{d,k}$. A design-based approximation for $cov(\hat{\bar{Y}}_{d,k}, \tilde{\bar{Y}}_{d,k'})$ is proposed by Van den Brakel, Buelens, and Boonstra (2016). They also derived a design-based approximation for $MSE(\tilde{\bar{Y}}_{d,k'})$ to avoid mixing design-based estimators for $Var(\hat{\bar{Y}}_{d,k})$ and $cov(\hat{\bar{Y}}_{d,k}, \tilde{\bar{Y}}_{d,k'})$ with a model-based estimator for $MSE(\tilde{\bar{Y}}_{d,k'})$.

As an alternative, the domain estimates of the regular and alternative approach can be modeled simultaneously in a bivariate random-effects model. See Benevant and Moralis (2016) and the reference therein for details. This approach improves the precision of both the estimates observed with the regular and the alternative approach using cross-sectional correlation and the correlation between the two domain parameters.

For the computation of small area estimates, several R packages are available, for example, the package *hbsae* (Boonstra, 2012) has been used. This package can be used to fit area level and unit level models in a Hierarchical Bayesian and a frequentist framework using maximum likelihood.

## 7.6 The Introduction of a Hybrid Mixed-Mode Design in the Dutch Crime Victimization Survey

### 7.6.1 Design

In the Netherlands, information on crime victimization, public safety, and satisfaction with police performance is obtained by the Dutch Crime Victimization Survey (CVS). This is a long-standing survey that has been redesigned several times. The CVS was based on a stratified simple random sampling of people aged 15 years or older residing in the Netherlands. The sampling frame is the Municipal Basis Administration, which is the Dutch government's registry of all residents in the country. The 25 police districts were used as the stratification variable in the sample design. In a regular yearly sample about 750 respondents were observed in each police district, resulting in a total net sample size of about 19,000 respondents. With a response rate that varies between 60% end 65% (in the years 2005–2008) this required a gross sample size of about 30,500 persons. Since police districts have unequal population sizes, inclusion probabilities varied between police districts. The estimation procedure was based on the GREG estimator, where the weighting scheme contains socio-demographic categorical variables like gender(2), age class(11), marital status(4), urbanization level(5), province(12), police region(25), and household size(5), where the number of categories is specified in brackets.

Until 2008 the data collection mode of the CVS was based on an adaptive mixed-mode design using CATI, if a telephone number of a landline telephone connection was available, and CAPI for the remaining sampling units. In 2008 the data collection mode of the CVS changed to a hybrid mixed-mode design that started with web interviewing. Respondents could also respond through PAPI on their own request, as a concurrent mixed-mode first phase. After three reminders nonrespondents were approached via CATI or CAPI, depending on the availability of a landline telephone number or cell phone number, i.e., the former adaptive mixed-mode design. The change-over to different data collection modes also required a redesign of the questionnaire. This new design was implemented in the regular survey in 2008. In order to maintain consistent series, the old approach was conducted in parallel on a considerably smaller scale in 2008 and is further referred to as the alternative approach. With this sample reliable design-based estimates for discontinuities can be obtained at the national level but not for the most important planned domains of this survey, the police districts. Therefore, model-based estimators were applied to construct estimates of sufficient accuracy for these domain discontinuities.

Discontinuities are analyzed for the following key variables of the CVS:

- Nuisance: the perceived nuisance in the neighborhood on a ten-point scale; this includes nuisance by drunk people, neighbors, or groups of youngsters, harassment, and drug-related problems;
- Unsafe: the percentage of people feeling unsafe at times;
- Propvict: the percentage of people saying to have been victim to property crime in the past 12 months;
- Offtot: the total number of offences per 100 people;
- Satispol: percentage of people satisfied with police at their last contact (if contact in past 12 months);

Police regions are the most important publication domains and were for that reason used as the stratification variable in the sample design. A natural choice for the parallel run is to set up a randomized experiment where police regions were used as the block variable in a randomized block design.

In planning a parallel run, a clear definition is needed about the treatments to be tested and the number of factors to be included in the experiment. In this case it was decided to estimate the overall difference between the old and the new design and not to explain the separate effects of a different data collection method and a different questionnaire. Confining the purpose of the experiment to estimating the net effect of a new data collection mode and new questionnaire requires a two-sample experiment, where the old and new approaches are compared. If, on the other hand, the purpose of the experiment was to explain the individual contributions of the two factors, then a 2 × 2 factorial design would have been required. This would have resulted in a smaller sample size for estimating the overall discontinuity of the new design. An important step in the design of an experiment is to decide in advance about which variables' hypotheses are to be tested and which treatment effect must be observed at a pre-specified significance and power level. Based on such considerations, the minimum required sample size of an experiment can be derived. Table 7.1 specifies the minimum detectable differences for Propvict and Satispol at a 5% significance level and different power levels. The first four rows present the obtained precision if the sample size of the regular survey is kept at its original size of 19,000 responses and the alternative sample gradually increases to 19,000 responses. The next three rows present the precision if with similar total sample sizes a balanced allocation is used; i.e., the subsample sizes are equal.

The available budget for parallel data collection allowed collecting additional data for about 3,750 respondents. The above-described power calculations resulted in a decision to reduce the sample size of the regular survey with 2,250 responses in order to obtain budget for a parallel run of about 6,000 responses. The precision obtained with this set-up is presented in the last row of Table 7.1.

For the sample assigned to the alternative survey, proportional allocation is used. With a sample of about 6,000 responses, sufficiently precise direct estimates at the national level are obtained, using the methods described in Subsection 7.3. This sample is, however, too small to obtain reliable direct estimates for the 25 police regions. Therefore, small area estimation methods, described in Subsection 7.5, are applied to obtain more precise estimates for the domain discontinuities.

**TABLE 7.1**

Minimum Detectable Difference Propvict and Satispol for Different
Sample Sizes at a Significance Level of 5% and Different Power Levels

| Propvict (13.73% in 2007) | | | | | |
| Respondents | | | Differences | | |
| Regular | Alternative | Total | Power 50% | Power 80% | Power 90% |
|---|---|---|---|---|---|
| 19,000 | 5000 | 24,000 | 1.47 | 2.11 | 2.43 |
| 19,000 | 10,000 | 29,000 | 1.14 | 1.64 | 1.89 |
| 19,000 | 15,000 | 34,000 | 1.01 | 1.45 | 1.67 |
| 19,000 | 19,000 | 38,000 | 0.95 | 1.36 | 1.57 |
| 12,000 | 12,000 | 24,000 | 1.19 | 1.71 | 1.98 |
| 14,500 | 14,500 | 29,000 | 1.09 | 1.56 | 1.80 |
| 17,000 | 17,000 | 34,000 | 1.00 | 1.44 | 1.66 |
| 16,750 | 6000 | 22,750 | 1.39 | 2.00 | 2.30 |

| Satispol (56.9% in 2007) | | | | | |
| Respondents | | | Differences | | |
| Regular | Alternative | Total | Power 50% | Power 80% | Power 90% |
|---|---|---|---|---|---|
| 19,000 | 5000 | 24,000 | 3.29 | 4.71 | 5.43 |
| 19,000 | 10,000 | 29,000 | 2.56 | 3.66 | 4.22 |
| 19,000 | 15,000 | 34,000 | 2.26 | 3.24 | 3.73 |
| 19,000 | 19,000 | 38,000 | 2.12 | 3.04 | 3.51 |
| 12,000 | 12,000 | 24,000 | 2.67 | 3.83 | 4.41 |
| 14,500 | 14,500 | 29,000 | 2.43 | 3.48 | 4.02 |
| 17,000 | 17,000 | 34,000 | 2.24 | 3.22 | 3.71 |
| 16,750 | 6000 | 22,750 | 3.11 | 4.46 | 5.14 |

## 7.6.2 Results

With the regular survey a response rate of 59.9% was obtained. This resulted in a sample of 16,964 responses. Under the alternative approach a response rate of 66.9% was realized, resulting in a sample of 6,113 responses. GREG estimator (7.6) is used to obtain direct estimates for the target variables under the regular and alternative approach. Variances are estimated with formula (7.10). Point and variance estimates for the five key variables of interest are summarized in Table 7.2. At the national level, discontinuities are clearly significant at a two-sided 5% significance level. The new data collection approach, in combination with the new questionnaire, resulted in systematically higher estimates for all five target variables.

The purpose of this parallel run was also to estimate discontinuities for the 25 police districts. As a first step the GREG estimator is applied to these 25 domains. Table 7.3 summarizes the results averaged over the domains. Standard errors for the variables under the regular and the alternative approach and thus also the standard errors for the discontinuities clearly increased. As a result, the average values of t-statistics are smaller than the critical value for a two-sided 5% significance level. The last column of Table 7.3 specifies the number of domains where the null hypothesis of no discontinuity is rejected at a two-sided 5% significance level.

**TABLE 7.2**

GREG Estimates Regular and Alternative Survey Approach at the National Level

| Variable | Regular | | Alternative | | Discontinuity | | |
|---|---|---|---|---|---|---|---|
| | $\hat{\bar{Y}}_{k;r}$ | $\sqrt{d_{k;r}}$ | $\hat{\bar{Y}}_{k';r}$ | $\sqrt{d_{k';r}}$ | $\hat{\Delta}$ | SE($\hat{\Delta}$) | t-statisitc |
| Offtot | 43.79 | 1.07 | 34.09 | 1.04 | 9.70 | 1.49 | 6.51 |
| Unsafe | 25.07 | 0.44 | 20.48 | 0.52 | 4.59 | 0.68 | 6.75 |
| Nuisance | 1.67 | 0.02 | 1.34 | 0.02 | 0.33 | 0.03 | 11.00 |
| Satispol | 59.88 | 0.92 | 55.10 | 1.25 | 4.78 | 1.55 | 3.08 |
| Propvict | 13.02 | 0.36 | 10.32 | 0.39 | 2.70 | 0.53 | 5.09 |

**TABLE 7.3**

GREG Domain Estimates Regular and Alternative Survey Approach Averaged over Districts

| Variable | Regular | | Alternative | | Discontinuity | | | |
|---|---|---|---|---|---|---|---|---|
| | $\hat{\bar{Y}}_{k;r}$ | $\sqrt{d_{k;r}}$ | $\hat{\bar{Y}}_{k';r}$ | $\sqrt{d_{k';r}}$ | $\hat{\Delta}$ | SE($\hat{\Delta}$) | t-statisitc | N.S.D. |
| Offtot | 42.29 | 4.73 | 33.28 | 5.73 | 9.01 | 7.69 | 1.17 | 3 |
| Unsafe | 24.38 | 2.03 | 19.86 | 2.87 | 4.52 | 3.57 | 1.30 | 8 |
| Nuisance | 1.61 | 0.11 | 1.28 | 0.13 | 0.33 | 0.17 | 1.91 | 12 |
| Satispol | 60.61 | 4.23 | 55.58 | 6.88 | 5.04 | 8.21 | 0.67 | 4 |
| Propvict | 12.55 | 1.60 | 9.78 | 2.19 | 2.78 | 2.77 | 1.00 | 2 |

N.S.D.: the number of domains with discontinuity significantly different from zero at a 5% significance level.

To improve the precision of the estimates for the domain discontinuities, the Fay-Herriot model proposed in Section 7.6 is applied. Covariates are selected in a step-forward selection procedure using social-demographic variables from the Municipal Basic Administration, the Police Register of Reported Offences, and related estimates from the regular survey. The conditional Akaike Information Criterion is used as a comparison measure to select the most suitable models (Vaida and Blanchard, 2005). The models obtained with this approach are summarized in Table 7.4. Covariates selected from the Municipal Basic Administration start with the prefix ADM, covariates from the Police Register of Reported Offences with PR, and covariates selected from the regular survey with REG.

Variables from the regular survey are potential covariates for the Fay-Herriot models. Indeed, for four out of the five models, variables observed with the regular survey are selected. Domain predictions obtained with the Fay-Herriot models for the variables under the alternative survey with their standard errors are summarized in Table 7.5. Comparing the standard errors of the GREG estimates under the alternative survey in Table 7.3 with the standard errors of the Fay-Herriot predictions in Table 7.5 shows that the precision of the domain predictions is substantially improved. Estimates for the discontinuities are obtained by taking the differences between the GREG estimate of the regular survey and the Fay-Herriot prediction under the alternative survey. These estimates, including the standard errors, are summarized in Table 7.5. The standard errors for the discontinuities that are based on the Fay-Herriot model are substantially smaller compared to the GREG estimates in Table 7.3. This is the result of a substantial improvement of the precision of the small domain predictions under the alternative survey with the Fay-Herriot model. Moreover, the use of direct estimates of the target variable and related variables from the

**TABLE 7.4**

Covariates for the Fay-Herriot Models

| Variable | Model |
|----------|-------|
| Offtot | 1 + REG_victim |
| Unsafe | 1 + REG_nuisance + ADM_benefit + PR_propcrim + PR_drugs |
| Nuisance | 1 + REG_nuisance + ADM_old |
| Satispol | 1 + REG_funcpol |
| Propvict | 1 + PR_propcrim + ADM_old |

- 1:                   *intercept*
- ADM_benefit:   *percentage of social benefit claimants*
- ADM_old:        *percentage of elderly people (aged over 65)*
- PR_drugs:        *reported drugs crimes*
- PR_propcrim:   *reported property crimes*
- REG_funcpol:   *opinion on functioning of the police on a 10-point scale, esti-mated with the regular survey*
- REG_nuisance: *perceived nuisance in the neighborhood, estimated with the regular survey*
- REG_victim:    *percentage of people saying that they have been victim to a crime, estimated with the regular survey*

**TABLE 7.5**

GREG Domain Estimates (Regular Survey) and Domain Predictions Fay-Herriot Model (Alternative Survey Approach) Averaged over Districts

| Variable | Regular | | Alternative | | Discontinuity | | | |
|----------|---------|---------|-------------|------------------------|----------------|----------------|---------|--------|
| | $\hat{\bar{Y}}_{k;r}$ | $\sqrt{d_{k;r}}$ | $\tilde{\bar{Y}}_{d,k'}$ | $SE(\tilde{\bar{Y}}_{d,k'})$ | $\hat{\Delta}$ | $SE(\hat{\Delta})$ | t-stat. | N.S.D. |
| Offtot | 42.29 | 4.73 | 33.21 | 2.90 | 9.08 | 3.92 | 2.24 | 17 |
| Unsafe | 24.38 | 2.03 | 19.83 | 1.64 | 4.55 | 2.46 | 1.85 | 11 |
| Nuisance | 1.61 | 0.11 | 1.29 | 0.08 | 0.33 | 0.07 | 4.65 | 24 |
| Satispol | 60.61 | 4.23 | 55.09 | 2.54 | 5.52 | 4.72 | 1.20 | 8 |
| Propvict | 12.55 | 1.60 | 9.85 | 0.84 | 2.70 | 1.83 | 1.41 | 6 |

N.S.D.: the number of domains with discontinuity significantly different from zero at a 5% significance level

regular survey as auxiliary variables in the Fay-Herriot model results in a positive correlation between $\hat{\bar{Y}}_{k;r}$ and $\tilde{\bar{Y}}_{d,k'}$, which further reduced the standard error of the discontinuity estimates. A model with estimates from the regular survey results in a substantial improvement of the precision of the domain estimates compared to models where covariates are only selected from register variables, see Van den Brakel, Buelens, and Boonstra (2016) for a comparison.

For Offtot, the point estimates for the regular and alternative survey are plotted against each other in Figure 7.2 for the GREG estimator and the Fay-Herriot model. The improved precision with the Fay-Herriot model is illustrated with the increased linear relationship between the point estimates. The deviation from the solid line with a slope equal to one illustrates the size of the discontinuity. For Offtot the discontinuities estimated with the GREG estimator and the Fay-Herriot model are plotted in Figure 7.3. The 95% confidence interval is clearly smaller for the Fay-Herriot model compared to the GREG estimator.

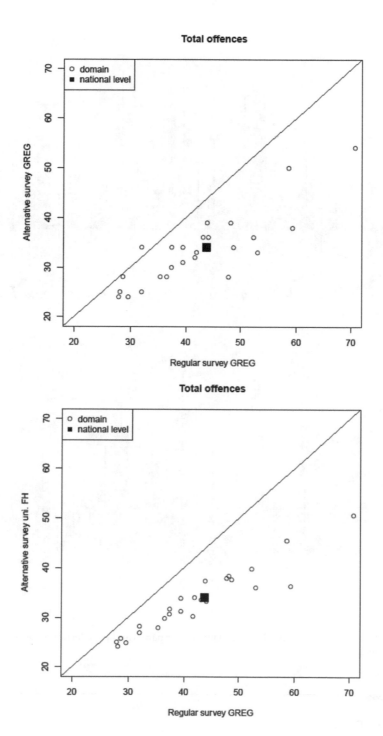

**FIGURE 7.2**
Domain estimates of Offtot alternative survey versus regular survey. Estimates at the national level are based on the GREG estimator.

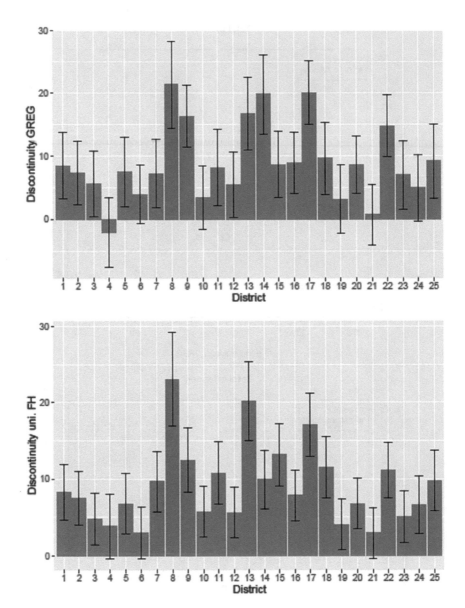

**FIGURE 7.3**
Discontinuities of Offtot based on the GREG estimator (upper panel) and Fay-Herriot model (lower panel), with a 95% confidence interval.

## 7.7 Discussion

In this chapter statistical methods to test the effect of new data collection methods on the outcomes of a sample survey are discussed. A safe and straightforward approach is to conduct a so-called embedded experiment. This implies that first a probability sample is drawn from a finite target population. Subsequently the sampling units are randomized over the different treatments, in this case two or more data collection procedures. To test hypotheses about systematic differences between finite population parameter estimates observed under different treatments or survey implementations, a design-based inference frame work is developed. This approach can be applied to experiments designed as completely randomized designs and randomized block designs embedded in generally complex sample designs. It is briefly discussed how these methods can be extended to factorial designs, to test the main effects and interactions of two or more factors, such as the effect of different data collection modes and different questionnaire designs. This kind of field experiments are typically conducted during the final stage of developing optimal mixed-mode data collection strategies and questionnaire designs, as discussed in Chapters 5, Mixed-Mode Data Collection Design, and 6, Mixed-Mode Questionnaire Design, and can provide additional quantitative insights that complement the cognitive and usability testing strategies treated in Chapter 6.

Large-scale field experiments are frequently applied to quantify the effect of the introduction of a new data collection approach in a repeated survey. Such experiments avoid confounding of real period-to-period change of the population parameters with the difference in measurement bias due to the introduction of a new field work strategy. Due to budget and fieldwork capacity limitations, sufficiently large field experiments that meet pre-specified precision requirements are often not tenable. In the most extreme case there is no parallel data collection at all. In that case, time-series methods can be used to quantify the effect of the change-over to a new fieldwork strategy. This comes at the cost of a higher risk that relevant discontinuities are not detected, estimates for discontinuities are subject to major revisions as new observations under the new design become available, and the final estimates for the discontinuities are not timely available.

In many cases there is a budget or capacity for an experiment with a reduced sample size. This information can be combined with the aforementioned time series method. It is also discussed how small area estimation methods can be used to improve the precision of domain estimates under the alternative survey, conducted with a small sample size. It is explained how reliable direct estimates obtained with the regular survey can be used as covariates in small area estimation models to improve the precision of domain discontinuities.

The methods are illustrated with an application to the Dutch Crime Victimization Survey, where in 2008 a new data collection approach in combination with a new questionnaire was introduced. There was a budget to conduct a parallel run with a sample of one-third of the size of the regular survey. This sample is large enough to analyze discontinuities with the design-based inference approach at the national level. To estimate domain discontinuities for the planned domains of the regular survey, small area estimation methods are applied. This results in more precise and stable estimates for the domain discontinuities. In the case of the Dutch Crime Victimization Survey, the change-over from a CATI-CAPI mixed-mode design to a hybrid mixed-mode design using web interviewing and PAPI with a follow-up with CATI and CAPI, in combination with a new questionnaire, resulted in systematically higher estimates for five key variables related to crime victimization, safety feelings, and satisfaction with police performance.

## 7.8 Summary

The main take-away messages from this chapter are:

- Different methods for quantifying relative mode effects are experiments embedded in sample surveys or time series analysis methods in case of repeated samples.
- Randomized experiments are timely and have a lower risk but are costly.
- Time-series methods are cost effective but have higher statistical risks.
- Key points to be considered in the design phase of an experiment are:
  - Clear definition of the purpose and hypothesis to be tested;
  - Minimum sample for a minimum detectable treatment effect at a pre-specified significance and power level;
  - Different types of randomized experiments (CRDs, RBDs, and factorial designs);
  - How to use survey design structures in efficient randomized experimental designs;
  - Fieldwork restrictions.
- Design-based inference for field experiments:
  - Formulating hypotheses for testing differences between finite population parameter estimates due to differences in data collection;
  - Design-based testing procedures for CRDs, RBDs, and factorial designs based on a Wald test; and
  - Special cases (where standard F-tests and t-tests are a special case of the proposed Wald test).
- Model-based inference for experiments with small sample sizes based small area estimation theory.

# 8

# Re-interview Designs to Disentangle and Adjust for Mode Effects

## 8.1 Introduction

Mode effects are the net result of multiple sources of non-sampling errors in the data collection phase of a survey process and can be divided into mode-specific selection effects and mode-specific measurement effects, see Section 2.3. In the first part of this chapter repeated measurement experiments are considered to disentangle the overall relative mode effects in relative coverage biases, relative nonresponse biases, and relative measurement biases, and to adjust the sample estimates for these effects. In the second part of this chapter re-interview designs and estimation methods are described that adjust for relative measurement effects in sequential mixed-mode designs.

Quantitative insight in mode effects and the decomposition in relative selection bias and relative measurement bias are crucial first steps in designing data collection strategies that are more robust against mode effects. For example, it is essential to know the coverage bias and nonresponse bias of telephone interviewing in a data collection approach procedure where only a sample frame for non-secret landline telephone numbers is available. In the case of large coverage bias, the survey organization could decide to allocate more budget to increase the coverage, e.g., through random digit dialing or contacting cell-phone numbers. In the case of large nonresponse bias, more budgets could be allocated to increase response rates, e.g., by increasing the number of contact attempts. This information is also crucial for adaptive survey designs that apply different data collection strategies within survey modes. Contrary to nonresponse bias, coverage bias cannot be altered by changing the timing and number of reminders or calls.

As explained in Chapter 2, Designing Mixed-Mode Surveys, survey methodology offers three options to make data collection strategies robust against mode effects. The first option is to reduce or minimize measurement effects by good questionnaire design, see Chapter 6, Mixed-Mode Questionnaire Design, and e.g. Dillman et al. (2014). The second option is to avoid mode effects by adapting the choice of survey modes to the sample units, see Chapter 11, Adaptive Mixed-Mode Survey Designs, and e.g. Schouten, Peytchev, and Wagner (2017). Adaptive survey designs may be applied to both reduce selection effects and avoid measurement effects between modes. The third option is to adjust mode effects by some form of weighting or matching afterward, see Chapter 9, Mixed-Mode Data Analysis, and e.g. Buelens and Van den Brakel (2015) and Pfeffermann (2017). In order to decide which options need most attention and are most promising, it is crucial that mode effects are decomposed into their underlying components.

DOI: 10.1201/9780429461156-8

Selection and measurement effects are typically strongly confounded when survey outcomes obtained under different modes are compared, e.g., using methods as discussed in Chapter 7, Field Tests and Implementation of Mixed-Mode Surveys. Separation of selection effects from measurement effects in empirical studies requires carefully designed experiments in combination with weighting or regression-based inference methods to control for selection effects, see e.g. Jäckle et al. (2010). As an alternative, Vannieuwenhuyze et al. (2010, 2012) propose to disentangle measurement and selection effects using instrumental variables; they use a single-mode comparative survey that is assumed to be equally representative of the population and that has the same measurement effects. Vannieuwenhuyze and Loosveldt (2013) compare this method to the generally applied backdoor method that assumes ignorable mode selection given a set of auxiliary variables. Biemer (2001) proposed a test-retest study and assumed a latent class model to separate selection bias and measurement bias in face-to-face and telephone modes. All these methods require strongly related auxiliary information to separate mode-dependent selection effects from mode-dependent measurement bias. Available auxiliary information generally concerns standard socio-demographic variables, which are only weakly related to the target variables of a survey. To overcome this problem a repeated measurement experimental set-up is described in Sections 8.2 and 8.3 with the purpose of collecting strongly related auxiliary information in the re-interview to construct regression estimators that correct for selection bias as good as possible. An application is described in Section 8.4.

This idea of collecting strongly related auxiliary information through a re-interview is extended in the second part of this chapter, where a statistical adjustment of measurement effects is proposed using a research design called the mixed-mode re-interview. Sequential mixed-mode surveys combine multiple modes of data collection in sequential order to maximize survey response while optimizing on data collection costs (De Leeuw, 2005; Groves et al., 2009; Lynn, 2013). Usually, a sequential design starts with a cost-efficient mode (e.g., web data collection) and, subsequently, nonrespondents to the first stage are approached by another mode (e.g., face-to-face). This second stage, typically, strongly improves survey response. When a face-to-face follow-up is used, for example, the combined mixed-mode design often reaches response rates comparable to those of single-mode face-to-face survey designs, but at lower costs (Klausch, Hox, and Schouten, 2015). The increase in survey response may be an indication of a reduction in survey nonresponse bias and may lead to more balanced response samples. However, any mode has particular measurement error properties. In particular, a mixed-mode design combining self and interviewer administration can, therefore, increase the measurement bias of sample estimates and may outweigh the increase in response. Therefore, statistical methods that correct for measurement bias can potentially improve the accuracy of statistics derived from sequential mixed-mode designs.

There is a growing body of literature that discusses statistical adjustment of mode-specific measurement effects (Kolenikov and Kennedy, 2014; Suzer Gurtekin, 2013; Vannieuwenhuyze, 2015; Pfeffermann (2017)). Statistical adjustment seeks to convert measurements obtained under different modes to the level of a common measurement benchmark mode. The measurement benchmark mode is assumed to be the desirable way (combination of mode and question format) to measure a target variable. Disentangling measurement and selection effects requires additional auxiliary data. Previous literature has often applied relatively weak auxiliary data for estimating measurement effects, such as socio-demographic sampling frame information, leading to the potential bias of unknown size in effect estimates (Vannieuwenhuyze, 2015; Vannieuwenhuyze and Loosveldt, 2013). In particular, estimates will be biased when mode-specific nonresponse does not occur at

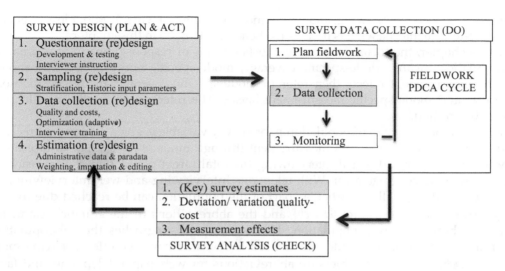

**FIGURE 8.1**
The plan-do-act cycle for surveys.

random in the mixed-mode design (Little and Rubin, 2002), as may be indicated by weak relations of auxiliary information and response mechanisms.

In Sections 8.5, 8.6, and 8.7, a mixed-mode re-interview design is described, including estimators to adjust for measurement bias in sequential mixed-mode designs. In the re-interview, respondents to the first stage of the mixed-mode design are re-approached under a second mode, where relevant questions from the main survey are repeated. This additional information can be expected to be strongly correlated with the benchmark variables and is exploited in the proposed estimation approach.

The methods described in this chapter to disentangle mode effects and to correct for measurement bias through special research designs involve the data collection phase and questionnaire (re)design and the three steps under survey data collection mentioned under 'do' in Figure 8.1. After data collection, approximately unbiased estimates for relative selection, and measurement bias, the key survey estimates are obtained. Therefore steps 1 and 3 in the survey analysis of Figure 8.1 are also involved.

This chapter requires a more advanced background in statistics, in particular knowledge of design-based and model-assisted survey sampling (Särndal et al., 1992).

## 8.2 Decomposition of Relative Mode Effects

Without external information it is very difficult or even impossible to estimate absolute mode effects from survey data, even under randomized experiments. Therefore, the focus of this chapter is to estimate relative mode effects. Most redesigns of surveys go from a single-mode telephone or face-to-face design to a mixed-mode design. As a consequence, there is a strong interest in measurement effects for respondents relative to these interviewer surveys. An experimental design is developed to disentangle relative mode effects into mode-specific coverage effects, mode-specific nonresponse effects, and mode-specific measurement effects (see Chapter 2 for definitions) using four modes, i.e., face-to-face interviewing, telephone interviewing, web interviewing, and paper-and-pencil interviewing. Without loss of generality, face-to-face interviewing is used as the benchmark mode in this chapter.

In this chapter, the focus is on mode effects on bias of response means as estimators of population means. Mode-specific coverage, mode-specific nonresponse, and mode-specific measurement effects then refer to mode-specific coverage, mode-specific nonresponse, and mode-specific measurement biases. The reference to 'mode-specific' will mostly be omitted.

Let $y_{p,m_1,m_2}$ be the mean of population $p$ for survey variable $y$, given respondent recruitment is based on mode $m_1$ and measurement through mode $m_2$. With face-to-face interviewing and paper-and-pencil interviewing the total target population can be reached, since there is no coverage error. With telephone interviewing and web interviewing, on the other hand, typically a particular part of the population can be reached due to coverage error. Therefore $p \in \{tot, tel, web\}$ and the abbreviations in the equation stand for total, telephone, and web population. This subscript distinguishes the subpopulation covered with a particular mode and is used to measure coverage effects. Furthermore $m_1, m_2 \in \{web, pap, tel, f2f\}$ which are abbreviations for web, paper, telephone, and face-to-face. The second subscript $m_1$ refers to the selection mechanism and is used to measure mode-specific nonresponse effects, while the third subscript refers to mode-specific measurement effects. For example, $y_{tel,f2f,tel}$ is the population mean for persons that can be reached through telephone, e.g., the population with a registered phone number, given they participated following a request by a face-to-face interviewer and given that their answer was administered by phone. As a mnemonic, it may serve that measurement follows response and response follows coverage.

Since $y_{tot,f2f,f2f}$ serves as the benchmark, mode effects are evaluated with respect to the face-to-face response mean; i.e., face-to-face was used both as a selection mechanism and as a measurement technique. Now the total relative mode effects for telephone, web, and paper are defined as:

$$M_{tel}(y) = y_{tel,tel,tel} - y_{tot,f2f,f2f} \tag{8.1}$$

$$M_{web}(y) = y_{web,web,web} - y_{tot,f2f,f2f} \tag{8.2}$$

$$M_{pap}(y) = y_{pap,pap,pap} - y_{tot,f2f,f2f} \tag{8.3}$$

The mode effect for telephone interviewing in (8.1), e.g., is defined as the difference between the outcomes of the same survey administered through telephone ($y_{tel,tel,tel}$) and face-to-face ($y_{tot,f2f,f2f}$), where $y_{tel,tel,tel}$ is the population that can be contacted by telephone ($p = tel$), is willing to respond by telephone ($m_1 = tel$), and completes the questionnaire by telephone ($m_2 = tel$) and $y_{tot,f2f,f2f}$ the population that can be contacted face-to-face ($p = tot$), is willing to respond by face-to-face ($m_1 = f2f$), and completes the questionnaire face-to-face ($m_2 = f2f$). The mode effects for web interviewing in (8.2) and paper-and-pencil interviewing (8.3) are defined in an equivalent way.

The relative mode effects in (8.1), (8.2), and (8.3), or more precisely the mode biases, are decomposed into coverage bias, nonresponse bias, and measurement bias. These decompositions can be made in six ways, resulting in alternative definitions of coverage, nonresponse, and measurement bias. The two decompositions that result in estimable components under the experimental set-up proposed in Section 8.3 are considered. The decomposition of the total mode effect is elaborated for telephone relative to face-to-face only. Mode effect decompositions for web and paper can be made analogously.

The first decomposition is defined as

$$M_{tel}(y) = \left(y_{tel,tel,tel} - y_{tel,f2f,tel}\right) + \left(y_{tel,f2f,tel} - y_{tel,f2f,f2f}\right)$$

$$+ \left(y_{tel,f2f,f2f} - y_{tot,f2f,f2f}\right) \tag{8.4}$$

$$\equiv NR_{tel}(y) + ME_{tel}(y) + CO_{tel}(y)$$

Where $NR_{tel}(y)$, $ME_{tel}(y)$, and $CO_{tel}(y)$ are the nonresponse, measurement, and coverage bias, respectively, for telephone relative to the face-to-face.

The second decomposition is

$$M_{tel}(y) = \left(y_{tel,tel,tel} - y_{tel,f2f,tel}\right) + \left(y_{tel,f2f,tel} - y_{tot,f2f,tel}\right)$$

$$+ \left(y_{tot,f2f,tel} - y_{tot,f2f,f2f}\right) \tag{8.5}$$

$$\equiv NR_{tel}(y) + CO_{tel}^{*}(y) + ME_{tel}^{*}(y)$$

The nonresponse bias in both decompositions is defined as the difference between the population that can be contacted via telephone ($p = tel$), is willing to respond as if they are recruited by telephone ($m_1 = tel$), and completes the questionnaire by telephone ($m_2 = tel$), and the population that can be contacted via telephone ($p = tel$), participates as if they are recruited in a face-to-face mode ($m_1 = f2f$), and answers the question as if the questionnaire is completed in a telephone interview ($m_2 = tel$). In (8.4), the measurement bias is defined for the population restricted to persons that can be contacted by telephone (e.g., persons with a registered telephone number) ($p = tel$) and participate as if they are recruited face-to-face ($m_1 = f2f$). The measurement bias is defined as the observed difference if this population completes the questionnaire by telephone ($m_2 = tel$) instead of face-to-face ($m_2 = f2f$). In (8.5), the measurement bias is defined for the full population, i.e., including those persons that cannot be contacted by telephone ($p = tot$) and participate as if they are recruited face-to-face ($m_1 = f2f$). Consequently, the coverage bias is different as well. In both decompositions the coverage bias is defined as the difference between the population which can be contacted by telephone ($p = tel$) and participate as if they are recruited in a face-to-face mode ($m_1 = f2f$) and the population that can be contacted face-to-face ($p = tot$) and participate as if they are recruited in a face-to-face mode ($m_1 = f2f$). In decomposition (8.4), the coverage bias is defined in terms of face-to-face answers ($m_2 = f2f$), while decomposition (8.5) defines the coverage bias in telephone answers ($m_2 = tel$).

$M_{web}(y)$ and $M_{pap}(y)$ can be decomposed analogously. For the paper survey mode there is no undercoverage, so it always holds that $CO_{pap}(y) = CO_{pap}^{*}(y) = 0$. In the following, for convenience, decomposition (8.4) is called MEMOD (measurement bias for respondents with access to mode) and (8.5) as MEFULL (measurement bias for all respondents).

## 8.3 Estimating Components of the Relative Mode Effects

### 8.3.1 Experimental Design

Variables in the mode decompositions (8.4) and (8.5), where the mode used to recruit respondents ($m_1$) is equal to the mode used to administer the questionnaire ($m_2$), can be observed directly. The mode decompositions contain also several variables where different modes are used to recruit the respondents and administer the questionnaire. These variables are not directly observable. An experimental design, in combination with an estimation procedure, is proposed that allows for the estimation of these conceptual variables.

The experiment consists of two waves. First, a probability sample of size $n$ is drawn from the target population. In the first wave of the experiment, the sample units are randomly assigned to one of the four survey modes: face-to-face, telephone, web, or paper. The data collection strategies for the four modes must be equal to the standard strategies of the survey to which results apply, e.g., length of data collection period and number of visits/calls/reminders. The questionnaire used in the first wave is typically equal to the questionnaire of the survey to which results are generalized. Observations obtained with these four subsamples can be used to estimate the total mode effects defined in (8.1)–(8.3), and the variables in (8.4) and (8.5) for which recruitment ($m_1$) and administration ($m_2$) are defined for the same mode.

The variables in (8.4) and (8.5) for which recruitment and administration are defined for different modes are estimated with the general regression (GREG) estimator. This estimator uses auxiliary information to correct for selection bias (Särndal et al., 1992). This property is used in this experimental set-up to separate measurement bias from nonresponse bias and coverage bias. The stronger the correlation between the auxiliary variables and the target variable, the better the GREG estimator corrects for selection bias. The main purpose of the second wave of this experiment is to collect auxiliary information for the sample units in the first wave that is strongly correlated with the target variables. This is achieved through partial repetition of the questions from wave 1 that concern the target variables to be analyzed in this experiment.

The full sample, excluding administrative errors and some exceptional nonresponse types like language problems, is approached once more in the second wave using the benchmark mode, which is the face-to-face mode. The questionnaire used in the second wave is different from the one used in the first wave. The purpose of the second wave is to collect variables to construct the ideal nonresponse adjustment variables in the weighting model of a GREG estimator. The questionnaire for wave 2, therefore, typically contains a repetition of the key statistics of the survey but also other potentially related variables might be added. Answers to these key statistics may of course show some structural change due to the time lag between the two waves. This is not a problem since the only requirement for the auxiliary variables in the GREG estimator is that they are strongly correlated with the target variables. Also other questions that might assist the interpretation of the results can be added to the questionnaire of the second wave.

The time lag between contacts in the two waves is a compromise between minimizing memory or recall effects and maximizing the predictive power of the auxiliary variables. The shorter the time lag, the more likely it is that respondents remember the answers provided in the first wave, but the stronger the predictive power as covariates in the GREG estimator. The longer the time lag, the less likely it is that respondents remember the answers provided in the first wave, but the predictive power of these variables becomes weaker since the real values of the variables change over time. In case of volatile variables a shorter time lag might be necessary compared to variables that are stable over time. Determining the optimal time lag is a topic of further research, in the application in Section 8.4. The time lag varies ranges from 4 to 8 weeks, depending on the survey mode of wave 1 and the time of contact in both waves.

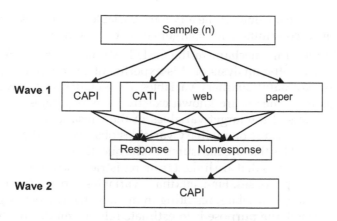

**FIGURE 8.2**
Experimental design.

## 8.3.2 Estimation Strategies

Two estimators are used to disentangle coverage, nonresponse, and measurement biases based on the experimental design: the response mean and the general regression (GREG) estimator. The response mean is used for the variables in (8.4) and (8.5) that can be observed directly, which implies that recruitment mode ($m_1$) and administration mode ($m_2$) are the same. Variables where the recruitment mode is not equal to the administration mode are not observed directly but can be estimated with the GREG estimator, since this estimator can be used to correct for selection bias.

GREG estimators, see chapter 6 of Särndal et al. (1992) for an introduction, are widely applied in survey sampling for the estimation of the unknown population variables of interest. The GREG estimator can be expressed in different ways. Let $x_i$ denote a q-vector containing the auxiliary variables in the GREG estimator. Following the notation of Chapter 7, the GREG estimator can be expressed as:

$$\hat{y}^{GR} = \hat{y}^R + \hat{\beta}^t \left( x - \hat{x}^R \right) \tag{8.6}$$

with $\hat{y}^R$ the response mean of the target variable $y_i$, $\hat{x}^R$ the response mean of the sample units of the vector with auxiliary variables $x_i$, and $x$ the corresponding true population mean of the auxiliary variables, available from registers or a census. Finally $\hat{\beta}$ is an estimator for the regression coefficients of the linear model

$$y_i = \beta^t x_i + e_i \tag{8.7}$$

which describes the relation between the target variable and the auxiliary variables. Note that $e_i$ in (8.7) is a residual for the unexplained variation of $y_i$. The linear model (8.7) is used to motivate the GREG estimator in (8.6), see Särndal et al. (1992) for details.

If model (8.7) explains the variation of the target variable $y_i$ in the population reasonably well, then the use of this auxiliary information reduces the sampling variance and corrects, at least partially, for selective nonresponse. This can be seen as follows: groups distinguished in the set of auxiliary variables, with low response rates, which have a large difference between the real values of the auxiliary variable and the response mean due to selective nonresponse, contribute with a larger correction to the response mean $\hat{y}$ in (8.6) via $\hat{\beta}^t\left(x - \hat{x}^R\right)$. Through this mechanism the GREG estimator corrects for selection bias. This property is used in the analysis of this experiment to separate measurement bias from selection bias. Normally, auxiliary variables are included in the weighting scheme for which the population means are known exactly from, e.g., registers. In this application the response means of the target variables in the subsamples of the first wave are calibrated to the sample means of auxiliary variables observed in the second wave. This holds for auxiliary variables that are measured in the questionnaire of the second wave, as well as the auxiliary variables that are linked from registries and administrative data. This modification serves two purposes. First, auxiliary variables can be obtained that are very closely related to the target variables, resulting in models that correct as good as possible for selection bias. Second, the purpose is to estimate relative mode effects where face-to-face interviewing is the benchmark. Therefore selection bias correction is conducted with respect to a standard face-to-face interview. This is the reason that the second wave is conducted face-to-face.

Let $\hat{y}_{p,m}^R$ be the response mean for mode $m$ for respondents in population $p$ of a variable $y$. Again, $p$ and $m$ are elements of $p \in \{tot, tel, web\}$, where abbreviations stand for total, telephone, and web population, $m \in \{web, pap, tel, f2f\}$, short for modes web, paper, phone, and face-to-face. Furthermore, $\hat{y}_{p_1,m}^{GREG,p_2}$ denotes the GREG estimator of wave 1 respondents in population $p_1$ calibrated to wave 2 respondents in population $p_2$. This means that the vector with the benchmark variables $x_i$ in (8.6) is the response mean of these auxiliary variables of the respondents in wave 2 belonging to population $p_2$. For example, $\hat{y}_{tel,tel}^{GREG,tel}$ is the GREG estimator for wave 1 respondents in the telephone subsample calibrated to all respondents in wave 2 that have a telephone.

In the GREG estimator, the respondents in wave 1 that did not respond in wave 2 are always ignored. This is a consequence of the GREG estimator, since auxiliary variables are required for all sample units. This group is also omitted since wave 2 is viewed as a regular face-to-face survey. The wave 1 respondents that do not respond to this survey are considered not to be part of the response to a regular face-to-face survey. For example, part of the web response may consist of persons that would not participate in a face-to-face survey.

The resulting estimators are again elaborated for telephone. Similar estimators can be constructed for web and paper survey modes. The telephone mode effect components of the two decompositions MEMOD and MEFULL are now estimated as follows:

$$\hat{y}_{tel,tel,tel} = \hat{y}_{tel,tel}^R \tag{8.8}$$

$$\hat{y}_{tot,f2f,f2f} = \hat{y}_{tot,f2f}^R \tag{8.9}$$

$$\hat{y}_{tel,f2f,f2f} = \hat{y}_{tel,f2f}^R \tag{8.10}$$

$$\hat{y}_{tel,f2f,tel} = \hat{y}_{tel,tel}^{GREG,tel} \tag{8.11}$$

$$\hat{y}_{tot,f2f,tel} = \hat{y}_{tel,tel}^{GREG,tot} \tag{8.12}$$

In (8.11), the face-to-face respondents of wave 1 with a telephone are used in the GREG estimator using auxiliary information from wave 2. Similarly, in (8.12) all face-to-face respondents of wave 1 are used in the GREG estimator using auxiliary information from wave 2.

The first three estimators, i.e. (8.8)–(8.11), are response means from wave 1 and do not use information from wave 2. Alternatively, the estimates of (8.9) and (8.10) can be improved with the GREG estimator, i.e.,

$$\hat{y}_{tot,f2f,f2f} = \hat{y}_{tot,f2f}^{GREG,tot} \tag{8.13}$$

$$\hat{y}_{tel,f2f,f2f} = \hat{y}_{tel,f2f}^{GREG,tel} \tag{8.14}$$

Combining the two decompositions MEMOD (8.4) and MEFULL (8.5) with these two estimation strategies results in four different analysis procedures, which are labeled as follows:

- MEMOD-R: Measurement bias for respondents with access to mode and response mean for variables where mode of recruitment and mode of measurement are the same, i.e., using estimators (8.8)–(8.12);

- MEMOD-GREG: Measurement bias for respondents with access to mode and GREG estimator for some variables where mode of recruitment and mode of measurement are the same, i.e., using estimators (8.8) and (8.11)–(8.14);

- MEFULL-R: Measurement bias for all respondents and response mean for variables where mode of recruitment and mode of measurement are the same, i.e., using estimators (8.8)–(8.12); and

- MEFULL-GREG: Measurement bias for all respondents and GREG estimator for some variables where mode of recruitment and mode of measurement are the same, i.e., using estimators (8.8) and (8.11)–(8.14).

Analytic approximations to the standard errors for the various estimators (8.8)–(8.14) are straightforward, see Särndal et al. (1992) for variance expressions of the GREG estimators. However, the approximation of standard errors for the coverage, nonresponse, and measurement biases is less straightforward due to covariances between the estimators. One approach to approximate these standard errors is using bootstrap resampling.

### 8.3.3 Assumptions

The proposed estimation strategies provide unbiased estimators for the relative coverage, nonresponse, and measurement biases with respect to face-to-face, when three assumptions are met:

1. The response to wave 2 is similar to the response to a regular face-to-face survey. This assumption can be broken down into two subassumptions:
   a. The face-to-face response to wave 1 resembles a regular face-to-face survey.
   b. The face-to-face respondents that drop out between wave 1 and wave 2 are similar to the wave 1 face-to-face nonrespondents that enter the response to wave 2.
2. The answering behavior to wave 2 is face-to-face, i.e., the answers to wave 2 are not affected by the mode in wave 1.
3. The nonresponse to wave 1 relative to wave 2 is missing at random for the CVS and LFS key survey variables, given the wave 2 variables and register variables.

Assumption 1 assures that the benchmark that is used to estimate relative mode effects resembles the population observed with a general face-to-face survey. This assumption provides the generalization of the results of the mode decompositions to this population. Historical survey data provides strong evidence that assumption 1 is valid. This assumption corresponds to the continuum of resistance assumption. The use of auxiliary information through the GREG estimator is based on the assumption that the auxiliary variables are a priori available for all elements in the population.

Assumption 2 assures that the auxiliary data observed in the second wave are not systematically influenced by the different modes in wave 1. Under this assumption, the auxiliary variables measured in the second wave are consistent so that the response mean of the entire second wave can be used as an a priori known population value in the GREG estimator. This assumption is enforced by demanding at least a four-week time lag between the two waves.

Assumption 3 assures that the weighting model of the GREG estimator is suitable to correct the selection bias. If conditionally on the weighting model the nonresponse is missing at random (Little and Rubin, 2002), then the regression weights correct for the selection bias induced by under- and overrepresentation of the different subpopulations distinguished in the weighting model. This assumption cannot be tested directly. An indication to which extent this assumption is met can be evaluated through the strength of the regression models underlying the calibration. The GREG estimator corrects better for selective nonresponse as its underlying linear model better fits the target variables in the population and also explains the response behavior. If assumption 3 does not fully hold, then, depending on the decomposition, some selection bias will end up in the measurement bias or in the coverage bias.

The various possible decompositions of the total mode effect are conceptually different and should, thus, in general lead to different estimates for the mode effect component. However, the various mode effect components should have the same order of magnitude.

Four sets of estimators for mode effect components were introduced, i.e., MEMOD-R, MEMOD-GREG, MEFULL-R, and MEFULL-GREG. Now, the conceptual and statistical differences are briefly described. The decompositions that define measurement bias for all respondents offer the advantage of comparability over modes. Access to web, paper, and telephone represents different populations and, hence, conceptually different measurement bias for the other decompositions. However, the decompositions that extrapolate measurement bias to the full population require stronger missing-at-random assumptions as the GREG estimators need also to adjust for the selectivity due to undercoverage. Furthermore, one can debate the existence of measurement bias for persons without access to a mode. Using GREG estimators instead of response means for wave 1 face-to-face may lead to gains in precision. However, since the GREG estimators use only those wave 1 respondents that also respond to wave 2, this gain may be lost by the reduction of the number of respondents. Table 8.1 summarizes the strengths of each strategy. For the group of estimators that is used under MEMOD-R, it is felt that this is overall the best strategy. As will be seen in the application, the differences in estimates between the strategies are small.

## 8.4 Application

The approach described in Section 8.3 is applied to one of the case studies from Section 2.6: the Dutch Crime Victimization Survey (CVS). The CVS is based on a sequential

**TABLE 8.1**

Differences between Decompositions and Estimation Strategies

| Strategy | Strength |
|---|---|
| MEMOD-R | Conceptual: No statements about counterfactual measurement bias of persons without access to mode |
| | Statistical: Requires weaker MAR assumption |
| MEMOD-GREG | Conceptual: No statements about counterfactual measurement bias of persons without access to mode |
| | Statistical: Requires weaker MAR, potential gain in precision |
| MEFULL-R | Conceptual: Measurement bias is comparable over modes |
| MEFULL-GREG | Conceptual: Measurement bias is comparable over modes |
| | Statistical: Potential gain in precision |

mixed-mode design. It appeared that variations in mode effects between the subsequent editions compromised the data quality of this survey. To obtain quantitative insight into the different components of these non-sampling errors, the experimental set-up proposed in Sections 8.2 and 8.3 is applied to the Dutch Crime Victimization Survey (CVS), with the purpose of disentangling mode effects for the most important key variables of this survey. The insights obtained with this experiment are used to improve the mixed-mode strategy of the CVS.

The sample size of the first wave amounts 8800 sample units that are randomly assigned over the four survey modes: face-to-face, telephone, web, and paper. The subsamples are equally sized, so the subsample size equals 2200 sampling units.

The field strategies per mode are the following: For web, an advance letter is sent with a login code to a secure website and two reminders are sent with time lags of one week. For paper, an advance letter is sent with the questionnaire as an attachment and two reminders, including again the questionnaire, are sent with time lags of one week. For telephone, an advance letter is sent announcing that an interviewer will contact the household over the phone. A strategy of three time slots on different days is applied where in each slot three calls are attempted until contact is established. Soft (a certain day) and hard (a certain day and time) appointments are possible. For face-to-face also an advance letter is sent, announcing contact by an interviewer at the door. Interviewers pay up to six visits to an address. The full sample is approached once more in the second wave.

Since the face-to-face mode is treated as the benchmark mode, the second wave was planned to be administered through the face-to-face mode. Due to budget constraints it was, however, necessary to conduct a small part through telephone. Approximately 77% of the sample persons are administered by face-to-face in the second wave. The remaining 23% of the sample persons is interviewed through telephone. When a sample person has a registered telephone number, then the allocation to telephone or face-to-face was random, with probabilities 67% to face-to-face and 33% to telephone. When a sample person does not have a registered number, then he/she is always allocated to face-to-face.

In the first wave, the CVS questionnaire is used. The second wave of the experiment employs a new questionnaire consisting of a repetition of the key statistics from the CVS, which are used to construct the ideal nonresponse adjustment variables in the weighting scheme of the GREG estimator. In addition, questions for further analysis and interpretation of the results are included, like general attitude toward safety, politics, and surveys; evaluation of survey participation in wave 1 and survey design features; evaluation of the CVS questionnaire; and access to web and mode preferences. The time lag between contacts in the two waves (if a contact is established) varies between four and eight weeks.

Table 8.2 contains an overview of the key variables from the CVS which are analyzed in this experiment. They include facts, attitudes, and opinions. These variables are repeated in wave 2. These variables are labeled as indicated in Table 8.2a but with a prefix 'w2'. A set of related variables from the CVS questionnaire were also repeated in wave 2. Those who were finally selected in the GREG estimator are listed in Table 8.2b Table 8.3 contains an overview of the auxiliary variables, which are available from registers. Finally, Table 8.4 contains an overview of the additional questions asked in wave 2.

The sample of 8,800 persons was randomly divided into four equal groups. The response within the face-to-face, telephone, paper, and web groups for the two waves is listed in Table 8.5. The fact that the effective sample sizes of the face-to-face and web groups are smaller than 2,200 is due to the omission of population frame errors (unknown/wrong addresses). As expected, in wave 1, the face-to-face response is the highest, 61%, and the web response the lowest, 29%. The relatively low response for telephone, 45%, is to some extent caused by undercoverage. Around 25% of the sample units do not have a registered phone number. Around 90% of Dutch households have access to the web at home. Hence,

**TABLE 8.2A**

Key Variables of CVS

| Variable name | Variable content |
| --- | --- |
| Offtot | Total number of offences per 100 inhabitants in the last 12 months |
| victim | Percentage of population victim of crime in the last 12 months |
| Nuisance | Scale score related to nuisance in neighborhood |
| Unsafe | Percentage of people feeling unsafe at times |

**TABLE 8.2B**

Selected Auxiliary CVS Variables from Wave 2 (other than key variables from Table 8.2a)

| Variable name | Variable content |
| --- | --- |
| w2_contpol | Percentage of people having contact with police in the last 12 months |
| w2_victviol | Percentage of population victim of violence crime in the last 12 months |
| w2_offviol | Total number of violence offences per 100 inhabitants in the last 12 months |

**TABLE 8.3**

Auxiliary Variables from Registers

| Variable name | Variable content |
| --- | --- |
| age | Age in categories of 3, 6, or 7 age classes |
| gender | Male/female |
| ethnic | Ethnicity (Dutch, foreign (Western), foreign (non-Western)) |
| urban | Degree of urbanization |
| hhsize | Household size |
| hhtype | Household type |
| hhpos | Position in household |
| income | Income category |
| Incometype | Type of income (employed, social benefits, …) |
| regunempl | Registered unemployed |

**TABLE 8.4**

Additional Wave 2 Variables

| Variable Name | Variable Content |
| --- | --- |
| betr1 | Opinion on Participation in surveys on safety is important |
| betr2 | Opinion on Safety should be on the top of the political agenda |
| betr3 | Opinion on Safety in the Netherlands is worrisome |
| betr4 | Opinion on The government should do more to improve safety |
| vmatt2 | Opinion on It is interesting to think about safety |
| vmatt3 | Opinion on Politics should take opinions on safety seriously |
| vmatt5 | Opinion on Surveys about safety concern issues that I would like to keep to myself |
| intpol | Extent to which a person is interested in politics |
| satt1 | Opinion on It is fun to participate in surveys on paper or the web |
| satt2 | Opinion on It is fun to be interviewed in a survey |
| satt3 | Opinion on In general surveys are interesting |
| satt4 | Opinion on Surveys are important for society |
| satt5 | Opinion on The government can learn a lot from surveys |
| satt6 | Opinion on Surveys are a waste of time |
| satt7 | Opinion on There are too many requests for surveys |
| satt8 | Opinion on Surveys are an invasion of privacy |
| satt9 | Opinion on It is tiring to fill in survey questionnaires |
| satt10 | Opinion on People in my environment generally participate in surveys |
| satt11 | Opinion on When asked for a survey, it is one's duty to participate |
| satt13 | Opinion on My time is too precious to use on a survey |
| satt14 | Opinion on There is too little time to participate in surveys |
| vmatt6 | Opinion on CBS protects the confidentiality of survey data |
| vmatt7 | Opinion on CBS statistics are reliable |

**TABLE 8.5**

Sample Sizes and Response Rates to Wave 1, Wave 2, and Both Waves

|  | F2F | Telephone | Paper | Web | Total |
| --- | --- | --- | --- | --- | --- |
| Sample | 2182 | 2200 | 2200 | 2199 | 8781 |
| Wave 1 response | 1338 | 993 | 1076 | 631 | 4038 |
|  | 61% | 45% | 49% | 29% | 46% |
| Wave 2 response | 1077 | 1036 | 1099 | 1084 | 4296 |
|  | 49% | 47% | 50% | 49% | 49% |
| Response to both waves | 933 | 700 | 726 | 444 | 2803 |
|  | 43% | 32% | 33% | 20% | 32% |

for web the response rate given access would be around 32% while the response for persons with a registered telephone number is about 60%. In wave 2, response percentages are comparable between the four groups; they are all around 50%. This is an important finding as it induces that response rates were not affected by the mode of wave 1. This is verified formally by modeling the binary response variable of wave 2 using a logistic regression model, with the mode in wave 1 as independent variable. The mode used in wave 1 is found to be insignificant (p = 0.24).

Table 8.5 also provides the proportion of sample persons that responded to both waves. These respondents form an important subset, as they are used in decomposing the mode effect into different components; it is this group that is used in the GREG estimator to calibrate wave 1 to the auxiliary variables observed in wave 2. The telephone and paper groups consist of approximately 700 respondents, while there are only 440 respondents in the web group. From these response rates, it can be conjectured that the variance reduction obtained with the use of auxiliary variables with the GREG estimators will not compensate for the reduced sample size. As a result, standard errors for the GREG estimators may at best be similar to those of the simple response means. Or in other words, the confidence intervals are not made smaller by weighting to auxiliary variables observed in wave 2.

In order to produce estimates for the various mode-specific bias terms, weighting models are developed for each of the selected variables for which the decomposition is to be performed. Table 8.6 shows the optimal models. Auxiliary variables are selected following a step-forward selection in the regression of a target variable on all potential auxiliary variables using AIC as the selection criterion. Potential auxiliary variables are the repeated key variables in wave 2 as listed in Table 8.2a, related CVS variables (selected ones are listed in Table 8.2b), the auxiliary register variables listed in Table 8.3, and the additional variables from wave 2 listed in Table 8.4. Interaction terms have not been considered. Auxiliary variables in the models are listed in the order in which they were selected. The first variables provide the strongest explanation.

As expected, the repeated variables in wave 2 are the most important ones. Some register variables appear; age, type of income, and urbanization degree are selected. The register variables are never the strongest variables, but in some cases do provide strong additional explanation. The special wave 2 attitudinal question, 'Opinion on Safety in The Netherlands is worrisome', is also selected, but provides only little extra explanation.

Using the models of Table 8.6, mode effects are quantified and decomposed. Table 8.7 provides the mode decomposition using MEMOD-R. All total effects that are significant are due to measurement bias. When not significant, the measurement bias remains the largest of the three components. The coverage bias generally contributes the least to the total mode effect. In some instances the bias terms have different signs, indicating that these effects counteract each other. For web interviewing, for example, offtot is found to be lower due to nonresponse bias, and at the same time higher due to measurement and coverage bias. The total mode effect is – in absolute value – smaller than the measurement bias.

For web interviewing, the total mode effect with respect to face-to-face interviewing is negative. Under web interviewing, more offences, more victimization, feeling less safe, and scoring higher for nuisance are observed. These differences seem to be mainly caused by measurement bias. The opposite applies to the total mode effect of telephone interviewing, which is positive compared to face-to-face interviewing.

The mode effect decompositions are provided for an additional set of CVS variables, see Schouten et al. (2013). From these mode effect decompositions, several implications for mixed-mode survey methodology were deduced. Following the measurement effects,

**TABLE 8.6**

Optimal Regression Models

| Variable | Model |
|---|---|
| offtot | w2_offtot + w2_nuisance + age6 + w2_contpol + w2_offbike + w2_victviol |
| victim | w2_victim + w2_nuisance + w2_contpol + inctype + w2_offtot + w2_offviol |
| nuisance | w2_nuisance + w2_unsafe + urban + w2_victviol |
| unsafe | w2_unsafe + w2_nuisance + age3 + betr3 |

**TABLE 8.7**

Decomposition of Mode Effects Using MEMOD-R, with
Significance Levels Indicated by $^*$(p < 0.05), $^{**}$(p < 0.01)

| Variable | Mode | Mean | F2F | NR | CO | ME | Total |
|----------|------|------|-----|-----|-----|-----|-------|
| Offtot | Tel | 35.3 | 41.6 | −2.0 | −1.4 | −2.9 | −6.3 |
| | Paper | 50.8 | 41.6 | −3.6 | – | 12.8$^{**}$ | 9.2$^*$ |
| | Web | 56.2 | 41.6 | −3.6 | 3.6 | 14.6$^{**}$ | 14.5$^{**}$ |
| Victim | Tel | 22.7 | 26.5 | 0.0 | −0.5 | −3.3$^*$ | −3.8$^*$ |
| | Paper | 28.2 | 26.5 | −1.8 | – | 3.5 | 1.7 |
| | Web | 32.0 | 26.5 | 0.4 | 1.5 | 3.6 | 5.6$^*$ |
| Nuisance | Tel | 1.27 | 1.47 | −0.01 | −0.05 | −0.15$^*$ | −0.21$^{**}$ |
| | Paper | 1.41 | 1.47 | −0.01 | – | −0.06 | −0.07 |
| | Web | 1.68 | 1.47 | 0.01 | 0.01 | 0.16 | 0.20 |
| Unsafe | Tel | 18.3 | 22.4 | −1.1 | −0.4 | −2.6 | −4.1$^*$ |
| | Paper | 23.5 | 22.4 | −0.1 | – | 1.2 | 1.1 |
| | Web | 28.5 | 22.4 | −0.6 | 0.4 | 6.3$^{**}$ | 6.1$^{**}$ |

a set of questions was derived that needed a redesign of the questionnaire, most strikingly the block of questions about victimization. The CVS questionnaire went through a major redesign. From the small selection effects in the CVS, it was concluded that the current nonresponse adjustment is sufficient and requires no further attention. However, the large measurement effects between interviewer and non-interviewer modes were a reason for concern; they threaten comparability over time when the shares of modes to the overall response vary between the subsequent editions of this survey. For this reason, an adaptive survey design was investigated that attempted to minimize measurement effects by allocating different subpopulations to different sets of modes given budget constraints. Since it turned out that measurement effects between modes are generally large over most investigated subpopulations, it was decided to avoid a mixed-mode design with both interviewer and non-interviewer modes. Therefore, the CVS design was changed to a combination of web and paper.

In Section 8.3, four sets of estimators were introduced. In Table 8.8 the mode decompositions for these four strategies are compared, where strategy 1 is MEMOD-R, strategy 2 is MEMOD-GREG, strategy 3 is MEFULL-R, and strategy 4 is MEFULL-GREG.

The differences between the four strategies lay in the coverage and measurement bias estimators. By definition, the nonresponse bias is the same for all four strategies. None of the differences found between the strategies is statistically significant at a 5% significance level (standard errors not provided). Generally, the impact of the definition of the mode effect components (strategies 1 and 2 versus strategies 3 and 4) is smaller than the impact of the use of auxiliary information through calibration (strategies 1 and 3 versus strategies 2 and 4). The largest difference based on the definition was found for the number of offences (offtot) per 100 inhabitants, a difference of 1.7% points. It is concluded that the definition of the mode effects is not influential for the selected target variables and that the differences are acceptable. The use of auxiliary information from wave 2 through the GREG estimator leads to differences in more variables. The largest difference, 2.6% points, is found for the 0–1 indicator for feeling unsafe (unsafe). There is a 2.2% jump in measurement bias from strategy 1 (MEMOD-R) to strategy 2 (MEMOD-GREG). Also, the sign of the measurement bias estimate changed. However, in none of the cases did the calibration significantly change conclusions.

**TABLE 8.8**

Mode Effect Decompositions for the Four Strategies (strategy 1 – MEMOD-R; strategy 2 – MEMOD-GREG; strategy 3 – MEFULL-R; and strategy 4 – MEFULL-GREG)

| Variable | S | NR | | | CO | | | ME | | |
|---|---|---|---|---|---|---|---|---|---|---|
| | | tel | pap | web | Tel | Pap | web | tel | pap | Web |
| offtot | 1 | −2.0 | −3.6 | −3.7 | −1.4 | – | 3.6 | −2.9 | 12.8 | 14.6 |
| | 2 | −2.0 | −3.6 | −3.7 | −1.4 | – | 3.6 | −4.6 | 11.0 | 12.8 |
| | 3 | −2.0 | −3.6 | −3.7 | 0.3 | – | 2.6 | −4.6 | 12.8 | 15.7 |
| | 4 | −2.0 | −3.6 | −3.7 | −1.7 | – | 2.2 | −4.4 | 11.0 | 14.2 |
| victim | 1 | 0.0 | −1.8 | 0.4 | −0.5 | – | 1.5 | −3.3 | 3.5 | 3.6 |
| | 2 | 0.0 | −1.8 | 0.4 | −0.5 | – | 1.5 | −4.0 | 2.8 | 2.9 |
| | 3 | 0.0 | −1.8 | 0.4 | 0.0 | – | 1.3 | −3.8 | 3.5 | 3.8 |
| | 4 | 0.0 | −1.8 | 0.4 | −0.4 | – | 0.9 | −4.2 | 2.8 | 3.5 |
| nuisance | 1 | −0.01 | −0.01 | 0.01 | −0.05 | – | 0.01 | −0.15 | −0.06 | 0.16 |
| | 2 | −0.01 | −0.01 | 0.01 | −0.05 | – | 0.03 | −0.15 | −0.06 | 0.16 |
| | 3 | −0.01 | −0.01 | 0.01 | −0.04 | – | −0.02 | −0.16 | −0.06 | 0.21 |
| | 4 | −0.01 | −0.01 | 0.01 | −0.05 | – | 0.03 | −0.15 | −0.06 | 0.16 |
| unsafe | 1 | −1.1 | −0.1 | −0.6 | −0.4 | – | 0.4 | −2.6 | 1.2 | 6.3 |
| | 2 | −1.1 | −0.1 | −0.6 | −0.4 | – | 0.4 | −4.8 | −1.0 | 4.1 |
| | 3 | −1.1 | −0.1 | −0.6 | −0.2 | – | 0.4 | −2.8 | 1.2 | 6.3 |
| | 4 | −1.1 | −0.1 | −0.6 | −1.2 | – | 0.8 | −4.0 | −1.0 | 3.7 |

There is a preference for strategy 1 (MEMOD-R) for the decomposition of mode effects for two reasons. First, this strategy requires a weaker MAR assumption when estimating the coverage bias. Second, the use of auxiliary information from wave 2 through the GREG estimator did not lead to an increase in precision. In fact, for some target variables the precision after calibration to wave 2 was smaller due to the reduced sample size.

## 8.5 Adjusting Measurement Bias Using Re-interview Designs

In the second part of this chapter, an estimation approach is explored to adjust for measurement bias in sequential mixed-mode designs. The method is based on the re-interview approach. The idea behind this is to collect auxiliary information that is strongly correlated with benchmark variables to correct as good as possible for mode-dependent selection bias. The method is introduced for a mixed-mode design with two modes in this section. A generalization to three modes is given in Section 8.2.

The true scores for the target variable in a sequential mixed-mode design are denoted by $y$. Furthermore, $y^{m_x}$ denotes the measurement of target variable $y$ using mode $m_x$. In the case of a mixed-mode design with two different modes, we have $y^{m_1}$ and $y^{m_2}$ if $y$ is measured using mode $m_1$ and $m_2$, respectively.

The missing data pattern in a standard sequential mixed-mode design with two different modes is depicted in Figure 8.3. Response is characterized by shaded areas and missing data by white areas. Sampling units are first approached using mode $m_1$. The response obtained under $m_1$ is depicted with the shaded box (A). For these respondents

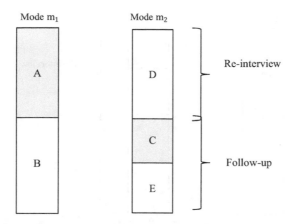

**FIGURE 8.3**
Standard sequential mixed-mode design. Grey area A represents $m_1$ response, white area B represents $m_1$ non-response, grey area C represents the follow-up $m_2$ response of the $m_1$ nonresponse, re-interview $m_2$ response (C), and follow-up $m_2$ response (D). White areas represent $m_1$ nonresponse (B) and re-interview $m_2$ nonresponse (D and E).

measurements $y^{m1}$ are observed. The nonresponse of mode $m_1$ is depicted with the white box (B) and has a follow-up using mode $m_2$. The response obtained with the follow-up under $m_2$ is depicted with the shaded box (C). For these respondents measurements $y^{m2}$ are observed. The nonresponse in the follow-up is the respondents of the first approach (white box D) and the nonresponse of the follow-up (white box E).

In the following, a sequential re-interview design is introduced with the purpose of estimating the true missed-mode response mean $y$ by adjusting for the measurement bias for one the modes. This estimator is denoted as $\hat{y}_{r_{mm}}$. To obtain strong auxiliary information for measurement bias adjustment, a part of the response under mode $m_1$ is re-interviewed in the follow-up using mode $m_2$. If, e.g., due to budget constraints, only a subset of the response under mode $m_1$ can be re-interviewed, then a random sample from the $m_1$ response is selected. This re-interview design is depicted in Figure 8.4, where it is shown that a part of the $m_1$ response (shaded box A) is re-interviewed using mode $m_2$. The response obtained with the re-interview is depicted in the shaded box (F). For these respondents measurements $y^{m2}$ are observed in addition to the measurements $y^{m1}$ observed under the first approach. The nonresponse of the re-interview is depicted with the white box D. It contains $m_1$ respondents that were not selected for a re-interview and sampling units that refuse to participate in the re-interview.

Since observed variables $y^{m1}$ and $y^{m2}$ are employed in estimation, the measurement error of both variables may bias an unadjusted estimator of $\hat{y}_{r_{mm}}$ that results when simply pooling the mixed-mode data by taking the sample mean across observed data. Therefore an estimator is constructed whose mean squared error is lower compared to the unadjusted estimator.

In doing so, we make an important assumption, called the measurement benchmark assumption. In the absence of true scores $y$, it is impossible to correct the measurement error bias contributed by both modes to the estimator. Instead, we focus on the situation when one of the modes is assumed as a measurement benchmark, setting the observed scores of this mode equal to $y$. The choice of benchmark mode is based on practical consideration of which combination of mode and question format can be assumed to evoke the least or no measurement error.

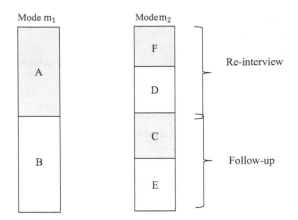

**FIGURE 8.4**

Re-interview design for a $m_1 \rightarrow m_2$ sequential survey design. Grey areas represent $m_1$ response (A), re-interview $m_2$ response (F), and follow-up $m_2$ response (C). White areas represent $m_1$ nonresponse (B), re-interview $m_2$ nonresponse (D), and follow-up nonresponse (E).

A major complication is the occurrence of non-random selection effects between modes; in the sense that the response mean of true scores $y$ under $m_1$ is not equivalent to the response mean under $m_2$. Such effects are desired in mixed-mode surveys, because they reflect that different modes reach different respondents. If both modes reached the same respondents, the second mode in the mixed-mode design would not have any additional value except for increasing the overall response rate. Unfortunately, selection effects are confounded with mode differences in measurement error (measurement effects). A difference in response means on $y^{m1}$ and $y^{m2}$ thus can denote a measurement or a selection effect or a combination. In a standard sequential design, there is insufficient information to determine the size of these effects. In adjusting measurement error it is therefore necessary to control for selection effects between modes.

Any estimator of $\hat{y}_{r_{mm}}$ that seeks to be superior to an unadjusted estimator necessarily needs to employ additional data in estimation. It may be an approach to use register (sampling frame) information for modeling response mechanisms and target variables. Generally, these variables are weak auxiliary information in many cases. Therefore the re-interview design is proposed to collect much stronger auxiliary information for measurement bias adjustment.

When adding the re-interview measurement to a sequential mixed-mode design, the repeated measurements in $m_2$ potentially may be influenced by the earlier measurement occasion. In this situation, $y^{m2}$ in the re-interview would not follow the same measurement model as standard responses in $m_2$. Therefore, it is assumed that the measurement error models in the re-interview and the regular $m_2$ model are identical. This assumption is called *measurement equivalence*.

Measurement equivalence is likely to occur in many practical situations. For inequivalence to occur, respondents first need to recall answers given at the first occasion ($m_1$) when the question is repeated under $m_2$. In addition, respondents need to be motivated to reproduce the answer they recall from the first occasion. The time lag between occasions, which in practice usually lies in the range of several weeks, plays a relevant role, because longer time lags increase the chances of answers being forgotten. Even if answers given earlier in $m_1$ are recalled, it is doubtful whether this causes the re-interview respondent to answer consistently with this answer in the response situation under $m_2$. Measurement equivalence, nevertheless, is an important assumption for the re-interview method.

A fixed response model is assumed that separates all units $i$ in a population of size $N$ for a given mixed-mode design into two response strata (units participating in either $m1$ or $m2$) and a nonresponse stratum (Cochran, 1977). Since we focus on the response mean, the nonresponse stratum is ignored in the following discussion. Let $N_{m1}$ and $N_{m2}$ denote the size of the strata that respond to mode $m_1$ and $m_2$, respectively. Furthermore, $y_{m1}$ and $y_{m2}$ denote the true population means of the strata that respond to mode $m_1$ and $m_2$, respectively. The mixed-mode response mean is then given by:

$$y_{r_{mm}} = \frac{1}{\left(N_{m_1} + N_{m_2}\right)} \left(N_{m_1} y_{m_1} + N_{m_2} y_{m_2}\right) \tag{8.15}$$

An unadjusted estimator of the mixed-mode response mean which simply pools the observed mixed-mode data without any measurement bias correction is given by

$$\hat{y}_{r_{mm}}^{Unadj} = \frac{1}{\left(\hat{N}_{m_1} + \hat{N}_{m_2}\right)} \left(\sum_{i=1}^{n_{m1}} d_i y_i^{m_1} + \sum_{i^i=1}^{n_{m2}} d_{i^i} y_{i^i}^{m_2}\right) \tag{8.16}$$

where $d_i$ denotes the design weight of the sample design, $n_{m1}$ and $n_{m2}$ the number of $m_1$ and $m_2$ responses, respectively. Furthermore, $\hat{N}_{m_1} = \sum_{i=1}^{n_{m1}} d_i$ and $\hat{N}_{m_2} = \sum_{i^i=1}^{n_{m2}} d_{i^i}$ are the estimated population sizes of the two response strata. Note that regression weights can be used as an alternative in (8.16) for the design weights.

Klausch et al. (2017) propose six different estimators that attempt to correct for the relative measurement bias between modes $m_1$ and $m_2$ using the information obtained with the re-interview data. Let $n_{m1m2}$ denote the number of respondents for which both an $m_1$ and $m_2$ response is obtained. Let $r_{i1}$ denote the indicator variable that is equal to one if a response from sampling unit $i$ is obtained under mode $m_1$ and zero otherwise. Similarly, $r_{i2}$ is the indicator variable that is equal to one if a response is obtained under mode $m_2$ and zero otherwise. Furthermore $r_{i12}$ denotes the indicator variable that is equal to one if sampling unit $i$ response under mode $m_1$ and also under mode $m_2$ in the re-interview and zero otherwise. Depending on which mode represents the measurement benchmark, we now seek to estimate the mean of either $y^{m1}$ or $y^{m2}$ over the full response sample. This objective presents us with the missing data problem illustrated in Figure 8.4. In doing so, we employ auxiliary data obtained from the subset of $n_{m_1m_2}$ re-interview respondents.

The first estimator is the propensity score estimator. This estimator estimates the propensity of respondents to reply under the benchmark mode and apply it for calibrating a selective sub-group (i.e. the response sample in the benchmark mode) to a reference group (i.e., the mixed-mode response sample). Let the propensity for unit $i$ to be observed in the benchmark mode be denoted as $\pi_i$ (Rosenbaum and Rubin, 1983, 1985). The propensity can be estimated using a model $f_\pi$:

$$\pi_i = P(r_{ib} = 1 | y_i^{m_j}, r_{ij} = 1 \cup r_{i12} = 1)$$

$$= f_\pi \left(y_i^{m_j}, r_{ij}, r_{i12} = 1, \theta\right) b, j = 1, 2; b \neq j \tag{8.17}$$

A typical choice for $f_\pi$ is the logistic or probit regression model with unknown parameter vector $\theta$. In the subscripts, $b$ refers to the benchmark mode and $j$ to the other mode. If $m_1$ is the benchmark mode, then $b = 1$ and $j = 2$ and the other way around. Now the propensity scores can be used in the following propensity estimator:

$$\hat{y}_{r_{mm}}^\pi = \frac{1}{\left(\hat{N}_{m_1} + \hat{N}_{m_2}\right)} \sum_{i=1}^{n_{mb}} d_i y_i^{m_b} \left( r_{i12} \frac{\left(1 - \hat{\pi}_i\right)}{\hat{\pi}_i} + 1 \right); b = 1, 2. \tag{8.18}$$

This estimator first estimates the total of the observed benchmark outcomes $y_i^{m_b}$ from the response in benchmark mode $b$. It then adds an estimate of the total of benchmark outcomes in the focal mode using benchmark outcomes from the re-interview response sample calibrated by weight $(1 - \hat{\pi}_i)/\hat{\pi}_i$.

Alternative estimators seek to find accurate predictions of the potential benchmark outcomes $y^{m_b}$ using a suitable model for $y^{m_b}$ and sum over observed and predicted scores. A general form of this type of estimators can be written as

$$\hat{y}_{r_{mm}}^{ypred} = \frac{1}{\left(\hat{N}_{m_1} + \hat{N}_{m_2}\right)} \left( \sum_{i=1}^{n_{mb}} d_i y_i^{m_b} + \sum_{i'=1}^{m_j} d_{i'} \hat{y}_{i'}^{m_b} \right); b, j = 1, 2; b \neq j. \tag{8.19}$$

In (8.19) $\hat{y}_i^{m_b}$ represents the estimated potential benchmark outcomes for the respondents in the focal mode $j$.

Estimator (8.19) requires specifying a model that describes the relation of benchmark responses to the alternative mode outcomes. It is then assumed that the model also holds in the response stratum to mode $j$ and can be used to transform observed $y^{m_j}$ to $y^{m_b}$. Three special cases of (8.19) are proposed by Klausch et al. (2017). The first one is the fixed-effect estimator, which corrects each focal mode outcome $y_i^{m_j}$ by the fixed observed mean difference in the re-interview sample as a simple estimate of measurement effect and is defined as

$$\hat{y}_{r_{mm}}^{fe} = \frac{1}{\left(\hat{N}_{m_1} + \hat{N}_{m_2}\right)} \left( \sum_{i=1}^{n_{mb}} d_i y_i^{m_b} + \sum_{i'=1}^{m_j} d_{i'} \left( y_{i'}^{m_j} - \left( \hat{y}_{re}^{m_j} - \hat{y}_{re}^{m_b} \right) \right) \right) b, j = 1, 2; b \neq j, \tag{8.20}$$

where $\hat{y}_{re}^{m_j}$ and $\hat{y}_{re}^{m_b}$ are the respondent means of focal and benchmark mode outcome in the re-interview. Estimator (8.20) omits scale differences between modes.

The second special case of (8.19) assumes a ration model to account for scale differences between modes, and is referred to as the ratio estimator:

$$\hat{y}_{r_{mm}}^{ratio} = \frac{1}{\left(\hat{N}_{m_1} + \hat{N}_{m_2}\right)} \left( \sum_{i=1}^{n_{mb}} d_i y_i^{m_b} + \sum_{i'=1}^{m_j} d_{i'} y_{i'}^{m_j} \frac{\hat{y}_{re}^{m_b}}{\hat{y}_{re}^{m_j}} \right) b, j = 1, 2; b \neq j. \tag{8.21}$$

The third special case of (8.19) is the regression estimator, which uses a standard or generalized regression estimator for the predictions of the potential benchmark observations and is defined as

$$\hat{y}_{r_{mm}}^{greg} = \frac{1}{\left(\hat{N}_{m_1} + \hat{N}_{m_2}\right)} \left( \sum_{i=1}^{n_{mb}} d_i y_i^{mb} + \sum_{i'=1}^{mj} d_{i'} \left(\hat{y}_{re}^{mb} - \hat{\beta}_{re}(\hat{y}_{re}^{mj} - y_{i'}^{mj})\right) \right) \Bigg| b,j=1,2;b \neq j. \quad (8.22)$$

where $\hat{\beta}_{re}$ denotes the slope of the linear regression of $y^{mb}$ on $y^{mj}$ in the re-interview stratum (Särndal and Lundström, 2005).

It can be seen that the fixed-effect estimator is a special case of the GREG estimator, where the slope is equal to 1. Furthermore, it can be shown that the ratio estimator is the special case of the GREG estimator, where the intercept is fixed at zero.

The fifth estimator uses an inverse version of a regression estimator (IREG). The idea of IREG is to use benchmark measurements $y^{mb}$ instead of $y^{mj}$ as auxiliary data for modeling focal mode outcomes $y^{mj}$ in

$$y_i^{mj} = v_0 + v_{re} y_i^{mb} + \varepsilon_i \quad (8.23)$$

Equation (8.23) is estimated by ordinary or generalized least squares from the re-interview response. Subsequently the inverse of $v_{re}$ is used in a GREG-type estimator. This gives rise to the so-called inverse regression estimator, which is defined as

$$\hat{y}_{r_{mm}}^{ireg} = \frac{1}{\left(\hat{N}_{m_1} + \hat{N}_{m_2}\right)} \left( \sum_{i=1}^{n_{mb}} d_i y_i^{mb} + \sum_{i'=1}^{mj} d_{i'} \left( \hat{y}_{re}^{mb} - \frac{1}{\hat{v}_{re}} \left(\hat{y}_{re}^{mj} - y_{i'}^{mj}\right) \right) \right) \Bigg| b,j=1,2;b \neq j. \quad (8.24)$$

Finally, Klausch et al. (2017) considered simultaneous multiple imputation (MI) for measurement error adjustment (Guo and Little, 2013) using the MICE algorithm (multiple imputation by chained equations; van Buuren, 2012). The procedure specifies the conditional distributions of benchmark and focal mode outcomes as normal regression models and initially completes the missing data by draws from the observed distributions (fields A for $y^{m1}$ and F and C for $y^{m2}$, in Figure 8.4, respectively). The procedure then alternates between the two variables predicting the potential outcomes by drawing from their predictive distributions, converging to the bivariate distribution like a Gibbs sampler. We evaluate this procedure with five imputed data sets pooled by Rubin's rules (Rubin, 1987).

---

## 8.6 Simulation Study

In Klausch et al. (2017) an extensive simulation study is described to evaluate under a wide range of conditions, which estimator performs best. To this end, data were simulated assuming the following measurement error model as the data generating process:

$$y_i^{mj} = \mu^{mj} + \lambda^{mj} \left( y_i + u_i^{mj} \right) \quad (8.24)$$

where $\lambda^{mj}$ is a scale parameter that is equal to 1 if $m_j$ measures on the scale of the true score, and $u_i^{mj}$ is an independently and identically distributed measurement error term with

$$u_i^{m_j} \sim iid\left(0,\left(\sigma_u^{m_j}\right)^2\right)$$                                                          (8.25)

Furthermore $\mu^{m_j}$ is a systematic measurement error common to all units, whereas $\left(\sigma_u^{m_j}\right)^2$ denotes the variance of random measurement errors in the population. It is assumed that true scores $y_i$ and random measurement errors $u_i^{m_j}$ are independent for elements $i$ in the population and all modes $j$.

Let $c_j = corr\left(y^{m_j}, y\right)$ denote the correlation between $y^{m_j}$ and $y$. The variance of the random errors $\left(\sigma_u^{m_j}\right)^2$ is dependent on $\sigma_y^2$ and $c_j$ since it is assumed that

$$\left(\sigma_u^{m_j}\right)^2 = \frac{1-c_j^2}{c_j^2}\sigma_y^2$$                                                       (8.26)

The correlation $c_j$ is sometimes referred to as validity or reliability coefficient (Biemer and Stokes, 1991). This correlation can be used to scale the error variance of $y^m$. Klausch et al. (2017) conducted simulations under different levels of $c_j$.

For the measurement benchmark mode it is assumed that $\mu^{m_b} = 0$, $\lambda^{m_b} = 1$, and $u_i^{m_b} = 0$ (and thus $c_b = 1$). This implies that in (9.24) $y_i^{m_j} = y_i$. Furthermore, the measurement equivalence assumption between re-interview and $m_2$ implies that $y_i^{m_{re}} = y_i^{m_2}$.

In the simulation the estimators are compared under the following settings:

- A range of values for selectivity between the response stratum of mode $m_b$ and $m_j$, with values from absent (0%) to strong selectivity (plus and minus 50% relative to $m_b$);

- A range of values for review selectivity, i.e., selectivity between $y_{m1}$ and $y_{m_{re}}$ with values from absent (0%) to strong selectivity (plus and minus 50% relative to $m_1$);

- Within each response stratum values are from a normal distribution with a range of values for the means of the two response strata and the nonresponse stratum and standard deviation equal to one;

- Using mode $m_1$ as well as mode $m_2$ as benchmark mode ($b = 1, 2$);

- A range of values for focal mode systematic error, i.e., different values for $\mu^{mj}$ in (8.24) and setting the benchmark mode systematic error equal to zero;

- A range of values for the focal mode scale parameter, i.e., different values for $\lambda^{mj}$ in (8.24) and setting the benchmark mode scale parameter equal to one; and

- A range of values for the correlation between $y^{mj}$ and $y$, i.e., $c_j$.

The levels of the aforementioned simulation settings are fully crossed with each other, giving rise to 1,800 different simulation settings. For each setting a population of 10,000 is generated. From each generated population 1,000 simple random samples are drawn without replacement and a sample size of 2,500. A response of 50% is assumed in the first approach for mode $m_1$. It is assumed that 50% of the $m_1$ nonresponse respond to mode $m_2$. Furthermore, 50% of the $m_1$ respondents are selected for a re-interview. It is assumed 60% responds to the re-interview.

The unadjusted estimator and the six estimators that adjust for measurement bias, discussed in Section 8.5, are applied to each of these samples. The root mean squared error across the $S = 1000$ simulation replications is calculated for each estimator;

$$RMSE\left(\hat{y}_{r_{mm}}\right) = \sqrt{\frac{1}{S}\sum_{s=1}^{S}\left(\hat{y}_{r_{mm}}^{s} - y_{r_{mm}}\right)^{2}} \qquad (8.27)$$

where $y_{r_{mm}}$ is the true population mean for the given condition.

In this section the main results of this large simulation study are summarized. For a detailed discussion of the results, see Klausch et al. (2017) and in particular the supplemental file of this paper.

If mode $m_1$ is used as the benchmark mode it is found that the IREG estimator (8.24) outperforms all other estimators. The variance of this estimator is large if the focal mode random error is high, indicated by low re-interview correlation. For re-interview correlations larger than 0.5 the RMSE of the IREG is smaller than 10% and if re-interview correlations larger than 0.7 the RMSE of the IREG is smaller than 5%. All other adjusted estimators as well as the unadjusted estimator perform worse under several conditions. This held for the regression, IPW, and multiple imputation estimator in particular. These estimators showed high (>10%) RMSE unless re-interview correlation was very high (>.90), a situation seldom expectable in practice. The ratio estimator performed only well in the absence of focal mode systematic error, i.e., if $\mu^{m_j} = 0$. The fixed-effect estimator performed only well if there is no focal mode scale parameter, i.e., $\lambda^{m_j} = 1$.

If mode $m_2$ is used as the benchmark mode IREG if re-interviewer correlation exceeds a moderate level (>0.4). It is observed that IREG leads to slightly smaller RMSE compared to $b = 1$. The finding that GREG, IPW, and MI have high RMSE was repeated, even though RMSE levels were lower than for $b = 1$. However, RMSE of GREG and IPW nearly vanished in the absence of a re-interview selection effects. Since it is hard to diagnose the size of the re-interview SE in practice, Klausch et al. (2017) recommend against the use of the estimators as in the case of $m_1$ as benchmark. The fixed-effect estimator had low RMSE under all re-interview selection effect conditions and all levels of scaling parameter $\lambda^{m_j}$. Also the ratio estimator showed low levels of RMSE under all levels of $\lambda^{m_j}$ and $\mu^{m_j}$ in case of strong re-interview correlation (>0.5).

---

## 8.7 Extension of Re-interview Design to Multiple Modes

In Sections 8.5 and 8.6, the number of modes in the sequential re-interview design was limited to two. However, many real surveys use three or four modes. Another of the case studies of Section 2.6, the Dutch Labor Force Survey (LFS), applies three modes. Households included in the sample first receive a letter with an invitation to complete the questionnaire via the web. Nonrespondents are approached by an interviewer by telephone if a registered phone number is available. Households without a registered phone number are approached by an interviewer at home. Extensions to other designs or designs with more than three modes can be done analogously.

The population under the sequential mixed-mode design outlined above is divided into five subpopulations:

- 1 = households with a registered phone responding to web;
- 2 = households without a registered phone responding to web;

- 3 = households with a registered phone responding to telephone after not responding to web;
- 4 = households without a registered phone responding to F2F after not responding to web; and
- 5 = households not responding to the MM design.

Subpopulation 5 is out of scope and no outcomes are estimated for this subpopulation. We label the remaining subpopulations as $l = 1, 2, 3, 4$.

Some additional notation is introduced first: Let $(l_1, l_2)$ be the union of subpopulations $l_1$ and $l_2$. Furthermore, let $\hat{N}_l$ be the estimated size of subpopulation $l$ based on the sample, and $\hat{N}$ be the estimated population size. Let $\hat{y}^u_{r_{mm}, t}$ be the type of estimator, with $t \in \{$unadj, prop. ratio, reg, ireg, fixed $-$ effect, multimp$\}$, applied to subpopulation $u \in \{(1,2), (1,3), (1,4), (2,3), (2,4), (3,4)\}$ of the population assuming $r_{mm} \in \{$web, tel, F2F$\}$ as the benchmark mode for measurement. Let $\bar{y}_u$ be the design-weighted mean of the outcomes for subpopulation $u$.

Under $r_{mm} = web$, estimators have the form

$$\hat{y}_{web} = \frac{\hat{N}_1 + \hat{N}_3}{\hat{N}} \hat{y}^{(1,3)}_{web, t_1} + \frac{\hat{N}_2 + \hat{N}_4}{\hat{N}} \hat{y}^{(2,4)}_{web, t_2} \tag{8.28}$$

where the estimators are allowed to be different per subpopulation given conjectures about the form of the measurement model.

The estimators for $r_{mm} = tel$ and $r_{mm} = f2f$ are conceptually the same; we give only the estimator for $r_{mm} = f2f$. Under $r_{mm} = f2f$, estimators have the form

$$\hat{y}_{f2f} = \frac{\hat{N}_1 + \hat{N}_4}{\hat{N}} \hat{y}^{(1,4)}_{f2f, t_1} + \frac{\hat{N}_2 + \hat{N}_4}{\hat{N}} \hat{y}^{(2,4)}_{f2f, t_2}$$

$$+ \frac{\hat{N}_3 + \hat{N}_4}{\hat{N}} \hat{y}^{(3,4)}_{f2f, t_3} - 2 \frac{\hat{N}_4}{\hat{N}} \bar{y}_{f2f} \tag{8.29}$$

where the term $\dfrac{\hat{N}_4}{\hat{N}} \bar{y}_{f2f}$ is included in each of the estimators and needs to be subtracted twice. Again, the estimators may be different per subpopulation. The simulation by Klausch et al. (2017), summarized in Subsection 8.7 is the only empirical evidence available to recommend the inverse regression estimator (8.24) to be used in (8.28) and (8.29).

## 8.8 Conclusions

In the first part of this chapter, an experimental design is presented that allows for a decomposition of relative mode effects. The definition of coverage bias, nonresponse bias, and measurement bias is not at all unambiguous, nor is the estimation of these components based on the experimental design. Several definitions and estimation strategies are compared. It is found that the various definitions and estimation strategies have relatively little impact on the estimates for the CVS. However, a preferred definition and strategy are proposed.

There is an important limitation to the proposed estimation strategy. The accuracy of the estimated coverage, nonresponse, and measurement bias depends on several assumptions. It is assumed that both response and answering behavior to the second wave are not affected by the mode and response of the first wave. Furthermore, it is assumed that nonresponse to wave 1, given response to wave 2, is missing-at-random conditional on the auxiliary variables measured in wave 2. See Schouten et al. (2013) for an extensive evaluation as to the extent to which these assumptions are met in the application to the CVS. It is, however, not guaranteed that these assumptions will hold in similar experiments. In the preparation of the experiment a lot of attention was paid to advance letters, interviewer instruction, and questionnaire design of the experimental second wave. A careful preparation is needed to avoid invalidity of the experimental assumptions.

The repeated variables in wave 2 were the strongest predictors in the mode effect decomposition. Hence, if standard frame data and linked registry data variables are used in the GREG estimator, then it can be expected that the estimates for coverage and measurement bias are partially obscured with nonresponse bias.

More generally, the application shows that mode biases may dominate the mean square error of survey estimates and that mode biases may be present even when careful questionnaire design and testing is applied and when nonresponse adjustment is performed. The experimental design and estimation strategy is independent of the survey context like the country in which the survey is conducted and the target population of the survey; it can generally be applied. Mode pilot studies like the study in this chapter may be well worthwhile, even if that means a relatively large investment.

A re-interview design is described in the second part of this chapter to correct for measurement bias in sequential mixed-mode data collection surveys toward a benchmark mode. This data is obtained from a subset of respondents to the first mode in a sequential design and it is employed as auxiliary data in a set of six adjusted candidate estimators. The performance of these estimators is evaluated by simulation and demonstrates how to adjust for measurement effects in the presence of non-ignorable selection effects. Earlier literature that attempts to estimate or adjust measurement effects can be criticized for potentially high bias, because researchers often had to assume that selection is ignorable conditional on weak auxiliary information (Vannieuwenhuyze, 2015; Vannieuwenhuyze and Loosveldt, 2013).

In summary, when mode 1 was the benchmark, only the IREG estimator performed reliably, whereas ratio and fixed-effect estimators performed only well under special circumstances. When mode 2 as the benchmark, IREG performed again well, but also the ratio and fixed-effect estimators showed low levels of RMSE when re-interview correlation was moderate, even in the most extreme scenarios considered here. GREG, IPW, and MI performed badly in the most considered scenarios due to their high bias.

## 8.9 Summary

The main takeaway messages from this chapter are:

- Mixed-mode designs potentially reduce nonresponse bias by improving coverage rates of a survey.

- Sequential mixed-mode designs potentially reduce administration costs and achieve response rates comparable to uni-mode face-to-face designs.
- The use of multiple data collection modes will potentially increase measurement bias.
- A decomposition of mode effects into measurement bias nonresponse bias and coverage bias provides important information to design mixed-mode surveys (field work strategies and questionnaire).
- Repeated measurement experiments allow a quantitative decomposition of relative mode effects through strongly auxiliary information that is observed in the re-interview, which is used to construct optimal weighting models to correct for selection bias in the analysis of the experiment.
- A decomposition of mode effects is based on several assumptions:
    - Response and answering behavior to the second wave are not affected by the mode and response of the first wave.
    - Nonresponse to wave 1, given response to wave 2, is missing-at-random conditional on the auxiliary variables measured in wave 2.
- A real-life application of a repeated measurement experiment showed that mode biases may dominate the mean square error of survey estimates and that mode biases may be present even when careful questionnaire design and testing is applied and when nonresponse adjustment is performed.
- The accuracy of statistics based on sequential mixed-mode designs can be improved by adjusting for measurement bias.
- Measurement bias adjustment in a sequential mixed-mode design requires strongly related auxiliary information that can be obtained via a re-interview design.
- An extensive simulation study showed that among a set of six adjustment estimators the so-called inverse regression estimator performs the best and is advised to be used in this context.
- Assumptions underlying re-interview design and the adjustment estimators are:
    - Measurement equivalence, i.e., the measurements in the re-interview and the regular follow-up of the $m_1$ nonresponse are identical.
    - The realizations of the benchmark mode represent the true values toward which focal mode measurements are adjusted.

# 9

# Mixed-Mode Data Analysis

## 9.1 Introduction

This chapter considers again the estimation design (see Figure 9.1). In doing so, it focuses on the survey estimates as they are produced in the analysis stage. The methods and techniques in this chapter do not require an experimental design or repeated observations, as was the case in Chapter 7, Field Tests and Implementation of Mixed-Mode Surveys, and Chapter 8, Re-interview Designs to Disentangle and Adjust for Mode Effects. However, they do require auxiliary variables that explain the mode-specific selection mechanism as explained in Chapter 4, Mode-Specific Selection Effects.

After data collection, survey response data are used for analysis, in particular, the estimation of quantities such as means and totals of survey variables. It is common practice to apply a weighting procedure to the response data to allow for the convenient application of calibration or generalized regression estimators. These techniques take unequal inclusion probabilities and selective nonresponse into account and are aimed at correcting selection bias that could otherwise be present. Researchers and analysts working with mixed-mode data should not proceed carelessly as if analyzing single-mode survey responses. Since multiple modes were used to collect data from sample units, not all units are measured in the same way, and their measurements may not be equivalent. When comparing survey estimates for population subgroups, for example, any found differences could be simply due – at least partially – to the fact that the groups were measured differently, in particular, using different combinations of data collection modes. When applying standard weighting methods to mixed-mode mode data, the resulting estimators consist of weighted combinations of observations collected through different modes. If there is indeed no measurement equivalence between the data collection modes, these traditional estimators contain components of differential measurement errors. These components are considered as bias if one of the modes is seen as the reference or benchmark mode.

After a literature review and the introduction of notation and definitions in Section 9.2, two types of approaches to mitigate the risk of biased estimates are discussed. Regression, imputation, and prediction techniques attempt to remove measurement bias from the observations, see Section 9.3. If no single mode can be identified as the reference mode, the measurement differentials can still cause instabilities in estimators, since different mixes of such errors can be present in different time periods or in different subpopulation groups. Balancing methods that address this issue are discussed in Section 9.4. Section 9.5 provides some advice for practitioners wishing to apply correction or balancing methods. Applications of both approaches are presented in Section 9.6 using the Health Survey, Crime Victimization Survey, and the Labor Force Survey of the Netherlands as examples. Section 9.7 contains the main take-home messages from this chapter.

DOI: 10.1201/9780429461156-9

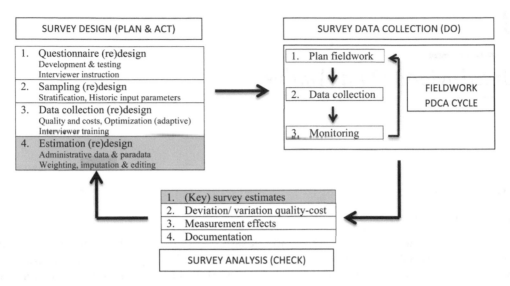

**FIGURE 9.1**
The plan-do-check-act cycle for surveys.

## 9.2 Analyzing Mixed-Mode Survey Response Data

In the following, the reference to 'mode-specific' effects is omitted, unless it could lead to confusion.

### 9.2.1 Literature

The emergence of web technology as a data collection mode has instigated renewed interest in research into mode effects in mixed-mode sample surveys. The confounding of mode-specific selection effects, see Chapter 4, Mode-Specific Selection Effects, and measurement effects, see Chapter 3, Mode-Specific Measurement Effects, is well known to form a core difficulty for estimation in mixed-mode data collection regimes. De Leeuw (2005) lists the advantages and pitfalls of mixing data collection modes. Voogt and Saris (2005) discuss the trade-off between improved selection and possibly hampered measurements in mixed-mode surveys. Dillman and Christian (2005) recognize the issue of differential measurement effects between modes and suggest preventing this issue through the design of questionnaires that prevent this phenomenon from occurring.

Mode adjustment methods are methods that adjust survey estimates obtained from mixed-mode designs to correct for mode effects induced by the use of multiple modes of data collection. In survey sample research, adjustment for selection effects due to coverage and nonresponse problems is typically conducted. Methods commonly used for this purpose include weighting, calibration, and regression methods (Bethlehem et al., 2011). If mixed-mode designs result in adverse selections of respondents, these common methods can be applied in the same way as they are used to correct selection effects due to coverage and nonresponse issues. In this respect, mixed-mode designs are not unlike single-mode designs in which selection effects are corrected in order to remove or reduce bias in survey estimates.

The adjustment methods discussed in this chapter are aimed at handling measurement effects, possibly in combination with the familiar selection effect adjustment methods. Common estimation methods in single-mode survey sampling do not handle measurement issues. These measurement adjustment methods are specific to mixed-mode designs. Section 9.5.3 will frame mode adjustment methods for mixed-mode surveys in single-mode settings, highlighting that neglecting measurement error issues in single-mode surveys is valid only under assumptions that remain mostly implicit in classic single-mode survey literature.

Adjustment to survey estimates in mixed-mode surveys is warranted and desirable when the point or variance estimates are biased compared to estimates from some benchmark design. Of course, it is assumed that the adjusted estimates are better – in mean square error sense – than the unadjusted survey estimates. Adjustment techniques are aimed at correcting survey estimates for the bias induced by one or several modes, or by the specific combination of several modes. Adjustment for bias requires the presence of a definition – or choice – of reference mode or design that serves as a benchmark, since bias of a certain design is only meaningful with respect to some other design.

Suzer-Gurtekin et al. (2012) presented some early results on estimation methods in the context of mixed-mode designs, expanding upon this work in her PhD thesis (Suzer-Gurtekin, 2013). In this work, mixed-mode measurements are regarded as treatments in a causal modeling framework of counterfactuals (Rubin, 2005), with potential outcomes defined as answers that would be given to survey questions through a mode that was not actually used for the respondent. Potential outcomes are obtained through regression modeling. Overall survey estimators of means and totals are proposed to be combinations of real answers and of potential outcomes. Uncertainty resulting from models to predict the counterfactuals adds to the total variance. Depending on the choice of benchmark, different mixes of counterfactuals can be produced; it is suggested to seek a mix that minimizes the mean square error.

Kolenikov and Kennedy (2014) compare the regression modeling approach with multiple imputation of non-observed answers, framing the problem rather as an imputation and missing data problem. In addition they studied a third approach, an imputation technique based on an econometric framework of implied utilities in logistic regression modeling. The multiple imputation method came out as the preferential technique.

Park et al. (2016) also propose an imputation approach to impute unobserved survey items with counterfactuals. They propose to use fractional imputation and obtain variance estimators using Taylor linearization. They present a limited real-world application in addition to a simulation study.

A recent application is discussed by Fessler et al. (2018) where they extend the potential-outcomes approach to distributional characteristics other than means and totals, to estimate measures of income inequality in Austria.

Another approach to mode adjustment is reweighting of the survey response. Buelens and van den Brakel (2015) address a situation where the composition of the survey response varies between population subgroups such as regions or age classes, or between editions of a survey in the case of regularly repeated surveys. Such variations hamper the comparability of survey estimates as the measurement effect in subgroups or editions is not constant due to the variability in the mode compositions. Their solution is to apply a calibration correction by reweighting the survey response to fixed-mode distributions. This method can be applied to non-experimental data assumed that there are no confounding variables that are not accounted for.

Buelens and van den Brakel (2017) compare their mode calibration method with the potential-outcomes approach and regression modeling (Suzer-Gurtekin, 2013). They discuss parallels of and differences between the two methods and give circumstances in which both methods are equivalent. They provide an example from the Labor Force Survey in the Netherlands and find that in this specific case no adjustments due to imbalances in mode distributions are required.

Vannieuwenhuyze et al. (2014) propose adjustments using covariates to correct for mode effects. While such methods are common to correct for selection effects, they propose to apply these methods to correct for measurement effects. Covariates must then be chosen not so that they explain selection differences between modes, but rather so that they explain measurement differences between modes. Which covariates can be used for this purpose remains an ad-hoc choice.

Pfeffermann (2017) described a unified approach to handle inference from non-representative samples. Section 8.1 of this article suggests an extension from the essentially Bayesian approach to include measurement error arising in mixed-mode designs. The target of inference is the posterior probability distribution of the variable of interest, which is free from measurement bias, and corrected for selection effects. This discussion is largely conceptual and has not been applied or simulated.

This chapter concentrates on the work of Suzer-Gurtekin et al. (2012), Suzer-Gurtekin (2013), Kolenikov and Kennedy (2014), and Park et al. (2016), on the one hand – in Section 9.3, and the work of Buelens and van den Brakel (2015, 2017), on the other hand – in Section 9.4. Before detailing the methods, the necessary formalism is introduced below.

### 9.2.2 Problem Statement, Definitions, and Notation

Based on a sample survey, the estimate of the total of a variable $Y$ is generally expressed as

$$\sum_{i=1}^{n} w_i y_i \tag{9.1}$$

with $n$ the number of respondents, $y_i$ the measurement of $Y$ for unit $i$, and $w_i$ a survey weight assigned to unit $i$. The survey weights typically take into account unequal inclusion probabilities and a correction for selective nonresponse. Generalized regression (GREG) is a commonly used method to compute the weights (Särndal et al., 1992).

In a mixed-mode survey, not all $y_i$ are obtained through the same data collection mode. For example, some may be obtained through web questionnaires while others are collected through telephone interviewing. When there is differential measurement bias between modes, expression (9.1) is contaminated with this error. In this chapter, two types of approaches addressing this problem are discussed.

The first approach seeks to adjust the estimator (9.1) by adjusting the measurements $y_i$. The goal of this type of adjustment is the removal of differential measurement bias from the measurements. The survey weights $w_i$ are left unaltered. This approach is detailed in Section 9.3.

The second approach leaves the measurements unchanged but consists of adjustments to the weights $w_i$ in expression (9.1). This method does not remove bias; rather, it counteracts adverse effects that differential bias may have by balancing differential mode-dependent bias between subpopulations or between time periods. This approach is discussed in Section 9.4.

The discussion of mode-dependent measurement bias necessitates distinguishing a variable $U$ from its measurement $Y$. More specifically, $Y_m$ refers to the measurement of $U$ through data collection mode $m = 1, ..., p$ for a mixed-mode design with $p$ different modes.

Expression (9.1) can be written more precisely, for the variable $U$, as

$$\hat{t}_U = \sum_{i=1}^{n} w_i u_i \tag{9.2}$$

which is the estimated total of $U$, with $u_i$ the values of $U$ for respondents $i$. Since these are not observed, the actual estimator that is commonly used is again like expression (9.2), but now in terms of $Y$,

$$\hat{t}_Y = \sum_{i=1}^{n} w_i y_{i,m} \tag{9.3}$$

with $y_{i,m}$ the measurement obtained from unit $i$ through mode $m$.

The relation between the $u_i$ and the $y_{i,m}$ can take various forms. In the present context a linear relation is assumed,

$$y_{i,m} = u_i + b_m + \epsilon_{i,m} \tag{9.4}$$

with $b_m$ an additive mode-dependent measurement bias and $\epsilon_{i,m}$ a mode-dependent measurement variance with an expected value equal to zero. Equation (9.4) is the measurement error model used in the remainder of this chapter, using $m$ to refer to data collection mode and $p$ to the number of modes.

In expression (9.4), the underlying true scores $u_i$ remain unknown, only the $y_{i,m}$ are observed. This is the case also in single-mode surveys. There, any mode-dependent measurement errors that might be present affect all respondents equally, in expectation. In a mixed-mode design, observations through some mode $m$ may measure $U$ differently than through a different mode $m'$.

Under model (9.4), the differential measurement error between modes $m$ and $m'$ is given by

$$y_{i,m} - y_{i,m'} = \left(b_m + \epsilon_{i,m}\right) - \left(b_{m'} + \epsilon_{i,m'}\right) \tag{9.5}$$

with expected value $(b_m - b_{m'})$, a quantity referred to here as differential measurement bias. Methods to handle this type of bias are presented in the following sections.

---

## 9.3 Correcting Differential Measurement Bias

### 9.3.1 Counterfactuals

When one of the survey modes in a mixed-mode survey is considered better or preferable compared to the other modes, this mode is taken to be the benchmark mode. In this case, ideally, every respondent had responded through that mode. In practice in mixed-mode designs, a certain proportion of the respondents responded through a mode other than the benchmark.

In expression (9.5), $y_{i,m}$ and $y_{i,m'}$ are not observed both, as each respondent $i$ gives only one answer. Introducing notation for this situation, $y_{i,m}^{m'}$ represents the response of unit $i$ through mode $m'$ while the true response in the survey was obtained through mode $m$. Since by definition $y_{i,m}^{m'}$ is never observed for $m' \neq m$, it is sometimes referred to as a counterfactual observation (Rubin, 2005), a term commonly encountered in the potential-outcomes framework.

The correction method discussed in this section proceeds by first estimating the counterfactual observations for the benchmark mode, and subsequently combining these in an estimator. Literature addressing such solutions includes Suzer-Gurtekin et al. (2012), Suzer-Gurtekin (2013), Kolenikov and Kennedy (2014), and Park et al. (2016). This section is largely based on their work.

Potential outcomes can be estimated or predicted using a wide variety of methods. In the present exposition a linear model as in expression (9.4) is used. Linearity and additivity of the measurement bias facilitate the discussion and the equations and will allow for a fair comparison with the balancing methods presented in Section 9.4.

Furthermore, it builds upon the linear model underpinning the GREG estimator. The classic GREG estimator does not cater to measurement errors. For the variable $U$, the GREG estimator is based on a linear model

$$u_i = \beta X_i + e_i \tag{9.6}$$

With $X_i$ a vector with covariates, $\beta$ a vector of regression coefficients and $e_i$ an error term. Combining this model with the error model (9.4) gives

$$y_{i,m} = \beta X_i + b_m \delta_{i,m} + \tilde{e}_{i,m} \tag{9.7}$$

with $\tilde{e}_{i,m} = \epsilon_{i,m} + e_i$ and $\delta_{i,m} = 1$ when respondent $i$ responded through mode $m$, and zero otherwise.

It can be seen from expression (9.7) that under the model assumptions the $b_m$ play the role of regression coefficients of data collection mode in a linear model.

Estimation of these coefficients will only be successful if the covariate *mode* in this model does not explain selection effects. In other words, the covariates $X$ must fully explain the selection effect. What remains is the mode-specific measurement effect, captured by $b_m$. If this requirement is not met, mode-specific selection effects will unduly be attributed to the estimate of $b_m$.

Seemingly strong, this assumption is also required for the GREG estimator (9.6) to result in unbiased estimates of $U$. Section 9.5.1 returns to this assumption.

Model (9.7) can be fitted using survey data, using, for example, least-squares regression. This results in estimates $\hat{\beta}$ and $\hat{b}_m$ of the regression coefficients. The fitted model can be used in a predictive manner to estimate the counterfactual observations:

$$\hat{y}_{i,m}^{m'} = \hat{\beta} X_i + \hat{b}_{m'} \tag{9.8}$$

which can be computed for every $m, m'$ in $1, \ldots, p$. The estimate $\hat{y}_{i,m}^{m'}$ is the predicted outcome of observing unit $i$ through mode $m'$ while it really was observed through mode $m$.

Alternatively, the estimated model coefficients can be used in a corrective rather than a predictive manner,

$$\hat{y}_{i,m}^{m'} = y_{i,m} + \hat{b}_{m'} - \hat{b}_m \tag{9.9}$$

which again can be computed for all units and for all modes. The terms $(\hat{b}_{m'} - \hat{b}_m)$ are an estimate of the expected measurement difference between modes $m$ and $m'$, as in expression (9.5). Essentially, this boils down to removing the estimated measurement error of the mode that was used to make the observation $m$, and adding the bias of the benchmark mode $m'$. Since the response mode is a categorical variable serving as a covariate in a regression model, only relative differences are estimated, with one arbitrary category (a certain mode) as the reference category.

## 9.3.2 Mode-Specific Estimators

Using the counterfactuals, a mode-specific estimate of the total is obtained as

$$\hat{t}_Y^{m'} = \sum_{i=1}^{n} \delta_{i,m} w_i y_{i,m'} + \sum_{i=1}^{n} (1 - \delta_{i,m'}) w_i \hat{y}_{i,m}^{m'} \tag{9.10}$$

which is the sum over measurements of units observed in mode $m'$ and of counterfactuals of units observed in other modes; $\delta_{i,m}$ is defined as in equation (9.7). This estimator is a potential outcome, in the sense that it is an estimate of the total of $Y$ had all units been measured through mode $m'$.

When adopting counterfactuals as in equation (9.9), expression (9.10) reduces to

$$\hat{t}_Y^{m'} = \sum_{i=1}^{n} w_i \hat{y}_{i,m}^{m'}. \tag{9.11}$$

Expressions (9.10) and (9.11) can be interpreted as the survey estimate that would result from a design with the selection mechanism of the mixed-mode design and the measurement mechanism of the benchmark mode. It combines the best of two worlds, since generally mixed-mode designs are assumed to have good selection mechanisms – using multiple data collection modes often improves response rates and representativity of the response – while the potentially adverse effects of measuring using different modes are corrected for.

The variance of the mode-specific estimator has two sources, associated with each of the two terms in equation (9.10). The first source is the design variance due to sampling. The second source is model based and is due to model uncertainty.

A viable approach is to use multiple imputation to estimate the model variance, and combining this with the design-based variance of the survey estimate. Alternatively, a bootstrap approach can be adopted, capturing both sources of variance simultaneously. This is achieved by sampling with replacement from the survey response. For each bootstrap sample one can compute the survey weights, fit model (9.7), compute counterfactuals (9.10), and finally estimators (9.11). The dispersion of the bootstrap distribution of the bootstrap $\hat{t}_Y^{m'}$ is a good estimator of the variance.

If there is no obvious benchmark mode, and all modes could potentially be valid, estimators consisting of a mix of counterfactuals can be applied. Rather than correcting observations conducted in non-benchmark modes toward the benchmark mode, now the different counterfactuals are combined,

$$\hat{y}_{i,m}^{combi} = \sum_{k=1}^{p} \propto_k \hat{y}_{i,m}^{k}$$
(9.12)

with the $\propto_k$ mixing coefficients, with the regularity condition $\sum_{k=1}^{p} \propto_k = 1$. The $\hat{y}_{i,m}^{combi}$ are mixes of estimated measurements in the different modes applied in the mixed-mode survey. Lacking good arguments about which mode to choose as the benchmark, this may be a good compromise.

Similar to expressions (9.10) and (9.11), the estimator for the total of the combined version can be written as

$$\hat{t}_Y^{combi} = \sum_{i=1}^{n} w_i \hat{y}_{i,m}^{combi}.$$
(9.13)

Applying some algebraic conversions,

$$\hat{t}_Y^{combi} = \sum_{i=1}^{n} w_i \hat{y}_{i,m}^{combi} = \sum_{i=1}^{n} w_i \sum_{k=1}^{p} \propto_k \hat{y}_{i,m}^{k}$$

$$= \sum_{k=1}^{p} \propto_k \sum_{i=1}^{n} w_i \hat{y}_{i,m}^{k} = \sum_{k=1}^{p} \propto_k \hat{t}_Y^{k}$$
(9.14)

from which it is seen that the combined estimator is in fact a linear combination of the mode-specific estimated totals.

Clearly, the mixing coefficients $\propto_k$ must be chosen or determined somehow. Optimal values for the $\propto_k$ can be determined, for example, by minimizing the mean squared error of $\hat{t}_Y^{combi}$. This requires an iterative numerical optimization, since the variance is obtained not from an analytical expression but from multiple imputation or from bootstrapping.

Rather than determining the mixing coefficients $\propto_k$ through some optimization routine according to some criterion, one could make an informed ad-hoc choice as well. Observe from (9.4) that the $\propto_k$ can be interpreted as the shares of contributions of the different modes to the total. If no mode is preferential over another mode, it may be sensible to choose mixing coefficients that are equal, or that are close to the actual shares of the modes in the total response realized in the survey. This idea is in fact fundamental to the approach presented in the next section. The choice of mixing coefficients is addressed again in Section 9.5.2.

## 9.4 Balancing Differential Measurement Bias

### 9.4.1 Mode Calibration

In the absence of a suitable or obvious benchmark mode, researchers could be tempted to simply use the response to a mixed-mode survey without considering adjustments for measurement errors. However, when the proportions in which the modes occur in the response vary between subpopulations or post-strata used for publication purposes, the differential bias varies also, compromising the comparability of survey outcomes. A similar situation occurs when a survey is repeated several times, and the composition of the mode mix changes between survey runs.

The method presented in this section is based on Buelens and van den Brakel (2015, 2017). The general idea is not to correct the mode-dependent differential measurement bias, but to balance the response to neutralize adverse effects of varying response mode compositions.

Combining the weighted survey estimates (9.3) with the assumed linear measurement error model (9.4), and taking the expectation over the measurement error model,

$$E\left(\hat{t}_Y\right) = E\left(\sum_{i=1}^{n} w_i y_{i,m}\right) = E\left(\sum_{i=1}^{n} w_i u_{i,m}\right) + E\left(\sum_{i=1}^{n} w_i \partial_{i,m} b_m\right) \tag{9.15}$$

since $E\left(\epsilon_{i,m}\right) = 0$, with $\partial_{i,m} = 1$ if unit $i$ responded through mode $m$, and zero otherwise. Since $U$ is the true score unaffected by measurement error, under the usual conditions of the GREG estimator, the first term of expression (9.15), $E\left(\sum_{i=1}^{n} w_i u_{i,m}\right) = \hat{t}_U$. The second term of expression (9.15) can be written as

$$E\left(\sum_{i=1}^{n} w_i \delta_{i,m} b_m\right) = \sum_{m=1}^{p} b_m \hat{t}_m \tag{9.16}$$

with $\hat{t}_m = \sum_{i=1}^{n} w_i \partial_{i,m}$, the weighted total number of respondents who responded through data collection mode $m$. This quantity depends on the selection mechanism of the mixed-mode design, not on the measurement mechanism.

Equation (9.15) can now be rewritten as

$$E\left(\hat{t}_Y\right) = \hat{t}_U + \sum_{m=1}^{p} b_m \hat{t}_m \tag{9.17}$$

stating that the expected total of the survey estimate for $Y$ consists of the estimated true total of $U$, plus a mixture of $p$ bias components $b_m$. The exact composition of the mixture is determined by the number of respondents whose data was collected through each of the modes.

Of the quantities in equation (9.17), only $\hat{t}_Y$ and $\hat{t}_m$ are observed, while $\hat{t}_U$ and $b_m$ are not. Hence, the usual GREG estimator for $Y$ in a mixed-mode survey is contaminated with a mixture of unknown mode-dependent biases.

When repeating a mixed-mode survey multiple times, which is commonly done, for example, to produce quarterly or annual estimates, the composition of the mode mixture might change, at least if the particular survey design does not explicitly controls for it. Since the $\hat{t}_m$ fulfill the role of mixing coefficients of the mode-dependent biases, the second term in equation (9.17) will generally be not constant between survey editions. Hence, estimates of change over time of $Y$ will be confounded with changes in the mode compositions in the different survey editions.

A similar problem arises when survey outcomes are published for population sub-groups, for example, by region, by gender, or by age groups. If the mode composition in these subpopulations is different, their estimated totals of $Y$ become incomparable because they incorporate unequal bias components.

The problem and the balancing solution are further developed using the case of repeating a survey twice; the case of subpopulation estimates can be treated analogously.

When repeating a mixed-mode survey twice, estimation of change over time occurs without bias if

$$E\left(\hat{t}_Y^{(2)} - \hat{t}_Y^{(1)}\right) = \hat{t}_U^{(2)} - \hat{t}_U^{(1)} \tag{9.18}$$

as seen from equation (9.17), with the superscripts in brackets referring to the first and second editions of the survey. If (9.18) holds, the expected change over time is equal to the change over time of the true score $U$. For this to hold, it is required that

$$\sum_{m=1}^{p} b_m \left(\hat{t}_m^{(2)} - \hat{t}_m^{(1)}\right) = 0 \tag{9.19}$$

where it is assumed that the systematic mode-dependent bias $b_m$ does not change between survey editions; see Sections 9.5.1 and 9.5.3 for details on this assumption.

Unless $b_m = 0$ for all $m$ – which would be the case if the modes have no differential bias – expression (9.19) requires that

$$\hat{t}_m^{(1)} = \hat{t}_m^{(2)} \tag{9.20}$$

which is the case only when the numbers of respondents using particular modes do not change between survey editions (1) and (2).

In general, in a mixed-mode survey where there is some flexibility in mode choice, there is no guarantee that equation (9.20) holds. This is where the balancing method comes in. The balancing method enforces equality (9.20) explicitly through a calibration approach.

The GREG is a calibration estimator in the sense that, for a covariate $X$ that is calibrated for, it holds that the weighted total is equal to the true, known total, $\hat{t}_X = \sum_{i}^{n} w_i x_i = t_X$. Since this is what must be achieved with response mode distributions, the usual GREG estimator can be extended to include response mode as a categorical variable in the model. Doing so results in an adaptation of equation (9.3), which, combined with expression (9.17), can be written as

$$E\left(\hat{t}_Y^c\right) = \sum_{i=1}^{n} w_i^c y_{i,m} = \hat{t}_U^c + \sum_{m=1}^{p} b_m \hat{t}_m^c \tag{9.21}$$

with the superscript $c$ referring to the mode-calibrated versions of the quantities.

Since calibration to mode has occurred, the $\hat{t}_m^c$ are equal to known calibration constraints, say $\Gamma_m$ for $m = 1, \ldots, p$. For typical socio-demographic variables, these known constraints are obtained from population or other administrative registers. In the case of response modes, the levels $\Gamma_m$ must be chosen or determined somehow. This choice is returned to below. The most important aspect is that $\sum_{m=1}^{p} b_m \hat{t}_m^c = \sum_{m=1}^{p} b_m \Gamma_m$. Hence, applying the same calibration to two editions of a survey renders requirement (9.20) fulfilled, since $\hat{t}_m^{(1)} = \hat{t}_m^{(2)} = \Gamma_m$. Consequently, expression (9.18) applies, stating that change over time of the measured variable $Y$ is estimated without bias.

Comparing equation (9.21) with (9.17) it is clearly desirable that $\hat{t}_U^c = \hat{t}_U$. Since mode calibration is meant to affect measurement errors only, the variable $U$, which is without error, should be unaffected. Section 9.5.1 returns to this requirement.

Under the mode calibration approach, the measurement errors $b_m$ are not quantified, nor is their total $\sum_{m=1}^{p} b_m \hat{t}_m^c$, nor is any kind of bias removed from the estimates. What is achieved is that the second term in expression (9.17) is rendered constant.

### 9.4.2 Choosing Calibration Levels

The aim of the mode calibration is to neutralize changes in the total mode-related measurement bias. It is not critical what the precise calibration levels $\Gamma_m$ are set to in a particular survey. Since the choice of these levels affects the survey weights, it is important to choose levels such that they do not incur negative weights or increase the dispersion of the weights, since this will inflate the variance of the mode-calibrated survey estimates. In the special case where the calibration levels are chosen equal to those obtained in the mixed-mode survey, the mode calibration will have no effect at all, and will consequently not affect the variance in any way. Intuitively, the $\Gamma_m$ levels should be chosen close to this special case.

If a survey has taken place more than once, an obvious choice for $\Gamma_m$ would be the average of the mode compositions obtained in the different editions.

Alternatively, levels can be chosen in anticipation of expected developments in the future, for example, an increase of the contribution of web interviewing, or a decline in the number of telephone interviews.

If a survey is conducted for the first time, levels can be chosen close to those realized. This will have little effect in that edition but prepares for future editions that can be balanced toward levels obtained in that first edition.

If large systematic fluctuations in the mode composition occur unexpectedly, it may become necessary to change over to new calibration levels in surveys running for many years. In that case, earlier editions that were calibrated with different levels can be recalibrated with the new levels, so as to maintain an uninterrupted series of survey estimates, without discontinuities due to changed calibration levels.

If interest is more in cross-sectional results than in change over time, mode calibration can be applied to population subgroups as required. Typically, categorical variables are used for this purpose, for example, regions, gender, or age groups. Comparability of survey estimates between such groups is only valid when their mode composition distributions are equal, or at least not very different. Crossing data collection mode with categorical covariates introduces additional degrees of freedom. To keep the model as parsimonious as possible, these covariates must be chosen with care. Classification variables are good candidates to use for this purpose if they are used in the main publication tables of the survey, and if the distributions of the modes over the respondents within the classes vary a lot. Conducting mode calibration in this case will render the publication cells comparable, with differences between cells no longer confounded with differences in mode composition of the cells.

Buelens and van den Brakel (2015) present a simulation study showing numerically how the choice of the calibration levels affects the point and variance estimates. As expected, they find that when a calibration estimator is applied with calibration levels that deviate a lot from those actually obtained in the mixed-mode survey, the variance of the mode-calibrated estimates becomes large. In such a situation, mode calibration will down-weight observations obtained in some mode(s) and up-weight observations in others, causing larger overall variance in the weights, resulting in an inefficient survey estimate, with large variance.

## 9.5 Handling Measurement Bias in Practice

### 9.5.1 Testing Assumptions

Both the correction and balancing methods require a few assumptions to hold. In fact, the assumptions for both methods are largely the same. Successful application of the methods includes tests or assessments of the validity of the assumptions. In this section the assumptions are addressed and ways to test them are outlined.

The most important assumption of both methods is that the covariates ($X$) other than mode explain the selectivity of the response. This is equivalent to stating that the variable $U$ – the underlying true score of $Y$ – is independent of mode, conditional on $X$. This assumption must hold for the regression coefficient of mode in equation (9.7) to represent the measurement bias, and for $\hat{t}_U^c = \hat{t}_U$ in equation (9.21) in the calibration approach. In both cases, if mode explains selection effects beyond those explained by the other covariates, the estimated measurement bias will unjustly include selection bias and would therefore be a biased estimate of the measurement bias.

A possibility to check this assumption is seeking an additional variable $Z$ known from an administrative register, in which case $Z$ is known for the entire population and is not affected by measurement error associated with the data collection modes applied in the mixed-mode survey. It is desirable that this variable $Z$ correlates with $Y$. In the correction approach, fitting model (9.7) with $Z$ as the dependent variable, one should find that none of the $b_m$ is significantly different from zero. In the calibration approach, one could compare $\hat{t}_Z$ and $\hat{t}_Z^c$, the standard weighted and the mode-calibrated estimates of the total of $Z$. If the assumption holds the two estimates should not be significantly different. In addition, since $Z$ is a register variable, the population total $t_Z$ is known, and should again not differ significantly from both estimated totals. If the conclusion would be that the proposed model does not explain all selectivity with respect to this variable $Z$, an obvious action would be to include $Z$ as an additional covariate in the model.

Another assumption, rather fundamental to the exposition in this chapter, is the specific analytical form of the measurement error model, expression (9.4). This model has been chosen because it has the advantage of being linear, consequently rendering the mathematical derivations relatively simple. The validity of this model assumption is very difficult to assess. Possibly, different model specifications could be tested against each other to select an optimal model. More advanced models could include measurement error terms that depend on the value of the underlying true score $U$, for example, terms like $\lambda_1 u_i$, $\lambda_1 u_i + \lambda_2 u_i^2$, or $\log u_i$. Such models are not considered in this chapter since they are less common in the literature.

### 9.5.2 Calibration Levels and Mixing Coefficients

The estimator (9.14) combines potential-outcome estimates for different data collection modes into a single estimate. Applying this combined estimator, the overall response consists of a mixture of observations conducted via the different modes, with the mixing coefficients $\propto$ characterizing the mixture. The mode-calibrated estimator (9.21) calibrates the overall response to levels given by the calibration constraints $\Gamma$, imposing a particular distribution of the weighted response over the modes.

Consequently, both estimators result in an overall response that is a mixture of measurements collected through the different data collection modes, and either $\propto$ or $\Gamma$ characterize the composition of this mixture.

When the regression model underpinning the GREG estimator used as a basis for the mode calibration is also used to model the measurement errors, the methods can be shown to agree if the mixing coefficients and the calibration levels are chosen in accordance, such that $\propto_m = \dfrac{\Gamma_m}{N}$ for all modes $m = 1, \ldots, p$, with $N$ the known population total. Expression (9.14) can be written using formula (9.9) as

$$\hat{t}_Y^{combi} = \sum_{i=1}^n w_i \left( \sum_{k=1}^p \propto_k \left( y_{i,m} - \hat{b}_{m(i)} + \hat{b}_k \right) \right) \tag{9.22}$$

with $\hat{b}_{m(i)}$ the estimated bias associated with the mode respondent $i$ actually responded through. Measurement error model (9.4) can be combined with equation (9.22) to obtain

$$\hat{t}_Y^{combi} = \sum_{i=1}^n w_i \sum_{k=1}^p \frac{\Gamma_k}{N} \left( u_i + b_{m,i} - \hat{b}_{m(i)} + \hat{b}_k + \epsilon_{i,m(i)} \right) \tag{9.23}$$

Taking the expectation with respect to the measurement error model gives

$$\hat{t}_Y^{combi} = \sum_{i=1}^n w_i \sum_{k=1}^p \frac{\Gamma_k}{N} \left( u_i + \hat{b}_k \right)$$

$$= \sum_{i=1}^n w_i u_i + \sum_{i=1}^p w_i \sum_{k=1}^p \frac{\Gamma_k}{N} \hat{b}_k \tag{9.24}$$

$$= \hat{t}_U + \sum_{k=1}^p \Gamma_k \hat{b}_k$$

where it is assumed that the weights are properly calibrated to the population total, $\sum_{i=1}^n w_i = N$. The last line of (9.24) is indeed equal to the mode-calibrated estimator, at least when $\hat{b}_k = b_k$, which will hold approximately.

Result (9.24) may come as a surprise since the first principles of the two estimation methods are quite different. The generic estimator (9.1) is adjusted by the correction methods by modifying observations $y_i$, while the calibration estimator modifies the weights $w_i$. However, since both methods are taken here to use the same GREG weighting model, and each expands this model with response mode as a covariate, it can be understood that (9.24) must hold. An example where both methods are used in the connection is given in Section 9.6.3.

### 9.5.3 Comparison with Single-Mode Designs

Some assumptions made earlier in this chapter seem strong and perhaps unlikely to hold. Single-mode survey designs, in which data collection occurs by means of one data collection mode, require a number of assumptions as well, although they are almost never made explicit. In this section some requirements for the models and mechanisms in mixed-mode designs are compared to their single-mode counterparts.

The measurement error model that is adopted throughout this chapter is given by equation (9.4). Mode-specific errors are assumed to consist of a systematic bias $b_m$ that does not depend on the individuals, and a zero-mean error variance component $\epsilon_{i,m}$ which does vary over individuals. This model states that the non-random, systematic bias of each mode equally affects all respondents. When only one mode is used to collect data from all respondents, a model like (9.4) can be assumed to be at play too. There, the question if that single-mode measures all respondents in a comparable way is generally never asked, although the consequences are the same. For example, when responses from women would be affected by a different bias than responses for men – using the same data collection mode – survey estimates for women and men would not be comparable since they would be affected by different biases. By positing model (9.4) this assumption is made visible, but it applies ordinarily to single-mode surveys too.

Another assumption regarding the measurement error model is made in Section 9.4.1, where it is stated that the mode-dependent systematic bias $b_m$ remains constant between survey editions. In single-mode surveys where the classic GREG estimator is used and measurement errors are ignored, the assumption of constant measurement error is made too, albeit implicitly. Changing measurement errors in repeated single-mode surveys would invalidate inference made about change over time of survey variables. Because in such settings measurement errors are not accounted for, confounding of varying measurement errors with true changes over time remains invisible. While in the present setting this assumption is made explicit, it is no stronger than what is typically assumed in survey sampling in general and in single-mode designs in particular.

## 9.6 Applications

### 9.6.1 Health Survey

The Health Survey (HS) in the Netherlands is conducted with a sequential mixed-mode design where sample units are initially invited to participate via a web questionnaire; the invitation is sent as a paper letter through the regular mail and contains a URL and login details. Sample units who do not respond via the web within four weeks are followed up for personal face-to-face interviewing.

In the context of an anticipated redesign of this survey it has been investigated what the effect on the survey estimates would be if a strategy could be designed where all face-to-face interviews could be converted to web interviews. The selection mechanism of the mixed-mode design should be maintained, as it was believed to result in a sufficiently representative response. This work is described in Buelens et al. (2015), which is not publicly available. This section summarizes the approach and the results. The analyses were conducted on HS data collected during 2014.

This research is an application of the correction method presented in Section 9.3. The web mode is considered as the benchmark mode: interest is in the survey estimate in the scenario where all responses were provided through the web mode, while respondent recruitment occurred by means of the mixed-mode design using the two modes.

In the HS, about 50% of the responses were actually collected through the web, the other half through face-to-face interviewing. Among the face-to-face respondents were relatively more people with low incomes, people in their early teens or aged 75-plus, immigrants from non-Western countries, singles and very large households. These characteristics are known from administrative registers. In addition, gender and region are available but no significant effects were found between web and face-to-face respondents. Clearly, the variables for which effects are found must be included in the covariates $X$ in model (9.7). Counterfactuals were obtained for the web mode, and equation (9.11) was used to obtain the potential outcome of the HS in the scenario that everybody had responded through web.

The potential-outcomes web estimator and the web-only estimator differ in terms of selection, but not in terms of measurement, as both use web interviewing as the single survey mode, be it in a counterfactual sense in the case of the potential-outcome estimator.

A total of ten survey variables were considered (see Figure 9.2). In this research no confidence intervals were computed for these alternative estimators, only for the original mixed-mode estimates. The potential-outcomes web estimator and the web-only estimator showed no large differences between them for any of these variables, suggesting that the GREG weighting model appropriately explains selection effects. The largest differences between the potential-outcome estimates and the mixed-mode estimates are seen in the percentage of smokers and in the number of contacts with GPs, specialists, and dentists.

It was concluded that the differences between the mixed-mode and potential-outcomes web estimator were predominantly due to relative measurement effects of the modes. Hence, comparisons of survey estimates between population subgroups in which the shares of the modes differ a lot are compromised. Differences found in such survey outcomes could in part be due to differences in the mode composition. The potential-outcome web estimator does not suffer from this problem as all answers are brought to the web level, and estimates for population subgroups are freed from relative mode-dependent measurement bias.

A recommendation based on these findings could be to use the potential-outcome estimates for the variables that exhibit large measurement differences between the modes. However, at the time of writing, the officially published results of the Health Survey are those here referred to as original mixed-mode estimates.

### 9.6.2 Crime Victimization Survey

An application of the mode balancing method of Section 9.4 is given in Buelens and Van den Brakel (2015) and summarized here. They studied the Crime Victimization Survey (CVS) in the Netherlands, an annual mixed-mode survey characterized by varying mode compositions during the years 2008–2011.

The CVS employed a hybrid mixed-mode design with four different modes. First, sample units were invited to respond through the web, in the same way as the Health Survey in the previous section. Together with a first reminder after one week, the sample units received a paper questionnaire that they could complete if they preferred. Nonrespondents of the

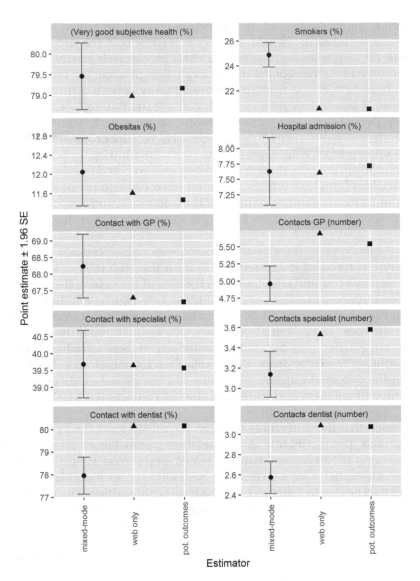

**FIGURE 9.2**
Health Survey estimates based on the original mixed-mode survey (circles), only the web respondents (tri-angles), and the potential-outcomes estimator for web (squares). Standard error bars are given for the original estimates only.

web-paper strategy, with a known telephone number, were approached by telephone, and other nonrespondents through face-to-face interviewing.

Every other year a large-scale oversampling was organized, to provide detailed esti-mates bi-annually. Unfortunately the sample units in the oversampling underwent a slightly different data collection strategy, in that the face-to-face mode was not used at all, due to budget constraints. As a result, the share of face-to-face interviews in the mode composition was 6–7% in the years 2008 and 2010, and only just over 1% in 2009 and 2011 when the oversampling took place. Web and paper together accounted for 65–70% in the even years and 80–85% in the odd years. For the purpose of the analysis, a distinction was made between the non-interviewer modes web and paper, on the one hand, and the interviewer modes telephone and face-to-face, on the other.

Extensive results and analyses are available in the article (Buelens and Van den Brakel, 2015). Here, a limited set of results is shown to illustrate the balancing calibration approach. The editions of the CVS in 2009 and 2010 resulted in mode compositions of, respectively, 26% and 36% of the response collected through interviewer modes. This means that equation (9.20) was not fulfilled, and that estimates of annual change in survey variables between 2009 and 2010 were potentially confounded with changed mode compositions of the response.

The response mode calibration method was applied, with calibration levels $\Gamma$ chosen to be 40% for the interviewer modes and 60% for the non-interviewer modes. This choice was based on the mode composition that was obtained in 2008, when no mode balancing calibration had been applied.

Table 9.1 combines results from tables 4 and 10 of the article cited above, for five survey variables, each referring to the last 12-month period: *victim* – the percentage of people been victims of crime, *offences* – the number of offences per 100 inhabitants, *unsafe* – the percentage of people feeling unsafe at times, *funcpol* – mean satisfaction with the functioning of police, *antisoc* – mean suffering from anti-social behavior. The last two are scale scores measured on a scale of 1–10. The table lists differences in estimates between the years 2009 and 2010. The column *uncalibrated* contains simply the differences of the original mixed-mode estimates. The columns with the fractions list differences of mode-calibrated estimates, with the fraction indicating the calibration levels that were applied. For example, in column *20/80* a mode calibration was applied with a level of 20% for interviewer modes and 80% for non-interviewer modes. In the column *50/50* both mode groups are calibrated toward equal levels.

The uncalibrated differences could be contaminated with differential mode bias, whereas the other estimates are not. For *victim* and *offences*, the mode-calibrated estimates are less extreme, in the same direction, and as significant as the uncalibrated estimates. For *unsafe* the direction of the change is different, albeit not significant in most cases. The change in *funcpol* is found significant based on the uncalibrated estimates, but calibration renders it insignificant, and smaller. For *antisoc* calibration changes the sign too, although none of the estimated changes are significant. Overall, it is clear that calibration does have an effect on some variables, mitigating the risk of drawing false conclusions based on the uncalibrated estimates. The differences seen in the table among the various calibrated estimators are not surprising, as different mode compositions correspond to different mixtures of bias, hence different point estimates.

**TABLE 9.1**

Annual Changes (2009–2010) in Five CVS Variables Estimated Using the Uncalibrated and Four Versions of Calibrated Estimators

| Variable | uncalibrated | 20/80 | 30/70 | 40/60 | 50/50 |
|---|---|---|---|---|---|
| Victim | −6.7 *** | −6.1 *** | −5.7 *** | −5.3 *** | −5.0 *** |
| Offences | −12.5 *** | −11.4 *** | −10.9 *** | −10.4 *** | −10.1 *** |
| Unsafe | −1.4 | +3.1 * | +2.6 | +2.0 | +1.4 |
| Funcpol | +1.5 *** | +0.3 | +0.3 | +0.3 | +0.3 |
| Antisoc | −1.3 | +1.2 | +1.4 | +1.6 | +1.7 |

*p < .05, **p < .01, ***p < .001, with unmarked changes not significant (p > .05).

### 9.6.3 Labor Force Survey

In many cases the methods of Sections 9.3 and 9.4 can both be applied. Buelens and Van den Brakel (2017) do so with the Labor Force Survey (LFS) and compare the results.

A mixed-mode design similar to the ones discussed in the previous sections is used for the LFS in the Netherlands. Web interviewing is applied first, with follow-up through telephone for people with known phone numbers, and face-to-face otherwise. At least, this is the case for the first wave. The LFS employs a rotating panel design consisting of five waves. Only the first wave is used in the analysis. Monthly estimates are considered for the period July 2012 through June 2015.

Figure 9.3 shows how the mode composition changes over time during the 36-month study period. The share of telephone interviews is fairly stable at 20–25%. Face-to-face interviews account for about 30% of the response at the beginning and increase toward 40%. The share of web interviews starts at 45–50% and generally decreases.

A key variable from the LFS is the estimated number of unemployed. The risk associated with the fluctuating pattern seen in Figure 9.3 is that this variation in the modes might affect the estimates of the unemployment numbers, rendering the estimates instable through time, and hence unreliable.

Both the correction method and the balancing method have been applied. In the balancing method calibration levels were chosen to correspond with the average over the study period: 44% web, 22% telephone and 34% face-to-face. Exactly the same proportions were used as mixing coefficients (see equations (9.12) and (9.14)) to combine the potential-outcome estimates for each of the three modes.

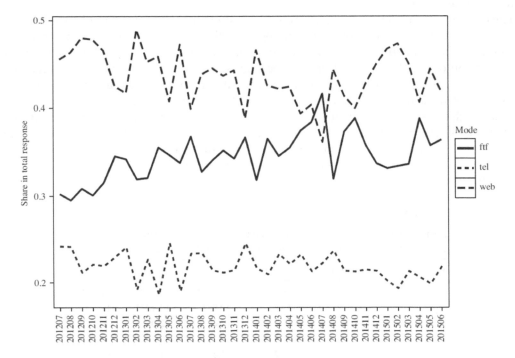

**FIGURE 9.3**
Response mode composition of the LFS during the 36-month study period. This figure is figure 1 from Buelens and van den Brakel (2017).

The conclusion of this analysis was that these alternative estimators result in estimated numbers of unemployed that are almost equal to the original estimates. Hence, the variations seen in the mode composition (Figure 9.3) do not have an adverse effect on the survey estimates. One might conclude that the relative mode-dependent measurement errors must be small. However, this turns out not to be the case. Averaging the estimated number of unemployed over the 36-month study period for each of the mode-specific potential-outcome estimators, resulted in estimates of 691 thousand unemployed for the web mode, 695 thousand for face-to-face, and 626 thousand for the telephone mode. The web and face-to-face modes measure unemployment about 10% higher than the telephone mode. This causes no instability in the series, since, coincidentally, the share of the telephone mode in the total response does not vary a lot. Substantial increases or decreases of the share of telephone in the mix would render the series of estimated unemployed less reliable. Applying the corrected or balanced estimators is therefore recommended, as a safeguard against potential instabilities caused by changing mode compositions in the future, in particular with respect to the telephone mode.

## 9.7 Summary

The main takeaway messages are:

- Estimates of population quantities that are based on survey data collected through a mixed-mode design are susceptible to differential measurement bias of the modes.

- Consequently, measurements in different population subgroups or at different moments in time may be incomparable due to differences in response mode compositions, and associated therewith, differences in measurement bias.

- The key difficulty is the confounding of measurement effects and selection effects, and the disentanglement of these two effects.

- The methods presented in this chapter require no additional data collection beyond the survey and require no experimental design. However, they do require the availability of covariates explaining the selection effects between the different modes.

- The first class of methods adjusts the survey measurements, using a model that predicts for each sample unit what the measurement would have been had the survey been conducted through another mode. These approaches are rooted in the potential-outcomes framework and can be conducted using regression or imputation techniques.

- The second method is a reweighting approach, leaving the survey measurements unaltered. Through weighting, the response mode composition is rendered constant among relevant population subgroups or survey editions. Rather than removing bias, this technique stabilizes the bias components and hence improves the comparability of survey estimates.

- When all models involved are linear, the correction and the balancing approaches can be shown to be almost equivalent under certain conditions.
- Some assumptions made in this chapter seem stringent at first but are in fact not stronger than what is commonly required for valid inference from single-mode surveys.
- Example use cases of the methods were shown in the context of the Crime Victimization Survey, Health Survey, and Labor Force Survey.

# Part V

# The Future of
# Mixed-Mode Surveys

# 10

## Multi-Device Surveys

### 10.1 Introduction

This chapter discusses the future of mixed-mode surveys looking into the possibilities and challenges of surveys using smart devices such as smartphones, tablets, wearables, and other sensor systems. The objectives of this chapter are to explain the potential functions of different devices in a survey, how these functions can be employed to create hybrid data collections in a mixed-mode context, and what the consequences for the total survey error are. This chapter does not address the questionnaire design across desktops, laptops, smartphones, and tablets, which is discussed in Chapter 6, Mixed-Mode Questionnaire Design. This chapter also does not focus on data collection strategies for the web survey mode accounting for different devices, which is part of Chapter 5, Mixed-Mode Data Collection Design. The chapter does, however, consider recruitment and measurement in surveys with new types of data that can be collected through smart devices.

Figure 10.1 displays the plan-do-check-act cycle. Adding smart devices to surveys affects the questionnaire design, data collection design, and estimation design. The only part of the design that can be left largely untouched is the sampling design.

Smart devices can add three new types of data to a survey. The first are *sensor data*, coming either from the device itself or from devices it can communicate to, such as wearables (see Link et al., 2014). Examples are location measurements and camera photos or scans. The second are *public online data* that can be downloaded and linked to respondent data. Examples are geographical points-of-interest data and pre-trained machine learning models (see Smeets, Lugtig, and Schouten, 2019). The third are *personal online data* that individual respondents are allowed to access under existing, applicable legislation, which is a form of data donation. (see Bietz, Patrick, and Bloss, 2019). The General Data Protection Regulation (GDPR), which was enforced by the EU in 2016, obligates all data holders to provide access to individuals of whose they data hold. Examples are data stored by commercial online personalized services and shop loyalty card data. Surveys employing any of these forms of data are examples of the so-called *smart surveys*.

These three types of data cannot be collected via paper-based methods. They could be collected through interviewer modes, but not longitudinally. This means that new types of data challenge mixed-mode surveys. In particular, they challenge the comparability of responses with and without such types of data.

There are two objectives to collect new types of data, even when they may lead to a risk of incomparability: reduction of respondent burden and improvement of data quality. Surveys that are demanding in time and/or cognitive effort, or are perceived or anticipated to be by respondents, naturally have lower response rates. Examples are diary surveys, such as time use surveys and budget expenditure surveys. Such surveys may also

DOI: 10.1201/9780429461156-10

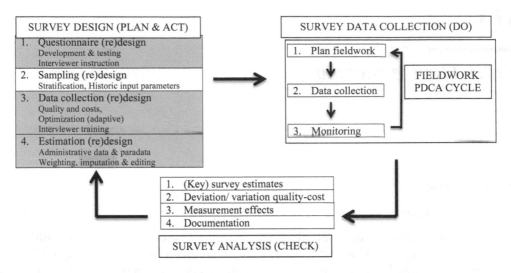

**FIGURE 10.1**
The plan-do-check-act cycle for surveys.

suffer from lower quality of survey data due to survey fatigue. Surveys that require recall or detailed knowledge may equally be affected by lower data quality. Respondents may simply not know answers to questions, even if they are fully motivated to do the survey. Examples are health surveys asking for health indicators and determinants, and travel surveys asking for traveled distances.

Replacing or supplementing questions by new forms of data requires a renewed look at total survey error (Groves and Lyberg, 2010; Biemer, 2010). This holds for both representation and measurement. In fact, the reference to 'survey' must be viewed more broadly as all data collected in a survey, not necessarily by means of a questionnaire alone. Sensor data have their own error properties, (Kreuter et al., 2018). However, just like the question-answer process, a sensor measurement passes through a number of steps that are more or less interactive, depending on how measurements are collected and validated by respondents. Existing external data linkage implies secondary data analysis, for which separate error frameworks have been developed (Zhang, 2011). In hybrid data collection, the different error frameworks for the different types of data need to be combined. Literature in this area is still very scanty and will most likely evolve over the coming decade.

So how do these smart surveys fit in a book on mixed-mode survey design and analysis? There are three reasons why it is important to discuss them in a mixed-mode setting. First, these surveys are likely to be mixed-mode surveys, although some modes will not support collection of all types of data. In general, response rates to web-assisted surveys are often deemed to be too low for official statistics; e.g. see Chapter 4, Mixed-Mode Selection Effects. Therefore, supplemental modes are employed to get a higher and more balanced response. This will also be true for smart surveys, as has been shown by various studies (see e.g. Keusch et al., 2019; Struminskaya et al., 2020). Second, telephone and/ or face-to-face recruitment, instruction, and motivation may be employed for more complex forms of sensor measurements. Respondents may need to be motivated and assisted. Furthermore, sensors may be handed to respondents in person, which typically is combined with a starting interview. Third, the new types of data may simply be viewed as alternative modes, even though they often imply a simultaneous conceptual change in the data that are being collected. For some respondents, answers are then provided through a questionnaire, while for other respondents they are derived from alternative data sources or measurements.

New error sources demand for new methodology. Data collection strategies will have to be adapted. Population coverage of devices and external data may be incomplete and selective. Willingness-to-participate mechanisms are different from interviews. An example is GPS (global positioning system) location data. Some respondents do not have adequate devices to collect location data. Other respondents may not be willing to consent to such measurements for privacy reasons. Questionnaire design becomes a much wider area, with more importance on user interfaces and user experience. Smart surveys may offer feedback of statistics to individual respondents, which is typically not included in traditional surveys. GPS location measurements may, for example, be shown to a respondent in order to request validation of detected tracks and stops. Traveled distances may be shown to respondents as a form of individual feedback. Finally, data that are being collected are much more diverse. Processing, adjusting, and combining data, as a consequence, change and tend to become more complex. Furthermore, part of the processing can be done on the device itself and only derived data may actually be submitted. GPS location data can be translated to tracks and stops on the device itself and only the derived traveled distances may be submitted.

Smart surveys are still in their infancy, especially in the context of official statistics. This chapter must, therefore, be seen as the first venture into this new area. Section 10.2 gives a taxonomy of smart surveys and provides examples. Section 10.3 extends error frameworks to include new types of data and provides simple criteria to judge whether smart surveys are promising. Section 10.4 focuses on the methodology in data collection, measurement, and estimation. The three case studies of this book, CVS, HS, and LFS introduced in Section 2.6, are discussed in Section 10.5. Finally, Section 10.6 gives takeaway messages.

## 10.2 Smart Surveys

This section provides more detail about surveys and a taxonomy on surveys that include smart features. These are termed *smart surveys*, when they use internal data, and termed *trusted smart surveys*, when they (also) use external data. See, for example, Ricciato et al. (2019).

### 10.2.1 A Taxonomy of Smart Surveys

A taxonomy of smart surveys distinguishes both different types of data and how these data are obtained. A smart survey starts from the viewpoint of a device on which the survey runs. This device will in practice usually be a desktop, laptop, tablet, or smartphone.

To a survey, the following features can be added:

1. Device intelligence: It can use the intelligence (computing and storage) of the device.
2. Internal sensors: It can employ the sensors that are available in the device.
3. External sensors: It can communicate through the device with other sensors that are close by.
4. Public online data: It can go online and extract publicly available data.

5. Personal online data: It can go online and request access to existing external personal data.

6. Linkage consent: It can ask for consent to link external personal data already in possession of the survey institute.

What makes a (trusted) smart survey? Computer-assisted surveys have the first feature, but do not go beyond asking questions and submitting answers. In its most rudimentary form, a smart survey is a computer-assisted survey on a smart device (i.e. desktops, laptops, smartphones, and tablets) with online access; i.e. it actively uses the local computing power and storage beyond merely asking questions. Computing power and storage can be used for various purposes such as plausibility checks, edit rules application, computation of new variables of interest, and gamification options. So a smart survey must have feature 1.

More advanced smart surveys use internal sensors, external sensors, and/or public online data. In order to do so, the survey usually demands explicit consent from respondents. In case of internal mobile device sensors, such consent questions are automatically posed to respondents whenever the survey tries to employ the sensor.

Trusted smart surveys have features 5 and/or 6 on top of feature 1; i.e. they use personal data either by request of the respondent herself/himself or by linkage after respondent consent.

A smart survey can add the following functions to a survey:

- It can initiate sensors and store the sensor data.
- It can go online and download data.
- It can communicate to nearby sensors and collect data (through Bluetooth).
- It can use internal sensor data and/or external sensor data and/or online data:
  - as supplements to survey data,
  - to execute decision rules in the questionnaire,
  - to perform plausibility/edit checks,
  - to facilitate respondents in answering questions, and
  - to provide new knowledge to the respondent.
- It can perform part of the interaction (reminders, motivation, and feedback) with respondents locally.
- It can perform part of the data processing locally instead of at the survey institute.

A trusted smart survey can add a different set of functions:

- It can ask for consent from respondents to connect to an external database.
- It can request data after consent through an external database API (application programming interface).
- It can perform linkage to existing data at the backend of the survey institute.
- It can use the downloaded external data:
  - as supplements to survey data,
  - to execute decision rules in the questionnaire,
  - to perform plausibility/edit checks,

- to facilitate respondents in answering questions, and
- to provide new knowledge to the respondent.

Since all the features can be combined, a simple taxonomy of smart surveys is not possible. If we combine all features and view feature 1 as a mandatory, default feature, then there are 32 (= $2^5$) possible types of trusted smart surveys. If we ignore the trusted smart features (features 5 and 6), then there are eight possible types of smart surveys. If, instead, we focus only on the trusted smart features, then there are four possible types. A simple taxonomy then is the following:

- Simple smart survey: Uses only the device intelligence (feature 1).
- Internal smart survey: Uses internal smart device sensors to add sensor data (feature 1 plus feature 2).
- External smart survey: Uses external data, either collected by external sensors or available in existing sources (feature 1 plus features 3 and/or 5).
- Full smart survey: Uses both internal sensors and external data (feature 1 plus feature 2 plus feature 3, 4, and/or 5).

In Section 10.5, examples are given of an external smart survey and a full smart survey.

Smart surveys and trusted smart surveys could be designed for any device. However, devices vary greatly in sensor availability. Desktops and laptops have only a limited set of sensors, whereas smartphones and tablets have a rich set of sensors. For some of the features, it is natural to use mobile devices only. Smart surveys are, therefore, often associated with mobile devices, but can in more simple forms also be run on fixed devices such as desktops and laptops.

Smart surveys could, strictly speaking, also be assisted by face-to-face interviewers, in case the data can also be collected on a device operated by the interviewer. This will only be natural in a few, very specific settings, where it does not matter who makes the measurements, such as living or working conditions.

An important technical distinction between smart surveys is whether the survey operates in a *browser* or in a *dedicated application*. Smart surveys and trusted smart surveys can be initiated within both, but the smart features that are included strongly determine which option is more practical. Dedicated applications demand a separate installing procedure and, as a consequence, create extra burden to respondents. Browsers support only a subset of sensor measurements and/or data linkage options, and typically perform no local processing or computations such as character recognition or image extraction.

The browser option is, thus, most natural when:

- the measurements and/or linkage can be done at a single time point, and
- the applied sensors are relatively simple, and
- no complex in-device operations or processing is performed.

The dedicated application option is most natural when:

- the measurements and/or linkage are repeated/longitudinally, or
- multiple sensors are applied, i.e. sensor fusion, or
- in-device operations or processing are needed.

### 10.2.2 Sensor Data

The first type of potential data in a smart survey is sensor data. Here, a distinction needs to be made between internal sensors, i.e. those available in the device itself, and external sensors, i.e. those available in other sensor systems that the device can communicate with.

The following elements and sensors are supported by many contemporary smartphones and tablets:

- 3D touch: This sensor measures the pressure exerted on the screen. Small objects up to 385 grams can be weighted.

- Accelerometer, gyroscope (motion sensors): A set of sensors measuring motion, acceleration, and position of the device. These sensors can be used for detecting movement patterns.

- Ambient light: This sensor measures the intensity of ambient light. More advanced versions also determine the light color or temperature. This is commonly used to adapt screen brightness and color to ambient conditions.

- Bluetooth: This sensor uses wireless communication protocol and can connect to small low-energy devices, such as wireless headphones, key fobs, smartwatches, or beacons.

- Camera: This device takes pictures or videos, or measures light intensity and is usable for image or pattern recognition, scanning of QR or barcodes and color analysis. It can also coarsely measure gamma rays, a form of radioactivity, and heart rate in combination with the camera flash.

- Cellular: The core of all cell phones, this is used to make and receive calls and text messages. The strength and ID of cell tower broadcasts can be measured. With the knowledge of tower positions, the user location can be determined with a precision of ~500 meters. More advanced cell phones – almost all cell phones today – can also connect to the internet. The presence of the internet connection as well as its speed (upload, download, and responsiveness/ping) can be determined.

- Fingerprint: Some devices are capable of detecting fingerprints. The raw data is not accessible, but it can be used as identification or simply as a button or – in some devices – as a small touchpad.

- GPS: Dozens of GPS satellites circle the earth and broadcast beacon signals. By measuring the time-of-flight of the satellite signals, the distance to that satellite can be calculated. With four or more satellites visible, the position on earth can be triangulated. The precision is ~5 meters outdoors under clear sky. Tall houses or a forest will reduce the accuracy. Indoor performance is poor.

- Heart rate: The heart rate can be measured usually optically on the finger (cell phone), wrist (smartwatch), or in ear (headphones).

- Humidity: These devices measure ambient humidity, but are not widely used yet.

- Magnetic field: Usually used as a compass, magnetic field sensors can also measure the strength of magnetic fields or can be used, within limits, as a metal detector.

- Microphone: This sensor detects speech and sounds that can be saved, streamed, or analyzed and can also determine loudness and detect ambient noise. Multiple microphones in one device allow for determining the directionality or distance of sound sources. This is used to filter out ambient noise in phone calls. Microphones can also be used to record the heartbeat or estimate lung function/spirometry.

- NFC (near field communication): The same technology as contactless payments, NFC can be used to pay with the cell phone or smartwatch, as 'contactless QR code'/'NFC tags' to change phone settings (muting) or trigger the start of certain apps. A phone can be a tag as well, so two devices can identify each other and initiate a data channel for communication.
- Pressure: This sensor measures ambient air pressure and functions as a barometer. The precision is so high that height differences of a few meters (ground floor versus first floor) can be detected. It is also used in combination with GPS for more precise height determination.
- Proximity: This sensor measures the presence of objects close to the screen, usually binary (object present or not) and can for example be used to switch the touchscreen off during phone calls.
- Thermometer: Usually placed in or near the battery to prevent overheating, the thermometer measures the battery/cell phone temperature which might be higher than ambient temperature.
- Vibration: Vibration sensor as a feedback mechanism induces vibration in the device and can be used as feedback or combined with other sensors such as the accelerometer.
- Wi-Fi: Usually used for internet access, Wi-Fi can detect the presence and strength of different wireless networks in different frequency bands. Measuring of connection speed (upload, download, and responsiveness/ping) is possible. Wi-Fi sensors are also used for location measurements with a precision of ~20 meters in the proximity of Wi-Fi routers/access points.

Some of these sensors cannot be initiated within browsers and demand for tailored applications. Browsers can, however, apply many of the most useful sensors such as location, microphone, and camera.

External sensor data may come from a great variety of sensor systems. The most commonly used are wearable sensors such as activity trackers and smartwatches. These are designed specifically to measure physical parameters such as type and intensity of activity, heart rate, and respiration. However, sensor systems do not have to be wearable, but can have a fixed location as well. Commonly used fixed sensor systems are indoor climate and indoor exposure systems. These contain a variety of sensors that can be modular, i.e. include sensors by demand. Examples of climate and exposure sensors are particulate matter of different sizes in micrometers (PM1, PM2.5, PM5, PM10, etc.), different gases such as $CO$, $CO_2$, $NO$, and $O_3$, air pressure, sound, light, temperature, and humidity. Figure 10.1 displays one day of measurements of such an indoor climate system.

External sensor data are, typically, made by commercial sensor systems. Some of the vendors of such systems have an academic profile and do allow for access to unprocessed, raw data, next to *stylized data*. However, most systems deliver stylized data only; i.e. they clean and process data using pre-trained machine learning models and only the predictions and outcomes of these models are made available to individual users. Figure 10.2 displays an example of raw sensor data, without any intermediate processing, but this is not true for all systems. For official statistics purposes, stylized data can be problematic as processing of data may be a black box and the accuracy of statistics becomes unknown. As a result, assuring comparability in time also may become problematic; data cleaning and processing procedures may change over time.

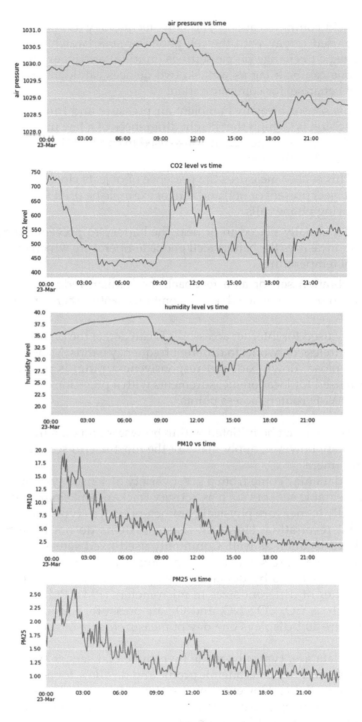

**FIGURE 10.2**
Indoor sensor measurements of air pressure, $CO_2$, PM1 and PM2.5, and humidity during a day.

**TABLE 10.1**

Participation Rates in a Housing Survey Study

|  | *n* | % |
|---|---|---|
| Invited | 2700 | 100 |
| Response |  |  |
|     Condition 1 (Photo) | 540 | 60.0 |
|     Condition 2 (Question) | 655 | 72.8 |
|     Condition 3 (Choice) | 564 | 62.7 |
|     Total | 1759 | 65.1 |
| Condition 3 |  |  |
|     Preferred photo | 320 | 56.7 |
|     Preferred question | 244 | 43.3 |

*Source:* Ilic, Lugtig, and Schouten (2020).
Condition 1 = photo, condition 2 = questions, condition 3 = choice between photos and questions.

In a mixed-mode survey setting, respondents may be offered the choice between answering questions and initiating sensor measurements. An example is given by Ilic, Lugtig, and Schouten (2020), where respondents in the LISS Panel of CentERdata, University of Tilburg, were offered the choice to answer questions about their housing condition or to take photos and submit these instead. The study included two control groups; one where respondents always provided photos and one where respondents always answered questions. Participation rates are presented in Table 10.1. Close to 57% of the choice group opted for taking photos instead of answering questions, demonstrating that even when offered a choice there is support for sensor data.

With commercial sensor data in mind, the step to data donation is a relatively small one. Commercial sensor systems usually provide dedicated applications that can extract sensor data directly from the sensor system itself or indirectly through an online database to which the system submits data on an on-going basis. Such data transfer is encrypted, so that other apps cannot read the data and the only access is through the vendor's application. Such applications do often offer the opportunity to download individual summaries of data, which may be linked by a respondent to a smart survey – a process called data donation.

## 10.2.3 Other Types of External Data

Two other types of data can be added: public online data and personal online data. Public online data are data that can be accessed by any online device. Personal online data can only be accessed by an online device through the household/person to which the data belongs.

Public online data come in many forms. The only prerequisites are that they are available online and can be linked to a survey.

One form is citizen science data, i.e. data that are collected on a voluntary basis by individual households/persons and made available to the wider community to serve a common good. A good example is data on outdoor air quality collected by households that installed outdoor sensor systems (see https://sensor.community/en and Figure 10.3).

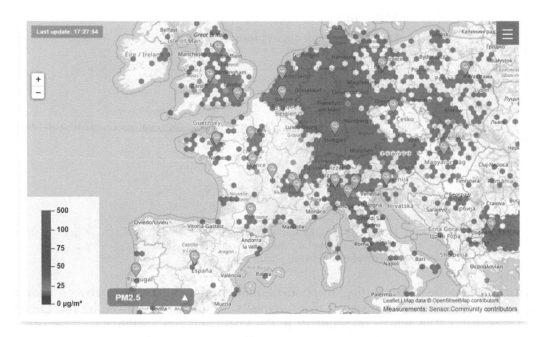

**FIGURE 10.3**
An example of public online data about particulate matter (PM2.5) taken from https://sensor.community/en on July 21, 2020.

Another form is points-of-interest data, where spatial and geographical data are made available based on latitude and longitude coordinates. One of the most commonly used options is open street map data; see https://openstreetmap.org. Both citizen science data and points-of-interest data can be linked based on location-time points; i.e. one would need to know the address of a respondent and/or the locations that a respondent has visited during a specified time frame.

Yet another form of data, not based on locations, are pre-trained machine learning models that are implemented in generally available applications. Examples are pre-trained optical character recognition (OCR) routines and pre-trained image recognition routines. The routines have been trained on large sets of annotated data in many countries and languages and can be employed in processing data submitted by respondents. An example of such pre-trained models is the extraction of products and prices from scanned shopping receipts.

Personal online data are equally diverse. The distinction of public online data is that only the owner or subject of the data can request access. In fact, legislation, such as GDPR in the European Union, may give the owners/subjects the right to have access and retrieve data and data holders the obligation to protect data against others. To increase utility of the data, often for commercial purposes, data holders may request owners/subjects to accept terms under which the data are shared with others. Typically, personal online data are voluntary and, as a consequence, are the result of self-selection. They are the results of households/persons buying certain products or using certain services. In the previous section, external sensor data was discussed. There is but a fine line between self-controlled measurements and automated measurements by online devices. This is best illustrated by two examples. A person may buy an activity tracker to evaluate physical movement and condition. Data of the tracker are collected because the person purchased the tracker by her/his own choice. But the person may have relatively little control over what is measured and what is not. A household may decide to use a shop loyalty card that will allow them to buy products with a discount. The household does so voluntarily but has to give consent to the shop to use the expenditure data for other purposes as well.

Including personal online data in a survey is another form of data donation. A responding household or person is asked to give consent to access and to link the personal data. This can be done actively by asking respondents to upload data themselves, but also passively by letting a dedicated application communicate with the data holder through APIs.

In a mixed-mode survey setting, respondents may be offered the choice between answering questions and consenting to link existing external data. At the time of writing, there was not yet much literature on willingness to opt for external data linkage. The future will have to tell.

## 10.3 Total Survey Error

The previous section presented examples of new types of data that can supplement or replace survey questions and lead to (trusted) smart surveys. They can do so even in a mixed-mode setting. When supplementing or replacing data, new sources of error arise. This section discusses these errors.

### 10.3.1 Representation and Measurement of Smart Surveys

New types of data affect both arms of the total error framework; see Figure 10.4. Representation depends on availability, access, and consent to these new data. Measurement

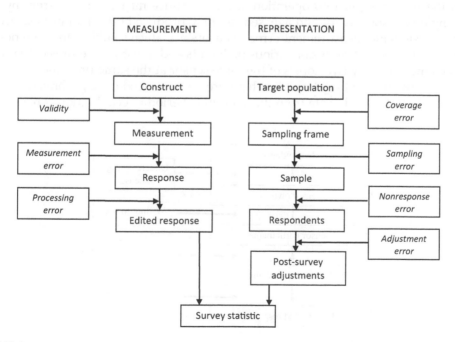

**FIGURE 10.4**
Survey errors for measurement and for representation taken from Groves et al. (2009).

may be partially performed by internal and/or external sensors. Some of these sensors need to be operated by the respondent, so that there is overlap in measurement errors between questions and sensors, but, in general, sensors have very different error properties.

Let us start by considering representation. Up to nonresponse, the errors for smart surveys coincide with those for traditional surveys. There is a target population that is represented by a sampling frame, which may be incomplete and contain ineligible units and/or contain double records. A sample is drawn at the hand of a sampling design from this frame leading to sampling error. Sample units are contacted and invited but may not respond, Respondents then proceed to perform the survey tasks. It is here that new causes of missing data may start to occur. In Figure 10.5 these are displayed for the case of sensor data. Respondents need to have access to sensors, they need to be willing to perform sensor measurements and/or provide access to the sensor data, they need to execute the sensor measurements, and the sensor data should be processed and transmitted. In all these steps, population units may drop out and cause representation to be selective and unbalanced across relevant population characteristics. Similar schemes can be made for data donation.

For measurement, changes are much more drastic as they involve the very concepts themselves. Figure 10.6 shows the measurement error again focusing on sensor data. The measurement that the survey institution intends to obtain from a respondent is the operationalization. For example, a survey aims to measure a number of air quality parameters over a specified time period in specific areas of dwellings. These are proxies for indoor climate, and, more generally, housing conditions.

The respondents need to be able and motivated to position and operate the sensors such that they measure these parameters, possibly following instructions given within the survey invitation. Respondents may, however, incorrectly initialize measurements or position devices in the wrong spots.

When the sensor is put into operation, it may produce measurement errors by itself. Depending on sensor quality and age, sensors may produce both random measurement errors and systematic measurement errors. Sensors are always subject to some noise. In the example, the concentrations of various pollutants and moisture are subject to imprecision. When measured by two copies of the same sensor at the same time, the two sensors will give slightly different readings. Furthermore, sensors need to be calibrated to known absolute levels, but may deviate from those in time. Periodic recalibration is needed, and

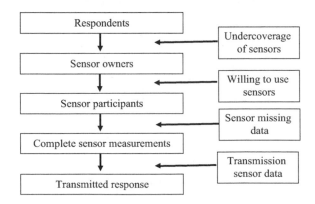

**FIGURE 10.5**
Representation errors in smart surveys.

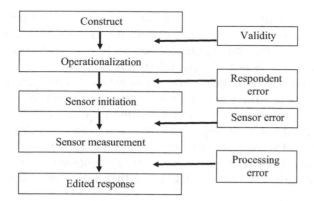

**FIGURE 10.6**
Measurement errors in smart surveys.

without such recalibration they will show a time-dependent systematic error. Even so, sensors may simply not be able to detect certain minimum levels or may not be able to differentiate between certain events. In the example, air quality sensors may not be sensitive enough to differentiate between particulate matter of different sizes.

Finally, sensor measurements need to be processed, e.g. different sensors may be combined, and errors may occur in doing so.

There is a close similarity between measurement and the cognitive process of answering questions (Tourangeau et al., 2001), with the main difference that some steps are now done by sensors. The four steps in answering questions are Interpretation, Information retrieval, Judgment, and Reporting. In sensor measurements, these are substituted by Interpretation of the sensor task, Initializing the sensor task, Performing the sensor task, and Transmitting the sensor data. Consequently, the same kind of deficiencies may occur. Respondents may hide or mask certain sensor measurements because of social desirability. Respondents may satisfice and produce low-quality sensor data. Respondents may also simply misinterpret or not understand the task they need to do. In the housing survey example of Table 10.1, after initial consent to take photos, some respondents did not submit photos, because they claimed they could not reach certain parts of the dwelling. Some other respondents submitted photos, but these displayed only part of the intended picture.

Figure 10.6 can be adapted to other types of data. Essentially, the same four steps in measurement arise and may lead to errors.

## 10.3.2 Criteria to Include New Types of Data

The mere possibility to include new types of data is not enough reason to also do so, as it requires a separate and new architecture and infrastructure, and respondents may not be willing to share the data. On the data collection side, new forms of data, especially when they are collected with the help of respondents (i.e. primary data collection), demand new data collection channels. These channels demand new and/or additional processing tools and skills, for expansion of existing monitoring and analysis tools and for a redesign of the survey estimation methodology. Such changes are costly and time consuming. On the respondent side, delivering the data may still be burdensome and/or may be privacy intrusive. Hence, a strong business case for alternative data types is needed and respondents need to benefit as well.

The previous section ventured into new error sources. From a theoretical perspective, classification and structuring of errors is very useful and helps to develop a methodology to reduce or adjust. However, from a practical perspective, with the business case in mind, such classification by itself is not sufficient. Virtually all errors are hard to quantify, but some quantification is needed to decide whether new types of data are to be introduced. In this section, criteria are given for making such decisions. Again the focus is on internal/external sensors, but similar criteria can be provided for other types of data. Three viewpoints are distinguished, each leading to a set of criteria: the survey measurement point of view, the sensor point of view, and the respondent point of view.

From the survey measurement point of view, survey topics may be candidates for enrichment or replacement with new types of data when they satisfy at least one of the following criteria:

- Burden: The survey topic(s) are burdensome for a respondent, either in terms of time or in terms of cognitive effort.
- Centrality: The survey topic(s) are noncentral to respondents; i.e. the average respondent does not understand the question or does not know the answer.
- Nonsurvey type: The survey topic(s) do not lend themselves very well to a survey question-answer approach to begin with.

Examples that satisfy the first criterion are topics that require keeping a diary for a specified time period, say a week or a month, and provide details about all time periods. Other examples are surveys that require consultation of personal information and archives, such as assets and finances. The second criterion is satisfied, for instance, for travel surveys where respondents need to provide exact coordinates of locations they have visited and health surveys where respondents need to describe sleeping patterns. The third criterion applies to complex socioeconomic or psychological topics such as happiness, health, or wealth, where many questions are needed to derive latent constructs.

From the sensor point of view, the main criteria are:

- Omnipresence: The sensor(s) are available in most, if not all, contemporary devices.
- Data access: Data generated by the sensor(s), as well as metadata about the properties and accuracy of the sensor data, can be accessed and processed.
- Quality: The sensor data is comparable, reproducible, and accurate.
- Costs: Any costs associated with the sensor(s) are affordable in most surveys.

The four criteria all link to the utility of the resulting sensor data. The omnipresence criterion refers to the coverage, and, hence, also price, of the sensors. In theory, tailored instruments can be developed that record complex phenomena and behaviors, but these are until now used only in lab settings. Smartphone sensors are examples of omnipresent sensors, whereas sensors in wearables have a much lower population coverage and pose challenges with regard to data access. The data access criterion means that sensor data can be stored, manipulated, and interpreted. For instance, location data can be stored and processed, but it is not always clear what sensor – GSM, Wi-Fi, or GPS – produced the data and how accurate the data are. The quality criterion originates from the statistical objective to derive the accuracy of statistics and to be able to compare statistics between persons and in time. For instance, location data can be used to estimate travel distances but are subject to missing data, measurement errors, and potentially also device effects. As such, sensors are just like other data collection instruments. The final criterion applies when sensors need to be provided to respondents and refer to costs associated with their use.

From the respondent point of view, sensors may vary in their attractiveness. Four criteria follow:

- Respondent willingness: Respondents are willing to consent to provide the sensor data.
- Data handling: Respondents can retrieve, revise, and delete sensor data on demand.
- Burden: Respondents are willing to devote the effort needed to collect and handle the sensor data.
- Feedback: Respondents may retrieve useful knowledge about themselves.

In order to employ sensors, respondents need to be asked for consent to activate sensors and to store and send data. Most mobile device sensors require consent by default. Exceptions are the various motion sensors that can be activated in Android without consent. However, even without the technical necessity to ask for consent, there are legal and ethical reasons why consent is imperative. Willingness to consent varies per type of sensor and depends on the context and purpose of the measurements. Obviously, the more intrusive a sensor measurement is, the more respondents will refuse and the larger the potential damage of missing sensor data. Legislation may require that respondents can get copies of their data and can request deletion of their data at any time[1]. This requirement puts constraints on the storage of and access to sensor data. Next, sensor measurement themselves, such as photos or sound recordings, require some respondent effort. This effort may be too great for respondents so that missing data and/or lower data quality result. Finally, the sensor data may be fed back to respondents in an aggregated form and may provide valuable information to them.

Summarizing, there are various criteria from the respondent, sensor, and survey measurement points of view that need to be confronted with costs and logistics of sensor data collection and processing.

## 10.4 Methodology for Hybrid Data Collection

Three types of methodology are discussed: data collection, measurement, and estimation. It must be noted that surveys mixing and combining different types of data are in their infancy, and the methodology still has many gaps and unknowns. Before the methodology is discussed, one important distinction needs to be made: active versus passive data collection. The section ends with a discussion on operational issues.

### 10.4.1 Active Versus Passive Data Collection

The new types of data can be collected passively or actively. Passive data are collected without respondent intervention or feedback, apart from consent to provide or link data. In active data collection, respondents are asked to check, revise, accept, and/or supplement data; i.e. the respondent gets previews of the data that are submitted. When opting for active data collection, the respondent also becomes involved in reducing and/or adjusting measurement errors.

Active and passive data collection have their advantages and disadvantages. Motives in favor of active data collection are increased motivation, increased engagement, and implementation of hybrid forms of data where questions depend on data that are being collected. Persons/households may be interested in obtaining new insights from the data, may want to control the data that are submitted, or may perform tasks more accurately when they see results. When survey questions depend on other types of data, then it is inevitable that the respondent is informed and becomes involved. Motives against active data collection are respondent burden, respondent inability to interpret accuracy of data, and complexity of the data collection instruments. Active data collection implies respondents have to invest some time. Also, they may not fully understand the data that are collected to begin with and presenting them with raw data could confuse them. Furthermore, survey interfaces that present data and provide options to respondents to alter or remove data are more demanding.

This can best be illustrated by an example. Consider again the housing survey example of Table 10.1. When taking a photo, the respondent could get instant feedback on the scanning circumstances such as insufficient light or contrast. The photo could also be shown to the respondent in order to judge whether it is of sufficient quality, and, if not, to retake the photo. A step further would be to apply computer vision approaches and show the result of image recognition to the respondent; e.g. Did we correctly identify the windows in this room or did we correctly detect the ground floor? Finally, the survey interface may allow for changing photos later, when deemed necessary by a respondent, before the whole survey is completed. Choosing between these options is a trade-off between burden and data quality.

### 10.4.2 Data Collection Strategies

When considering data collection strategies, at least two methodological areas need attention:

1. Recruitment and motivation strategies
2. Adaptive survey sensor designs

The first area is recruitment and motivation. These must be viewed from the perspective of mixed-mode surveys that adopt a push-to-web tactic (Dillman et al., 2014). In the context of multi-device surveys, this amounts to a push-to-smart-device tactic.

Despite the high population coverage of mobile devices, there is no guarantee of high participation rates in surveys that employ mobile devices. Field tests and surveys exploring willingness to participate typically find strongly varying consent rates for different types of sensor measurements and linkage of existing sensor data (see, e.g. Keusch et al., 2019; Struminskaya et al., 2020a,b). Some of the consent rates are hypothetical and were estimated within traditional surveys. It is possible that sensor surveys may attract respondents that would normally not participate when this is made more salient in the invitation. In the studies that have been done at Statistics Netherlands to date, real consent rates are surprisingly similar to those predicted from hypothetical consent rates estimated from surveys. Figure 10.7 shows consent rates for a number of measurements for a study fielded in the Dutch LISS Panel, reported by Struminskaya et al. (2020a). In the study, around 30%

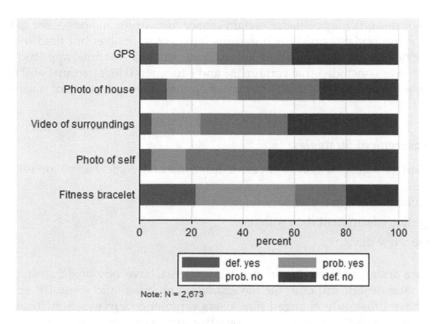

**FIGURE 10.7**
Consent rates for different sensor data estimated from a pilot survey in the Dutch LISS panel taken from Struminskaya et al. (2020a).

of the respondents replied that they would probably or definitely allow for GPS location measurements. In a travel app study in 2018 (McCool, Lugtig, and Schouten, 2020), roughly 30% of the sample participated and around 20% provided a week of travel data. These data were collected through a mix of location measurements and additional survey questions.

Strategies to lift participation rates look at recruitment strategies, i.e. invitation letters, instruction material, a dedicated webpage, incentives, personalized respondent overviews of statistics, interviewer assistance, and at motivation strategies, i.e. incentives, app tutorials, help functions, and again respondent overviews of statistics. Especially, the personal overviews are considered a potentially powerful option to recruit and motivate respondents: Respondents are provided overviews and insights about themselves at the end of data collection or gradually in stages during data collection. More generally, the data collection strategy aims at more than monetary incentives. Research in this area is oriented at evaluating whether respondent feedback alters the behavior of interest. In intervention studies, this is an intended objective, while in observatory studies, this should be avoided as much as possible. In all three case studies discussed in this book (Section 2.6), tailored strategies to recruit and motivate will play crucial roles.

The second area concerns an extension of adaptive survey designs to designs that include sensor measurements and/or existing sensor/big data after consent. Adaptive survey designs make explicit trade-offs between effort/costs and allocation of design features to different sample units. Four ingredients are needed: explicit quality and cost objectives, auxiliary data to distinguish different sample units, interventions/design adaptations, and optimization strategies. Typically, adaptive survey design focuses on overall response rates and balance of response rates across population subgroups that are relevant to the survey. However, measurement and data quality may also be included. In hybrid data collections, the number of design features and possible interventions increase. Adaptive survey sensor designs may consider certain sensor measurements or sensor data for certain types of respondents only. Such designs have yet to evolve, but field tests already revealed the need for such adaptation. For example, in the 2018 travel app study (McCool et al., 2020), it was concluded that certain age and educational level groups will need more attention or will need to be offered alternative options; their recruitment rates were relatively low.

### 10.4.3 Measurement Strategies

To reduce and/or adjust measurement error, three methodological areas are relevant:

1. User interface and experience
2. Integration of different data sources
3. Re-interview designs

User interface design and user experience, the first area, have, obviously, always been part of questionnaire design, but both the navigation on mobile devices and the employment of sensors have drastically changed how users experience surveys. Statistical institutes still have relatively little experience in cognitive and usability testing of such types of data collection (Gravem et al., 2019). This means that test plans often evolve during survey projects. An additional complexity is the two operating platforms (iOS and Android) and the wide range of operating systems that force developers to react quickly to new developments and to anticipate on technical errors. To date, commercial parties exist that simulate applications on hundreds of different devices and provide feedback within a few days. This area is still in need of best practices and guidelines, although first attempts have been made such as by Antoun et al. (2017a,b). Integration of sensor data in the mobility and consumption survey user interfaces is a prominent feature in application design. For physical activity, wearable activation and monitoring needs to be incorporated in survey designs.

The second area is the integration of difference data sources, which is at the heart of (trusted) smart surveys. Integration goes further than mere linkage and analysis of a single file including all variables. Different sources of data, such as mobile device sensor data and wearable sensor data, concern different concepts. Integration requires reconsideration and reconciliation of concepts. Although this is most likely specific to each form of hybrid data collection at hand, a general methodology is needed to model latent constructs that are underlying to all data sources. This area is most challenging for the CVS and HS case studies as new types of data may provide different proxies of the statistics of interest.

The last area is detection and, possibly, correction of device/mode-specific measurement biases. This is very similar to the methodology described in Chapter 9, Mixed-Mode Data Analysis. While the aim is for more accuracy and/or even reconsideration of survey concepts, hybrid data collections will affect estimates and may lead to breaks in time series as all mode changes do. In fact, it is anticipated that such risks are very real as data sources become more diverse and technological advances have (yet) not converged and may bring new changes. It is crucial that comparability between relevant groups in the population and comparability in time are preserved. Achieving such comparability is not an easy task. It may require that part of the budget be reserved to constantly evaluate changes, especially on measurement. An option to do so is a re-interview design in which respondents are surveyed or observed through two different measurement modes or devices. For the physical activity case study such re-interviews have indeed been conducted, comparing questionnaire data to different types of physical activity sensors. These will be discussed in Section 10.5.

### 10.4.4 Estimation Strategies

Estimation strategies have to consider at least the following two areas:

1. Planned missing designs
2. Active data collection checks and validation

The first area, planned missing design, is essentially an extension of sampling designs. Sensor measurements, but also data linkage, can be costly and/or time consuming. Hybrid data collections may invite subsamples of recruited respondents. Such subsampling may be viewed as a second stage, and the question is what inclusion probabilities are most efficient. The subsampling may be stratified based on auxiliary variables available at the start, e.g. the age of the sample person, and answers provided in the survey, e.g. yes/no smoking and self-reported activity profile in a health survey. Given the costs of sensor measurements, an optimization problem follows that requires extension of the existing sampling theory. The theory must include the statistical properties of sensor data. When sensors are expensive, such as the HS case study described later, then efficient subsampling designs are imperative.

The choice and design of active data collection, as discussed in Section 10.4.1, is a great challenge in terms of estimation strategies. This is best illustrated by two examples.

In an expenditure survey, respondents may submit scans of receipts. These receipts may be pre-processed, subjected to OCR, and classified using machine learning after the survey is completed. The resulting classifications may or may not be returned to respondents. A black box construction is desirable from a processing point of view, but undesirable from a respondent feedback point of view. Furthermore, classification accuracy may be around, say, 80% to 85%, meaning that 15% to 20% of classified scans are incorrect. As a consequence, manual intervention by trained coders, an approach called human-in-the-loop, or by respondent themselves may be needed. When respondents are involved, passive data collection gets an active component that needs to be organized and structured.

The last example concerns stop detection in travel apps. A 'stop' is usually defined as a respondent being in a location point within a certain perimeter for at least a certain time duration. Literature (Killaars, Schouten, and Mussmann, 2019) has shown that such a rigid definition is prone to false positives (falsely detected stops) and false negatives (missed stops). To some extent these can be identified afterward by linkage to public online data with points-of-interest. However, the best option is to let respondents check. Such an active component has an impact on the design of the app, the labeling of data, and the estimation of traveled distances between stops.

The implementation of respondent checks and feedback has a strong influence on estimation.

### 10.4.5 Logistics and Operations

Challenges in logistics and operations lie in a number of process steps:

1. Case management
2. Active versus passive data collection

3. Data processing

4. Privacy and ethics

The first challenge lies in structuring case management, i.e. the fielding, monitoring, and handling of sample units. App-assisted surveys apply tools that are new to statistical institutes. This is especially true when mobile device sensors are employed. Such sensor data need to be collected and checked and may form the basis for reminder and motivation strategies, feedback to respondents, and incentive payments. Sensor data are diverse by nature and, until now, atypical for survey data collection. The definition of a complete response becomes less clear, as sensor data may have missing data or measurement errors that are outside the influence of the respondent. For example, the Statistics Netherlands travel app (CBS verplaatsingen) uses GPS, Wi-Fi, and GSM for location sensing. However, under certain circumstances, the app may be put in hibernation mode when it uses too much processing/battery (McCool et al., 2020). This can to some extent be prevented by the respondent but is mostly a technical issue. Obviously, case management systems need clear definitions of complete response in order to act. Another setting that complicates case management is where multiple devices and/or modes are employed. Some devices support sensor data, while others do not or only partially. An integrated design where survey data can be submitted through different devices and where sensor data can be checked in all devices demands flexible case management and continuous handling of survey and sensor data at the end of the statistical institute. Apart from this, interviewer-assisted modes or paper modes may be added and imply clear definitions of response.

It must be explored how the collection of new forms of data affects case management and what a separate app channel should look like. For consumption and mobility case studies, monitoring is key as data is longitudinal and comes in on a continuous basis.

The second challenge is the choice and design of active versus passive data collection. Three examples are given: In an expenditure survey, respondents may submit scans of receipts. These receipts may be pre-processed, subjected to OCR and classified using machine learning after the survey is completed. The resulting classifications may or may not be returned to respondents. A black box construction is desirable from a processing point of view but undesirable from a respondent feedback point of view. Furthermore, classification accuracy may be around, say, 80% to 85%, meaning that 15% to 20% of classified scans are incorrect. As a consequence, manual intervention by trained coders or by respondent themselves may be needed. When respondents are involved, the passive data collection gets an active component that needs to be organized and structured. Another example is set in the same context. Many statistical institutes have access to price data of products identified by the so-called EAN/GTIN identifiers that are printed on products as barcodes. Respondents may scan such barcodes and the statistical institute may link the barcodes to these existing databases. At Statistics Netherlands, price data are available on a weekly basis and with some regional granularity. From a process point of view, it is again convenient to perform the linkage to price data after data collection, taking sufficient time to first process the price data. From the respondent point of view, this is not true, however, and it would be useful to be able to feedback the linked data to the respondent apps. The last example concerns stop detection in travel apps. A 'stop' is usually defined as a respondent being in a location point within a certain perimeter for at least a certain time duration. Literature has shown that such a rigid definition is prone to false positives (falsely detected stops) and false negatives (missed stops). To some extent these can be identified afterward by linkage to open databases with points-of-interest. However, the best option is to let respondents check. Such an active component has an impact on the design of the app and the labeling of data.

The third challenge is structuring data processing at substantive departments. Typically, hybrid data collections provide a mix of data, some of which have regular and straightforward data models and others have new and more exotic models. Substantive departments have separate teams for processing raw data to analysis of data sets that form the basis for publications and further linkage and integration into the statistical process. Two strategies may be adopted: One is where the data are manipulated and handled in such a way that they conform to traditional data models, i.e. from the output perspective they look similar. Another is where also the new types of data are delivered and need to be handled. An example again is stop detection in travel apps. The time-location data may be transformed within the app to a number of stops, at an average distance per track and the total distance traveled. These would then conform to the existing survey data model. Alternatively, the time-location themselves, being much richer, may be provided and processed during analysis and production of statistics. Such choices have a big impact on the tasks, roles, and expertise needed, and also on documentation and maintenance of software and methods. There is a strong parallel to the implementation of big data/trusted smart statistics studies that involve machine learning techniques.

Structuring data processing and analysis is to date still an open area and one that hampers implementation. All three case studies will imply major changes to the processing of data.

The last and fourth challenge concerns privacy and ethics. Although privacy and ethics conditions are two different aspects in data collection design, they are essentially based on the same underlying principles such as informed consent, public utility, and autonomy of the respondent. Again, privacy and ethics concerns are nothing new, but sensor measurements have two new features: The first is that they may be collected passively, i.e. as a black box, leading to uncertainty about what is collected and when. The second is that sensor data offer insights that may also be new to the respondent, much more so than do answers to questions. These two features alter respondents' perceptions of privacy, as studied by Keusch et al. (2019) and Struminskaya et al. (2020a). The GDPR put into operation in the EU in 2015 gives the right to respondents to access and remove data about them at any given time. In process terms, this adds extra importance to how data is stored and how respondents are informed and provided access, if requested. Statistical institutes yet have to find general procedures on how to handle such data collections. Location data and physical activity data are considered very sensitive.

## 10.5 Case Studies

In Section 2.6, three case studies are introduced: the Labor Force Survey (LFS), the Health Survey (HS), and the Crime Victimization Survey (CVS). These three surveys are evaluated from the smart survey perspective. One of the case studies, measurement of physical activity in HS, is elaborated.

### 10.5.1 Smart Survey Criteria

In Section 10.3.2, criteria are proposed to judge whether including new types of data is promising for existing surveys. There are three viewpoints: survey measurement, sensor/alternative data, and respondent. The criteria are discussed for the three surveys.

The survey measurement error criteria can be assessed regardless of alternative data sources. Table 10.2 flags the three criteria, burden, centrality of survey topics, and proximity of survey question operationalization to the concepts of interest.

The LFS is a relatively factual survey with the main statistic, the unemployment rate. Up to 2020, it has been implemented as a household survey, and detailed information on employment and education has been asked from all household members 16 years and older. The survey is, thus, considered burdensome, but central to respondents and survey questions are relatively good proxies of the statistics of interest.

The HS scores on all criteria. It is demanding in length and detail, ranging from perceived health, lifestyle, health determinants to medicine use, and hospitalization. Part of the questionnaire modules are noncentral. A good example is the type and duration of various forms of physical activity on an average day. Furthermore, the HS attempts to measure the health status of a person through a range of indirect questions, without any physical inspection or test.

The CVS is a relatively short survey with topics, such as victimization and perceptions of safety, that are topical to respondents. However, the underlying interest is in general well-being and safety, and the survey may be viewed as providing only indirect measures of these latent constructs through batteries of questions. The CVS scores on the last criterion.

So what would be useful new types of data for the three surveys? Table 10.3 gives some examples that are by no means exhaustive.

For the LFS, studies have suggested to employ passive sensor data to derive whether persons are employed or unemployed. One option to do so is to consider their daily lifestyle patterns such as the locations they visit and the times and types of social media that they use. Another option is to consider the content of their social media messages and

**TABLE 10.2**

Flags for Each of the Survey Measurement Criteria for the Three Case Studies

|      | Burdensome | Noncentral | Weak Proxy of Concepts |
|------|:----------:|:----------:|:----------------------:|
| LFS  | ×          |            |                        |
| HS   | ×          | ×          | ×                      |
| CVS  |            |            | ×                      |

A flag means that the risk of measurement error is very real.

**TABLE 10.3**

Potential New Data in the Three Case Studies

|      | Sensor Data | Personal/Public Online Data |
|------|-------------|------------------------------|
| LFS  | Time-location (GPS) <br> Mobile device use | - |
| HS   | Time-location (GPS) <br> Motion <br> Heart rate <br> Camera <br> Indoor climate sensors | Wearable sensor data <br> Outdoor climate data |
| CVS  | Camera <br> Heart rate <br> Time-location (GPS) | Wearable sensor data |

internet searches. While these sensor data may be related to employment and pursuits to get employed, they mostly serve to assist derivation of employment rate status and cannot fully replace survey questions. The LFS as a household survey would still require passive sensor data for all of its members. Furthermore, per the LFS wave, the sensor measurements would need to be collected for a longer time period that corresponds to the LFS reference periods. Essentially, sensor measurements would need to run for the full duration of the LFS waves, i.e. a year or longer. Hence, the sensor measurements do not necessarily reduce the burden.

The HS topics are physical and mental, and, thus, lend themselves to wearables. In fact, the most frequently used wearables aim at health, fitness, and lifestyle indicators. Activity trackers, smart scales, and smartwatches may be employed for at least a specified duration in order to derive physical conditions, and, possibly, also mental conditions. Wearable sensor data may replace parts of the HS questionnaire but may also provide important insight into the underlying concepts that the HS attempts to measure. Instead of wearables, mobile device sensors may also be activated that measure similar data such as activity, heart rate, and locations. Obviously, mobile devices are usually not attached to one's body, so that, in practice, the physical activity of the device is measured. As a supplemental type of sensor data, camera photos may be included that give an impression of a person's physical condition. In order to measure exposure, one may use indoor climate sensor systems and public online data on outdoor air population.

The CVS measures victimization and perceptions of safety. For the latter, questions are asked about neighborhood characteristics, neighborhood safety, police contacts, feelings of unsafety, and measures respondents take to improve safety. While perceptions are hard to measure, the mobile device camera may be used to take pictures of the neighborhood and dwelling that allow for a derivation of characteristics such as litter, graffiti, broken windows, etc. Heart rate sensors in combination with time-location measurements, and possibly motion sensors, may be employed to measure increased stress while moving through certain areas or neighborhoods. Wearable sensor data may be used for the same purpose to combine location with motion and heart rate.

The examples in Table 10.3 may be scored on the other two sets of criteria: sensor viewpoint and respondent viewpoint. Given that the HS examples seem most promising, this is done only for the HS case study. Tables 10.4 and 10.5 present flags for the HS examples for the sensor viewpoint and the respondent viewpoint, respectively.

**TABLE 10.4**

Assessment of the Sensor Criteria for the HS

| Survey | Sensor | Omnipresence of Sensor | Access to Sensor Data | Quality of Sensor Data | Costs of Sensor(s) |
|--------|--------|:----:|:----:|:----:|:----:|
| HS | Location | v | v | v | v |
| | Motion | v | v | v | v |
| | Heart rate | v | v | v | v |
| | Wearables | | | v | |
| | Indoor climate | | | v | |
| | Camera | v | v | | v |

A flag is a positive score on the criterion.

**TABLE 10.5**

Assessment of the Respondent Criteria for the HS

| Survey | Sensor | Willingness to Participate | Respondent Data Handling | Burden to Respondent | Feedback to Respondent |
|--------|--------|:-------------------------:|:------------------------:|:--------------------:|:----------------------:|
| HS | Location | ∨ | | ∨ | ∨ |
| | Motion | ∨ | | ∨ | ∨ |
| | Heart rate | ∨ | | ∨ | ∨ |
| | Wearables | ∨ | | | ∨ |
| | Indoor climate | | | | ∨ |
| | Camera | | ∨ | | |

A flag is a positive score on the criterion.

## 10.5.2 A Case Study Elaborated: Physical Activity

Physical activity is a promising survey topic to supplement and/or replace with sensor data. It consists of two types of statistics: the type of activity and the intensity of activity. The types of activity are a hierarchical classification with main classes, such as lying, sitting, sporting, standing, walking, and other activity, that can be detailed to very specific activities. The intensity of activity refers to the amount of energy/calories that a person uses for the activity. Typically, in survey questionnaires the two statistics are measured by asking respondents to estimate the average duration in a normal week or day that is spent on a range of activities. The resulting durations are multiplied by the so-called MET (metabolic rate) values that translate to energy use. Each type of activity has a different MET value that has been historically determined. MET values are average values; they are taken the same for all respondents, regardless of their characteristics and regardless of the actual intensity. It is clear that taking average values removes between- and within-person variation in the actual intensities. However, even apart from this smoothing of individual statistics, durations for different activities are known to suffer from both underreporting and overreporting.

Both type and intensity of activity can be predicted through sensor measurements. Useful sensors are motion sensors, heart rate sensors, and respiration sensors. Motion sensors detect shocks in three dimensions and are combined with a gyroscope and magnetometer. Commercial wearables, such as smartwatches, contain motion sensors and heart rate sensors, but typically only on one position on the body, usually the wrist. These wearables may have a display and/or may send data to a website, where summaries can be consulted and downloaded/saved. There are also more dedicated sensors that are to be worn in a different position on the body, such as the upper leg or waist, or even on multiple positions. Literature on activity trackers and the prediction of type and intensity is enormous, but almost entirely outside the survey and official statistics literature. Figure 10.8 shows examples of sensors.

Physical activity data may be included in three ways. One is to use activity trackers that respondents already have and to ask them to donate these data within the survey data collection period. Another is to provide respondents with a tracker that they can keep when they deliver data during the survey data collection period. Yet another is to provide respondents with a dedicated activity tracker, ask them to apply it during the survey data collection period, and then return it. There are advantages and disadvantages for each of the three options that relate to the total error components of Section 10.3.1.

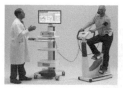    

**FIGURE 10.8**
Examples of activity sensors.

**TABLE 10.6**

Representation and Measurement for Activity Tracker Options

|  | User-Owned | Incentive/Gift | Institute-Owned |
|---|---|---|---|
| Representation |  |  |  |
| Coverage | – | + | + |
| Willingness | +/– | + | – |
| Sensor error | +/– | – | + |
| Transmission | + | +/– | + |
| Measurement |  |  |  |
| Validity | +/– | – | + |
| Competence | + | +/- | +/- |
| Sensor error | – | – | + |
| Processing error | – | – | + |

'+' = good performance, '+/–'= average performance, '–'= weak performance.

Table 10.6 presents sensor-specific representation and measurement errors for each of the three options.

Under the first option (user-owned), the weak features are population coverage, the diversity of incoming data due to the wide range of trackers on the market which leads to device effects, and the processing of data which is completely handled by the vendors of the trackers. The strong features are the familiarity of respondents with the trackers and the processing that is already prepared by vendors of the trackers. It is also the cheapest option.

Under the second option (incentive), the weak feature is that all errors are intrinsic to the sensors themselves. The reason is that trackers must be relatively cheap in order not to create a very expensive survey. The strong features are coverage and the likely positive impact on willingness.

The third option (institute-owned) is a middle ground option where sensors are provided, but they are used multiple times. Typically, black box sensors are used under this option which are robust to misuse, but give no direct feedback to respondents. A weak feature may be the willingness of respondents to wear such a tracker. The lack of immediate feedback is what attracts activity researchers, as it is suspected to reduce the experimental impact on the behavior of respondents. However, for respondents, this option gives no sense of control.

**TABLE 10.7**

Accuracy of Type of Activity Prediction for Sensors at Four
Different Positions, Without and with a Heartrate Sensor

| Feature | Wrist | Ankle | Upper Leg | Shirt |
|---|---|---|---|---|
| Motion | 58% | 67% | 84% | 71% |
| Motion + Heart | 65% | 65% | 84% | 63% |

Accuracy is based on fully labelled data.

Despite this potential impact on participation rates, the third option is further explored.

Given the risk of low participation rates, an important question is how to optimize recruitment and motivation strategies. Two strategies may be effective, but costly, in increasing participation rates and interviewer recruitment and incentives. Both strategies lift the costs of a survey. The interviewer recruitment option can be combined with existing face-to-face surveys. Respondents to these surveys are invited to participate in a follow-up sensor survey. Incentives may be added to reduce nonresponse and also to guarantee return of sensors. In an experimental Dutch study conducted in 2019 (Toepoel, Lugtig, and Luiten, 2020), incentive schemes were varied and randomized across sample subgroups. Four strategies were tried:

1. Unconditional = 0, conditional = 10: Response rate = 11%
2. Unconditional = 0, conditional = 20: Response rate = 15%
3. Unconditional = 0, conditional = 40: Response rate = 19%
4. Unconditional = 5, conditional = 10: Response rate = 16%

In another study, different sensors were tried and activities were fully labeled and recorded. Table 10.7 presents the accuracy of sensors at four different positions: ankle, shirt, upper leg and wrist, i.e. the percentage of time frames that was correctly classified. In addition, a heart rate sensor was included, which improved accuracy for some of the sensor positions. The sensor on the upper leg performed the best by far and had an accuracy of 84%.

Much more can be said about the case study. Both planned missing designs and adaptive survey designs may be important to consider. Furthermore, in order to increase response rates respondent feedback may be arranged through a dedicated app and through respondent statistics that are prepared at the end of data collection. Physical activity surveys are a good example of a smart survey combining questionnaires with external sensor data.

## 10.6 Summary

The main takeaway messages from this chapter are:

- Promising new types of data exist that may supplement and/or replace survey questions: mobile device sensor data, sensor data from other sensor systems, public online data, and personal online data.

- Surveys are termed smart surveys when they include internal or external sensor data and trusted smart surveys when they include public or personal online data.
- (Trusted) smart surveys are interesting when survey topics are demanding for respondents, when survey topics concern knowledge that most respondents do not have and/or when survey topics do not lend themselves for a question-answer format.
- (Trusted) smart surveys are promising from the respondent's perspective when they are not intrusive or burdensome, when respondents can access and control data and/or when respondents can learn about themselves.
- (Trusted) smart surveys are promising from the quality perspective when the new forms of data have high coverage in the population, when the raw data can be accessed and processed, when the data is relatively inexpensive to obtain and when accuracy is hig.
- (Trusted) smart surveys are likely to have a mixed-mode design as response rates to web invitation is often too low for official statistics.
- (Trusted) smart surveys require a renewed look at total error frameworks; they introduce new types of error.
- (Trusted) smart surveys demand for extension of existing methodology and development of new methodology.
- One of most crucial choices when including new types of data is between active (with explicit respondent involvement) and passive data collection (without a strong involvement of respondents).

## Note

1. See https://gdpr-info.eu/issues/right-of-access/ and https://gdpr-info.eu/issues/right-to-be-forgotten/

# 11

## Adaptive Mixed-Mode Survey Designs

### 11.1 Introduction

This chapter considers mixed-mode designs in which mode allocation may differ across sample units based on auxiliary information that is available before data collection or that becomes available during data collection. Such designs are specific forms of adaptive and responsive survey designs (Schouten, Peytchev, and Wagner, 2017), where survey modes are a design feature that may be adapted. Although literature points at differences between adaptive and responsive survey design, here, no explicit differentiation is made between the two and the designs are simply termed adaptive.

Adaptive mixed-mode survey designs alter the data collection strategy. In Chapter 5, Mixed-Mode Data Collection Design, mode strategy decisions were uniform, i.e. non-adaptive. In adaptive designs, decisions may be made differently for different subgroups within the sample. Adaptive survey designs are an example where mode effect reduction is pursued through the data collection design, see Figure 11.1. The decision if and how to adapt the mode strategy is, generally, made before data collection starts in the data collection design. Nonetheless, adaptation may depend on paradata, i.e. data collection process data, that comes in during data collection. Hence, adaptive mixed-mode survey design also affects the plan-do-check-act cycle during survey field work. Since multiple strategies may be applied simultaneously, more flexibility is demanded from survey case management, monitoring, and analysis.

Adaptive survey designs (ASD) have four key ingredients (Schouten, Calinescu, and Luiten, 2013): a set of quality and cost functions, linked auxiliary information, a set of design features/interventions, and an optimization strategy. Loosely said, ASD explicitly optimizes quality given cost constraints by assigning different design features to different sample units identified by auxiliary information. In order to do this, quality and cost functions need to be explicit and be accepted by the survey stakeholders, the auxiliary information should be complete and accurate, and the optimization strategy should be transparent and robust. In this book, survey modes are the design feature of interest and the focus will be on adaptation and intervention in the mode strategy. The fourth key ingredient, the optimization strategy, can be very explicit and result from the (approximate) solution to a formal mathematical optimization problem. The optimization strategy can also be implicit only and based on expert knowledge and experiences from historic survey data. In practice, therefore, this ASD ingredient may be absent and the ASD implementation may be the result of a relatively organic process. This does not mean that the ASD is less effective or efficient as will be discussed later in this chapter.

DOI: 10.1201/9780429461156-11

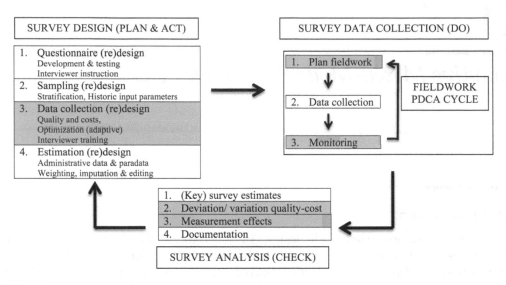

**FIGURE 11.1**
The plan-do-check-act cycle for surveys.

The aims of this chapter are to explain potential ASD design decisions in a mixed-mode context to explain the consequences of such decisions for design and analysis and to provide examples that may aid readers in exploring ASD for their surveys.

As has been discussed throughout this book, modes affect both representation and measurement. This means that mode strategy decisions should be considered from the perspective of all survey errors. To date, most ASD literature addresses survey nonresponse and ignores the impact on other survey errors, most notably measurement error. This is for understandable reasons, as estimation of mode-specific measurement effects is known to be hard without relatively costly and complex experimental designs. Chapter 8, Re-interview Design to Disentangle and Adjust for Mode Effects, addressed such decompositions in the context of survey modes. This chapter goes a step further and uses estimated or anticipated mode effects in ASD.

Literature on ASD in a mixed-mode context is (still) relatively thin (see Chapter 12 of Schouten, Peytchev, and Wagner (2017)). This may be surprising, given that the survey mode is the design feature that has the largest impact on both quality and costs. One reason for this may be that ASD is mostly about efficient rearrangement of resources and modes, especially face-to-face, which implies big differences in costs and interviewer workloads; differences that are too large are left to final decisions during field work. Another reason may be the required flexibility of survey case management systems to move sample cases around different modes. A third reason may be the possible confounding of mode-specific nonresponse and measurement biases, which makes ASD decisions across modes more complicated than ASD decisions within modes. A recent exception is given in Van Berkel, Van der Doef, and Schouten (2020), who present an ASD for the Dutch Health Survey in which face-to-face follow-up is applied only to subgroups of the sample during data collection.

This chapter employs the Health Survey and Labor Force Survey as case studies; see Section 2.6. In Section 11.2, the four elements of ASD are discussed in the context of mixed-mode surveys. In Section 11.3, the two case studies are presented and choices are motivated. The chapter ends with take away messages in Section 11.4.

## 11.2 Elements of Adaptive Multi-Mode Survey Designs

Schouten, Peytchev, and Wagner (2017) discuss in detail the main elements of ASD. ASD can be viewed as adjustment by design; rather than ensuring balance by weighting or calibrating afterwards, the response is balanced during fieldwork. Just like nonresponse adjustment, ASD requires auxiliary information that is relevant both in terms of response behavior and in terms of key survey variables. Adjustment in ASD terms means adapting survey design features before or during data collection. To make such decisions, quality and cost criteria are needed and a strategy to find the optimal balance needs to be constructed and applied.

Lo Conte, Murgia, Luiten, and Schouten (2019) modified the existing ASD checklist constructed by West and Wagner (2017) focusing on mixed-mode designs. It has the following steps:

1. Identify survey priorities;
2. Identify major risks and explicitly consider the following three risks:
   a. Incomparability in time;
   b. Incomparability between subgroups;
   c. Budget overrun and heavy interviewer workloads in follow-up modes;
3. Define quality and cost indicators;
   a. Consider nonresponse indicators;
   b. Consider measurement error indicators;
   c. Consider cost indicators;
4. Define decision rules from:
   a. Expert knowledge;
   b. Case prioritization;
   c. Sampling with quota;
   d. Mathematical optimization;
5. Modify the survey design and monitor the outcomes;
   a. Develop a dashboard for survey errors, both in representation and in measurement;
   b. Develop a dashboard for survey costs;
6. Compute estimates;
7. Document results, including detected deviations from expected quality and costs;

The four main ASD elements are part of the checklist steps. Steps 3 and 4 concern the quality-cost indicators and the optimization strategy, while the link to auxiliary data is included in steps 1 and 2.

The ASD checklist steps can also be mapped on the plan-do-check-act -cycle in Figure 11.1. Steps 1 to 4 correspond to the 'Plan' step. Step 5 is the 'Do' step. The 'Check' step consists of steps 5 and 6. Step 7 is the 'Act' step that feeds back to the 'Plan' step of the survey.

In the following subsections, the checklist steps and the ASD elements are discussed. In Section 11.3, the steps and elements are detailed for the two case studies.

### 11.2.1 Survey Mode as a Design Feature

In the context of this book, the survey mode is the design feature of interest. Potential ASD choices follow the taxonomy and terminology of mixed-mode designs as presented in Section 2.3. It must be remarked, however, that mode choice indirectly affects other design features such as the contact strategy and the training and assignment of interviewers. Furthermore, as explained in Chapters 2, Designing Mixed-Mode Surveys, and 5, Mixed-Mode Data Collection Design, there is considerable within-mode variation of design features and many choices are still open when adapting the mode. In practice, adaptation of the choice of modes goes hand in hand with adaptation of other design features such as the number of reminders, calls and visits, and the incentive strategy.

Possible adaptations in mixed-mode surveys are:

1. Some sample units get all follow-up modes in a sequential mixed-mode design, while others do not. For example, all sample units are invited for a web survey, while only a subset of the web nonresponse is re-approached through an interviewer mode;

2. Some sample units skip the starting mode in a sequential mixed-mode design, while others get the starting mode. For example, part of the sample is first called for a telephone survey and in case of nonresponse visited by an interviewer, while other sample units are only visited by the face-to-face interviewer;

3. Some sample units are offered a concurrent mixed-mode design (and, thus, a choice between modes), while others get a prescribed mode. For example, all sample units are invited for a web survey, while only part of the sample is sent a paper questionnaire as an alternative;

4. Sample units are assigned one mode out of a fixed set of modes. For example, one subset of the sample is invited to fill in a paper questionnaire, while others are called and yet others are visited;

5. Sample units are assigned one mixed-mode design out of a fixed set of mixed-mode designs. For example, one subset of the sample is first invited to a web survey and in case of nonresponse is contacted by a face-to-face interviewer, another subset is called, and yet another subset receives paper questionnaires followed by telephone calls.

The different ASD options are not exclusive and can be combined as well. Option 5 is the most general and encompasses all others when a single mode design is viewed as a special case of a mixed-mode design. Option 5 also includes exotic, but unrealistic, options where the sequence and the number of modes is completely open and varied. In practice, the first two options are the most realistic, since mixed-mode designs usually start with cheaper self-administered modes before moving to more expensive interviewer-assisted modes.

Two examples are introduced that will be elaborated in Section 11.3, one for option 1 and one for option 5. In the first example, linked to the Health Survey (HS), all sample units are invited to participate online and only some nonrespondents to the web mode receive a follow-up by a face-to-face interviewer. So the choice is between two mode strategies {web, web → F2F}. This example is shown in Figure 11.2. In the second example, linked to the Labor Force Survey (LFS), sample units get assigned one mixed-mode design out of a set of five designs {web, telephone, F2F, web → telephone, web → F2F}. So there are three single mode designs and two sequential designs with two modes, as illustrated in Figure 11.3.

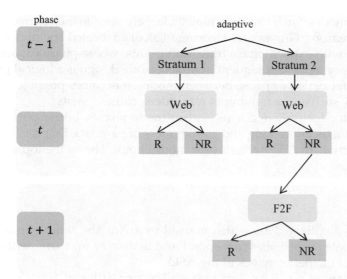

**FIGURE 11.2**
Schematic view of an adaptive mixed-mode design for the HS example.

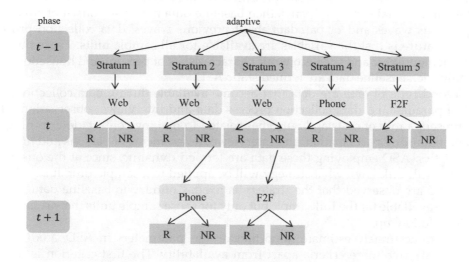

**FIGURE 11.3**
Schematic view of an adaptive mixed-mode design for the LFS example.

In an ASD with mode as a design feature, other design features may be adapted as well. For example, in the web mode the number of reminders may be varied, in the telephone mode the number and priority of calls may be adapted, and in the face-to-face mode the number and priority of visits may be chosen differently. Across modes, the advance and/ or invitation letters may be tailored to known characteristics of the sample units.

Obviously, some caution is needed when altering many design features at the same time. While the flexibility of the resulting design to meet quality-cost constraints grows, the complexity of logistics and case management grows as well. Furthermore, sampling variation and unpredictable circumstances cause variation in quality and costs that impact

ASD decision in follow-up modes. Such variation may be largely outside the control of the surveyor. Partly for these reasons, Groves and Heeringa (2006) advocated the division of data collection into phases with explicit phase transition points, where phase capacity is evaluated. A change of survey mode in a sequential mixed-mode design is a logical phase transition. By adding a buffer period or pause between modes, it becomes possible to add ASD choices within phases, such as the number of reminders, calls, or visits.

Apart from providing time for ASD decisions within mode phases, buffer periods in between modes also facilitate the processing and employment of paradata, i.e. process data collected during data collection, in ASD decisions for mode choice. This is the topic of the next subsection.

### 11.2.2 Population Strata

The second key element is auxiliary data, which is used to divide the sample into subgroups, to model and estimate propensities to respond and to display measurement error and to monitor and evaluate the performance of the ASD.

In ASD, there are two types of auxiliary data that lead to two different types of ASD. The first type of auxiliary data are data available at the start of the survey data collection, and, typically, consist of linked data from the sampling frame, administrative data, and contextual data about the areas of sample unit dwellings. When ASD is set within the context of a longitudinal survey or panel, then baseline data may also consist of survey data from previous waves and/or paradata about previous waves' data collection processes. A crucial feature is that the variables are available for all sample units. These data often describe demographic and socioeconomic characteristics of persons and households. ASD using only such baseline data are termed static ASD.

The second type of data are data that become available during data collection. These consist of paradata, i.e. data collection process data and interviewer observations. These data represent a mix of costs and outcomes related to making contact, determining eligibility and obtaining participation, and interviewer assessments of persons and household characteristics. ASD employing these data are termed dynamic, since at the onset of the survey, it is still unknown what strategy will be applied to a sample unit. It is only when the paradata are observed that the strategy is fixed. Contrary to baseline data, paradata need not be available for the full sample but only for those sample units that are candidates to receive a follow-up.

In order to accurately estimate important design parameters in ASD, auxiliary data need to satisfy two more criteria apart from availability. The first criterion is that they are complete and are not subject to item-nonresponse. If item-nonresponse occurs, then some sample units cannot be assigned to a mode strategy. The second criterion is that any measurement error on the variables is unrelated to response and answer behavior in the survey. If auxiliary variables are subject to varying measurement error that is predictive of survey behavior, then ASD decisions may become less effective or even ineffective. The three criteria are termed availability, completeness, and measurement error independence.

For ASD considering survey mode as a design feature, it has been argued in this book and in this chapter that next to nonresponse mode-specific answer behavior and measurement biases also need to be accounted for. This broader total survey error view does not change the distinction between the two types of auxiliary data, baseline data, and paradata. However, it does shift the attention from ASD auxiliary data to baseline data. This has two reasons.

One reason is that it is less common to return to a respondent and ask for new or supplementary survey data, when it is concluded after the interview that the respondent produced measurement errors. In establishment surveys, there is a tradition to do such follow-up, especially for the larger and influential sample units. In person and household surveys, such practice is rare, because of costs and the absence of very influential population units. This means that, even if paradata would be informative about the occurrence of mode-specific answer behavior, paradata is received too late to intervene or change strategies.

Another reason is that there are no (perceived) cost implications of measurement error as they exist for nonresponse. Rarely are survey targets set at some upper threshold to measurement error, e.g. include only responses without measurement error. The exception is item-nonresponse. Respondents with large proportions of item-nonresponse may be discarded and, obviously, item nonresponse by itself reduces statistical power in any multivariate analysis. Arguably, item-nonresponse may be viewed as an extreme form of measurement error, where the answer to the item is completely useless. However, in general, survey data can be less valid or reliable with little perceived consequence to cost. The absence of cost implications is, however, only indirect. Less reliable survey data would call for larger sample sizes and less valid survey data may reduce the efficacy of resulting statistics.

Nonetheless, there have been some attempts to introduce dynamic ASD based on paradata that are informative about answer behavior and have a potential risk of measurement error. An example is Conrad et al. (2017). They prompted respondents to slow down, when they had an average answering time below a pre-specified threshold. Such prompting may be implemented as a function of person/household characteristics and survey mode, e.g. based on age, and would then lead to a dynamic ASD. Another example is moderation of the speed and pace of the interview by interviewers. Interviewers are trained to create an interview atmosphere that keeps respondents motivated and concentrated. They tailor to respondents' characteristics in doing so. Interviewers may be instructed and trained to react differently to undesirable answer behavior for different types of persons/households. Again a dynamic ASD is introduced when this is done in a structured way.

However, given that dynamic interventions accounting for measurement error are less common in person surveys, mixed-mode ASD will often be static.

Let us first focus on nonresponse and distinguish auxiliary variables for static ASD and auxiliary variables for dynamic ASD.

Relevant auxiliary variables for nonresponse in the context of static mixed-mode ASD are the same as for any ASD. The variables need to relate to both key survey variables and to general nonresponse behavior. Auxiliary variables play the same role as in weighting or imputation after the survey is done, with the difference that in ASD the relation is to anticipated nonresponse patterns without adaptation. As explained by Little and Vartivarian (2005), including auxiliary variables with no relation to key survey variables may only lead to an increase in estimation variance and a loss of precision. Since key variables differ from one survey to the other, relevant auxiliary variables for ASD also change. The two examples of this chapter have very different topics, health and employment. For health surveys, for example, the size and type of household are less relevant, while they are important predictors for the employment status in labor force surveys. For health surveys, lifestyle and health determinants are very relevant, while for employment they are less relevant. Some auxiliary variables such as age and income are relevant for almost all surveys, including health and labor force surveys.

Relevant auxiliary variables to explain and predict nonresponse in dynamic mixed-mode ASD should also have the feature that to relate to the key survey variables. Paradata may indeed contain such variables. Paradata are often about contactability of sample units or about motives sample units may have to not participate (Kreuter, 2013). Contactability depends on lifestyle and at-home patterns which may be indirectly related to the key survey variables (Groves and Couper, 1998). Nonresponse arguments derived from paradata may obviously be directly related to the topics of the survey (Groves et al., 2002).

Paradata may be used to inform ASD decisions in subsequent modes. In sequential mixed-mode design phases, paradata may be collected in early modes. An example is a sequential design with telephone followed by face-to-face. Sample units that have not been contacted by telephone may be assigned to face-to-face or may be treated differently from those that refused or indicated they had no time. Another example is a sequential design with web followed by an interviewer-assisted mode. Sample units that broke off in web may be contacted by an interviewer or may be treated differently from those without a sign of life. In these examples, it must be decided, however, beforehand whether contact patterns, familiarity with online surveys, and lack of ability and/or motivation of sample units relate to the key survey variables.

Potentially useful paradata for follow-up modes are:

- In web: Occurrence of break-off, including the position in the questionnaire and the type of device that was used. Break-off may have different causes, ranging from technical performance of the questionnaire to a lack of motivation and/or inability to answer the questions. Lack of motivation and inability may be directly related to the survey variables;

- In telephone: Number and timing of calls, motives for refusal, and other interviewer observations like presence of other household members or the mood of the nonrespondent. Number and timing of calls may be indicative of at-home patterns and lifestyle. Motives for refusal may relate directly to the survey. Interviewer observations may be tuned such that they relate to survey topics;

- In face-to-face: Number and timing of visits, motives for refusal, impediments for contact, other interviewer observations like state of the neighborhood and dwelling, and characteristics of the nonrespondent and her/his household. Paradata are similar to telephone but generally richer, because of visiting the respondent's neighborhood and dwelling;

- All modes: Information on contacts through helpdesk calls by the sample units themselves. Sample units may themselves contact the survey institute to ask about the purpose of the survey and the data that are being collected. They may also express concerns about the survey, express belief that they are not eligible or relevant to the survey, or even announce their refusal to participate. All these may be related to the survey variables.

The three criteria of ASD auxiliary data, availability, completeness, and measurement error independence can, however, be restrictive for utility of paradata in hybrid mixed-mode designs. As process data is about survey data collection, paradata are mode-dependent. Paradata collected in interviewer-assisted surveys is much richer that paradata in web surveys. In mail surveys, paradata may be almost absent.

The mode dependence of paradata means that in a concurrent mixed-mode design phase no uniform set of auxiliary variables may be available for all nonrespondents. Consider, for example, a design where sample units are offered a choice between paper and online questionnaires. For the mail mode, there is no paradata for nonrespondents, except when sample units contact the survey institute. For the web mode, there may be paradata when a sample unit logs in but does not complete or submit the questionnaire. After this concurrent step, there are four types of sample units: Those that responded by mail, those that responded online, those that logged in at the online website but stopped, and those that gave no sign of life. For the last group, it is unclear whether they did read the invitation, whether they had access to web and/or were familiar with web questionnaires, whether they had a look at the web portal and decided to refuse, or whether they started the paper questionnaire but decided to stop. This uncertainty makes it hard to act in potential follow-up phases, which is crucial for ASD, while for the nonrespondents that broke off in web the paradata may be very informative. Another example is a concurrent design where households with a registered phone number are called and all others receive an invitation for an online questionnaire. The nonrespondents in the telephone mode may have relatively strong paradata about failed contact attempts and/or sample units' arguments to refuse participation. For nonrespondents in the web mode, again only paradata is available for sample units that stopped after logging in.

It may also be that measurement error independence of the paradata does not hold across modes. In a concurrent design with telephone and face-to-face interviewing, telephone interviewer observations may be less accurate than face-to-face interviewer observations due to the absence of nonverbal cues.

Let us, next, consider measurement error. Nonresponse and measurement error have common causes; less motivated or able sample units may produce more measurement error, but also have lower response propensities (e.g. Olson 2007). This means that relevant auxiliary variables for measurement error must and do overlap with those for nonresponse. As explained in Chapter 3, Mode-Specific Measurement Effects, measurement error is the result of deficiencies in the cognitive answering process, and these relate to motivation, ability, the survey topics and, importantly, the survey mode. ASD may differentiate the mode strategy based on the interaction between mode features, respondent features, and questionnaire features. For example, when a survey is cognitively demanding, an ASD may assign interviewer modes to those less able to answer the questions themselves. Another example is when survey topics are sensitive and an ASD assigns self-administered modes to sample units for which the topic may be more sensitive than for others.

The application of auxiliary data in ASD addressing measurement error is, however, different from ASD addressing nonresponse. Unlike nonresponse, measurement error cannot be adjusted for by means of weighting. Since ASD is essentially a form of adjustment by design, ASD aiming at measurement error use different quality metrics and, hence, different optimization strategies. As will be discussed in the next subsection, quality indicators are not aimed at a balanced response with respect to some set of auxiliary variables, but at predicting and minimizing the probability that measurement error occurs and at minimizing the size of the error when it occurs. Auxiliary variables are, thus, used to estimate propensities to display measurement error and the magnitude of such errors. As a consequence, it is not sufficient to just focus on explaining the key survey variables; relevant auxiliary variables for measurement error must also relate to the prevalence and magnitude of measurement errors.

Another difference to nonresponse is the multidimensionality of measurement error. Each survey question may have its own measurement error. Minimizing the probability and size of measurement error, thus, becomes a multidimensional problem. For example, measurement error caused by inability to perform a task may occur for different survey items than measurement error caused by a lack of motivation. Since the mode strategy applies to the questionnaire as a whole, ASD decision for some survey items may be in conflict with ASD decisions for other survey items.

Consequently, auxiliary variables are relevant for measurement error when they are associated with the survey topics, and/or with intrinsic motivation to do surveys in general, and/or with ability to do surveys.

Again baseline data and paradata can be distinguished. As explained, baseline data can be demographic and socioeconomic characteristics and sometimes survey data from previous surveys or earlier waves. Educational level, age, and ethnic background are obvious predictors of the ability to do a survey. Intrinsic motivation and/or survey-specific motivation are, however, much more volatile and harder to grasp with demographic and socioeconomic characteristics. In the second case study, the LFS, where registered employment and unemployment can be linked, it will be shown that even such characteristics can play a role to understand motivation. Data from previous surveys or waves can be relevant as they may signal answer behavior that implies an increased risk of measurement error. In a study including ten different panel surveys, Bais et al. (2020) investigated consistency in answer behavior across surveys. They found, however, only weak evidence of consistent 'risky' answer behavior.

Paradata on actual answer behavior, such as audit trails and time stamps, can be indicative of measurement error (Medway and Tourangeau, 2015). However, as mentioned, in social surveys it is very unusual to re-approach respondents to verify answers or to do the interview again. Hence, such paradata cannot be used for ASD purposes. The exception is break-off, where a nonrespondent may receive a follow-up. The break-off itself, as well as all available paradata on answered survey items before break-off, are informative of potential measurement error.

From the availability of sample frame data, administrative data and paradata for actual use in ASD decisions, for example by forming subgroups strata, are not trivial. The utilization of auxiliary data depends heavily on the specified quality and cost functions and will be discussed in the next subsection. For the two examples, a brief summary is given of the data that are employed.

Auxiliary data used in the Health Survey example:

- Sample frame data: gender, age, country of birth, marital status
- Administrative data: size and type of household, personal income, household income, country of birth of parents
- Geographical area data: province, degree of urbanization
- Paradata web: binary indicator for break-off
- Paradata F2F: number and timing of visits

Auxiliary data used in the Labor Force Survey example:

- Sample frame data: gender, age, country of birth, marital status
- Administrative data: size and type of household, personal income, household income, country of birth of parents, registered employment, registered unemployment, registered educational level
- Geographical area data: province, degree of urbanization

- Paradata web: binary indicator for break-off
- Paradata telephone: number and timing of calls
- Paradata F2F: number and timing of visits

In both cases, the number of potential auxiliary variables from administrative data is much larger but are weakly related to key variables or not timely enough.

### 11.2.3 Quality and Cost Objectives in Multi-Mode Surveys

The first two steps of the checklist for ASD implementation are the identification of priorities and the identification of risks. In ASD terms, priorities should be translated to quality and cost objectives and risks to quality and cost constraints. In order to perform optimization of the design and to make decisions, the objectives and constraints need to be formalized and made explicit. Without explicit consensus on how quality and cost must be measured, there can be no proof that one survey design outperforms another survey design. For example, survey stakeholders may want to obtain a minimum number of respondents, well-balanced across a number of relevant population strata, without overrunning the budget. This means that the budget, balance of response, and size of the response need to be translated into functions in which all parameters are known or can be estimated. The parameters in such objective and constraint functions are called *survey design parameters* and well-known examples are *response propensities* and *interviewer costs per sample unit*. The quality and cost functions depend on decision variables that can be manipulated before or during data collection. Such decision variables are called *strategy allocation probabilities* in ASD, as sample units are allocated to a strategy with a certain probability that is specified beforehand. In mixed-mode ASD, these allocation probabilities amount to mode strategy allocation probabilities, i.e. the probability that a certain unit is assigned to a certain set and order of modes.

Let us start with the metrics for nonresponse. Consider the most common quality metric, the response rate. Let $\rho_g(s)$ be the response propensity of a stratum $g$ to a strategy $s$ and $p_g(s)$ be the probability that the stratum is allocated to the strategy. Furthermore, let $q_g$ be the relative size of stratum $g$ in the population. The response rate then equals

$$RR(p) = \sum_{g \in G} \sum_{s \in S} q_g p_g(s) \rho_g(s), \tag{11.1}$$

where $G$ is the set of the strata and $S$ is the set of mode strategies. The $\rho_g(s)$ need to be estimated and so are the survey design parameters. The $p_g(s)$ can be manipulated and so are the decision variables.

The response rate is an example of the so-called type 1 or covariate-based quality metric; see Wagner (2012) and Schouten, Calinescu, and Luiten (2013). These are metrics that do not depend on the survey questionnaire variables of interest and can often be written as a function of the strategy response propensities and the strategy allocation probabilities. The covariate-based quality metrics often used for nonresponse are *R-indicators* and *coefficients of variation* of the response propensities. These are defined as

$$R(p) = 1 - 2S(p), \tag{11.2}$$

$$CV(p) = \frac{S(p)}{RR(p)},$$ (11.3)

where $S^2(p) = \sum_{g \in G} q_g \left( \sum_{s \in S} p_g(s) \rho_g(s) - RR(p) \right)^2$ is the variance of the response propensities.

Type 2 or item-based quality metrics focus on survey questionnaire variables and cannot be written solely as functions of the strategy response propensities and the strategy allocation probabilities. Survey variables are unknown for nonrespondents. Consequently, this type of metric, typically, makes assumptions about the nature of the missing-data-mechanism or, alternatively, adopts a sensitivity analysis across the range of missing-data-mechanisms that are considered plausible. An example is the *fraction-of-missing-information* (FMI) for a specific survey variable, $Y$, given a set of auxiliary variables, $X$. It is defined as

$$\text{FMI}(p) = \frac{(1 - CD_R(p))(1 - RR(p))}{RR(p) + (1 - CD_R(p))(1 - RR(p))},$$ (11.4)

where $CD_R(Y \mid X)$ is the coefficient of determination of $Y$ given $X$, i.e. the fraction of variance of $Y$ explained by $X$. The FMI is different for each survey variable, so that a choice about adaptation has to be based on one or a few survey variables.

In the adaptive mixed-mode survey design, the measurement error must be included as well, as has been argued in several of the preceding chapters. Since measurement error is a feature of single survey questionnaire variables, metrics accounting for measurement typically are type 2 or item-based.

Metrics for mode effects depend on a new survey design parameter, termed the *stratum mode effects*. Let $D_{Y,g}(s; BM)$ be the difference in the estimate for a survey variable $Y$ in stratum $g$ when strategy $s$ is applied relative to the estimate obtained using a benchmark strategy $BM$. As, in general, true values are unknown, or may even be considered not to exist without specifying a data collection strategy, mode effects can only be quantified in relative sense. See also Section 2.4 of this book. The benchmark strategy is a mode strategy that either has been used historically or that is considered to be least subject to survey errors. An example of the first setting are interviewer-assisted modes when they have been the traditional mode(s) in a survey for many years. An example of the second may be face-to-face in long surveys, where motivation and concentration are challenged and interviewers are needed to motivate and stimulate.

The choice of a benchmark mode strategy is not at all trivial. Questionnaires contain many variables, some of which may benefit more from one choice of mode and others more from another choice of mode. Chapter 3, Mode-Specific Measurement Effects, discusses the measurement features in detail. A straightforward option is to derive a questionnaire profile (e.g. as in Bais et al., 2019) that summarizes the characteristics of survey items and supports an assessment of measurement risk. Here, it is assumed that a benchmark strategy has been chosen, either based on historical motives or on questionnaire expert judgment.

The mode effects $D_g(s; BM)$ can be translated to aggregate metrics in various ways. One option is the *expected mode effect per stratum*:

$$M_g(p; BM) = \frac{\sum_{s=1}^{S} p_g(s) \rho_g(s) D_g(s; BM)}{\sum_{s=1}^{S} p_g(s) \rho_g(s)},$$ (11.5)

in which mode effects per strategy are weighted by their response propensities. The stratum mode effect (11.5) represents the expected mode effect in a stratum not adjusted for nonresponse using other variables than the variables deployed for the stratification. Accounting for such nonresponse adjustment would make it difficult or even impossible to express the moded effect explicitly in terms of the strategy allocation probabilities.

Instead of the stratum moded effect, the second option is to consider the *overall expected mode effect* as an alternative option:

$$M(p; BM) = \frac{\sum_{g=1}^{G} W_g \sum_{s=1}^{S} p_g(s) \rho_g(s) D_g(s; BM)}{\sum_{g=1}^{G} W_g \sum_{s=1}^{s} p_g(s) \rho_g(s)}. \tag{11.6}$$

The stratum mode effect and the overall mode effect are nonlinear and nonconvex functions of the strategy allocation probabilities. In optimization terms, such functions are harder, and often no closed forms exist for optimal solutions. See, for example, Schouten, Calinescu, and Luiten (2013). However, when they are included as constraints, then they can be rewritten to be linear functions of the allocation probabilities. This can be done as follows: Say (11.5) or (11.6) must be smaller than some threshold, $\varepsilon$, in absolute sense, then first both sides of the inequality equation may be multiplied by the denominator of (11.5) or (11.6) and next either the left side may be subtracted from the right side or vice versa. The resulting inequality is a linear constraint.

A third option to set a metric is the *expected difference in mode effects* over pairs of strata. Say $g_1$ and $g_2$ are two strata, then the difference is defined as

$$\Delta M_{g_1, g_2}(p; BM) = M_{g_1}(p; BM) - M_{g_2}(p; BM). \tag{11.7}$$

The mode effect difference is a natural indicator when it is important to compare survey variables over strata. Again, these strata can be defined separately from strata used for the allocation of strategies. They can, for example, be the strata of analytical interest. If such differences are large, then any cross-strata comparison is hampered by nonstructural differences caused by the data collection instrument and may lead to false conclusions. For example, one may find the difference between younger and older respondents that are partly the result of mode effects and not entirely of real differences.

Finally, as an overall metric, one may look at the *expected variance of stratum mode effects*:

$$VM(p; BM) = \sum_{g=1}^{G} w_g (M_g(p; BM) - \sum_{h=1}^{G} w_h M_h(p; BM))^2. \tag{11.8}$$

The variance of mode effects may be viewed as a component in the total variance created by the mode strategies. In general, it will reduce statistical power and complicate statistical tests of differences between population strata. In other words, multivariate analyses including both stratum characteristics and multiple survey variables are affected by noise introduced by the modes.

Like the mode effect metrics (11.5) and (11.6), the differences and variances of mode effects (11.7) and (11.8), are nonlinear, nonconvex functions of the allocation probabilities, but, unlike the mode effect metrics, as constraints they cannot be rewritten to linear expressions. This feature implies that optimization problems will become more difficult to handle when the differences and variances of mode effects are included. In the numerical optimization literature, several solutions have been proposed, but these are beyond the scope of this handbook. For an introduction see, for example, Nocedal and Wright (2006).

In order to avoid item-based metrics, which lead to multi-dimensional decision problems, a totally different approach to the inclusion of measurement error are metrics about answer behavior. Behaviors such as high or low interview speed, frequent do-not-know (DK) answers or refuse-to-say (REF) answers and apparent straightlining to grid questions may be signs of the lack of respondent ability and/or motivation to provide high quality survey data. Covariate-based metrics may be built around the prevalence of such behaviors, and constraints may be imposed in optimization problems. Survey design parameters then would be stratum propensities of an interview duration below a certain threshold and stratum propensities of providing at least one DK or REF answer. Example metrics would aggregate these parameters so that constraints can be imposed. An example is Calinescu and Schouten (2015b) where administrative data on jobs were used to detect motivated underreporting in the Dutch Labor Force Survey, i.e. respondents seem to accidentally or purposively forget to report one or more jobs and do not have to answer a subsequent list of questions on these jobs.

Since metrics on answer behavior only give very indirect evidence of measurement error, the actual application, to date, has been very limited. In the context of mode strategies, the additional complication is that paradata are a prerequisite to estimate the metrics and paradata are strongly mode-specific. The interview duration, for example, is mode-dependent, so that constraints would have to be mode-dependent as well.

The counterparts to quality metrics are cost metrics. Since survey modes differ hugely in their costs per sample unit and per respondent, cost metrics are imperative in adaptive mixed-mode survey design. Let $c_g(s)$ be the costs for a sample unit in stratum $g$ for a mode strategy $s$. The expected required budget is then equal to

$$B(p) = \sum_{g \in G} \sum_{s \in S} q_g p_g(s) \rho_g(s) c_g(s). \tag{11.9}$$

These costs can be written as costs per data collection phase. If $s$ is a sequence of modes $s_1 \to s_2 \to \ldots \to s_K$, where $s_p$, $k = 1, 2, \ldots, K$ may also be concurrent sets of modes, then costs are a mix of costs per phase, response propensities per phase, and allocation probabilities per phase. Notation quickly becomes very cumbersome, so that only an example is given here. Let the mode strategy consist of a sequence of three modes, say web, telephone, and face-to-face, of which the last two are optional. Let the costs be $c_g(web)$, $c_g(tel \mid web\ NR)$ and $c_g(F2F \mid web + tel\ NR)$, where 'NR' refers to nonresponse. Let the notation for stratum response propensities and for mode allocation probabilities be analogous, then (11.8) changes to

$$B(p) = \sum_{g \in G} q_g (c_g(web) + p_g(tel \mid web\ NR)(1 - \rho_g(web))$$

$$\rho_g(tel \mid web\ NR) c_g(tel \mid web\ NR) + p_g(F2F \mid web + tel\ NR) \tag{11.10}$$

$$(1 - \rho_g(web + tel)) \rho_g(F2F \mid web + tel\ NR) c_g(F2F \mid web + tel\ NR))$$

In the two case studies in this chapter, the Dutch LFS and the Dutch HS, the choice of metrics is different.

The main survey statistic for the LFS is the monthly unemployment rate. This makes any optimization problem one-dimensional, even for item-based metrics. Since absolute levels of the unemployment rate are important, and not just month to month changes, mode effects may play a dominant role. The above-mentioned metrics (11.5) to (11.6) are closely monitored and included in any optimization problem.

Contrary to the LFS, the HS has many key statistics, e.g. general health, hospitalization, medication, and health determinants, which come from different modules in the questionnaire. Inherently, any optimization problem is multi-dimensional when including measurement error. For the HS absolute levels are of far less importance than relative change over time and between population strata. This means that metrics (11.7) and (11.8) are more important, because comparisons between strata and multivariate analyses need to be conducted.

## 11.2.4 Optimization Strategies

Step 4 of the ASD checklist is the derivation of decision rules. Decision rules are the outcome of the imposed mode strategy optimization problem that includes quality and cost metrics, as discussed in the previous subsection. The decision variables in the optimization are the stratum mode strategy allocation probabilities. The implementation of the resulting probabilities leads to decision rules that prescribe mode allocation based on the auxiliary data available at the onset of the survey and paradata that are observed during data collection.

An outcome may be, for example: adolescents are sent to a sequential mode strategy with web followed by a face-to-face follow-up; middle-aged sample units receive the same strategy but only get a face-to-face follow-up when they break-off in web; and older sample units are assigned to face-to-face directly. In this example, the strategies for adolescents and older sample units are static, whereas for middle-aged sample units it is dynamic. The allocation probabilities are all either zero or one, i.e. there is no randomization. The latter can be generalized; one may randomly assign half of the older sample units to the sequential mode strategy and half to a single mode face-to-face design. Such allocation probabilities in the interval [0,1] may be efficient in terms of variances.

Before proceeding to optimization strategies, it must be stressed again that ASD may be the result of a relatively implicit and organic process stretching out over multiple years to improve the efficiency of survey resources. The optimization strategy, thus, can be replaced by a trial-and-error approach.

Schouten and Shlomo (2017) distinguish different optimization strategies and, hence, imply different decision rules:

1. Expert knowledge: Mode allocations are based on a mix of expert knowledge and historic survey data without an explicit model and link to quality and cost metrics. An example is Luiten and Schouten (2013), where web, paper, and telephone are allocated to different subgroups based on historic data from other surveys and experience with efficacy of different mode strategies. The advantage of this approach is its relative simplicity. The disadvantage is its less predictable outcome and more subjective nature. It leans heavily on the expertise of survey designers and may lack the possibility to replicate the decisions;

2. Case prioritization: Response propensities are estimated and sorted in ascending order during data collection. The lowest response propensities are allocated first to follow-up modes. This approach will be demonstrated in the Dutch HS case

study. The advantage of this approach is the direct link to response propensities. The disadvantage is the risk of allocating effort to unsuccessful follow-up and the lack of an explicit quality metric, although it, obviously, aims at a more balanced response across characteristics that are used in response propensity estimation. An example is Van Berkel, Van der Doef, and Schouten (2020);

3. Stopping rules based on quota: Follow-up in strata is based on quota, say 50% or 60% stratum response rates. When thresholds are met, any follow-up is aborted and no more effort is spent. Implicitly, the approach attempts to obtain equal stratum response rates. The advantage is again its relative simplicity. The disadvantage is the unpredictable fieldwork effort. This approach is simulated by Lundquist and Särndal (2013) for the Swedish EU-SILC;

4. Mathematical optimization: The most advanced but also most demanding approach is to formulate an explicit optimization problem in which mode allocation probabilities are decision variables. The optimization problem chooses a quality or cost metric as an objective function and optimizes this function subject to constraints on other metrics. The advantage of the approach is it transparency and link to metrics. The disadvantage is the requirement to estimate all components in quality and cost functions. This approach is demonstrated in the Dutch LFS case study;

A choice between the optimization strategies depends on the length of data collection periods, the operational flexibility of data collection staff and systems, the availability and richness of historic survey data, and the speed at which survey design parameters change. The less flexible processes are or the shorter the data collection period is, the stronger the need for a simple set of decision rules. Furthermore, the weaker the historic survey data, the less one will rely on estimated survey design parameters. Finally, the faster the change in survey design parameters, the less one would build decision rules on estimated survey design parameters that may be outdated.

Schouten, Peytchev, and Wagner (2017, Chapter 7) discuss optimization of adaptive survey designs in various settings. One extreme setting is repeated, large scale surveys by national institutes with access to rich administrative data. The continuous availability of rich auxiliary data and rich historic survey data lends itself to mathematical optimization problems of strategies with a relatively short data collection period. Mode strategies can still be sequential, but decisions have to be made quickly and through automated decisions rules. Changes to designs are made between waves of a survey and not during the wave of a survey. The other extreme setting is one-time-only surveys by statistical institutes with little access to administrative data. In such a setting, data collection periods tend to be longer, in order to allow for learning during data collection and to let modes reach their full potential, called phase capacity.

The two case studies of this chapter are set in the context of relatively rich administrative data and paradata. Decision rules are based on estimated survey design parameters that can be updated continuously.

## 11.3 Case Studies

We illustrate the various steps of ASD for the two case studies, the Dutch Labor Force Survey (LFS) and the Dutch Health Survey (HS), described in Section 2.6.

### 11.3.1 The Dutch LFS

At the time of writing at Statistics Netherlands, a redesign of the LFS is prepared in which adaptation is also considered. Since 2012, the first wave of the Dutch LFS uses a mixed-mode design with web, CATI and CAPI. Web is offered as the first mode and web non-respondents are sent either to CATI or to CAPI. Households for which a phone number is available and that have at most four registered persons between 15 and 65 years are allocated to CATI. All other households go to CAPI. Hence, the LFS has a simple form of adaptation based on household size. In the redesign, more variables are included and stratification will be more advanced. The design features of interest are CATI and CAPI follow-up.

The Dutch LFS is a monthly household survey. It has a rotating panel design with five waves with time lags of three months. The second to fifth wave use only CATI interviews based on phone numbers provided in the first wave. We consider only the first wave in this example. The sampling design of the LFS is a stratified simple random sample without replacement of addresses. Addresses with persons registered between 15 and 25 years, with persons registered unemployed and persons registered as non-western non-native are oversampled, while addresses with persons registered as 65 years and older are undersampled.

In 2012 and 2013, extensive research was conducted into an adaptive survey design for the LFS (Calinescu and Schouten, 2013). Since the main survey statistic of the LFS is the unemployment rate, the research was focussed on this statistic. At the end of the study, an adaptive survey design was recommended, but implementation was postponed due to demands on logistics and case management. We describe again the five steps as they were taken in the study: priorities, risks, indicators, decision rules, and monitoring.

The main priorities of the Dutch LFS are high precision of estimates, high response propensities across all relevant population subgroups, low bias towards true population values, and a specified budget. LFS statistics is published monthly for six age and gender groups and annually for municipalities. Required precision is very high, around 0.5% on monthly statistics. Since statistics is used for various policy reasons, a high coverage of the population is imperative, i.e. high overall response propensities are requested. Both absolute values and changes in the time of LFS statistics are deemed very important. With the high precision in mind, these priorities demand a very high accuracy (mean square error) and comparability. Mode effects need to be avoided at all times and need to be quantified if they are conjectured to exist.

Given these priorities, the LFS faces the risk of mode effects in time series due to mode-specific selection and/or measurement biases. Furthermore, when such biases occur, the comparability of age and gender groups and of regional estimates are also at stake. As for the Health Survey in Section 3.2, the LFS may suffer from varying CATI and CAPI workloads due to variation in the web response. Unlike the Health Survey, the LFS produces monthly statistics and a problem in a particular month may immediately affect statistics.

In the 2012 and 2013 study, the main quality indicator was the absolute weighted mode effect of the unemployment rate relative to a single mode CAPI LFS. This means that the weighted unemployment rate was estimated for a candidate design and compared with the CAPI estimate. This was possible as various parallel runs had been conducted in 2010 and 2011. As a first constraint, a maximum was set to the overall budget. Another constraint concerned precision; lower limits were set to the numbers of respondents per age and gender group. As a third constraint, in order to ensure comparability between subgroups, absolute mode effects between relevant subgroups were allowed to differ at a specified value, say 1%. This was operationalized as follows: For a candidate design, the subgroup unemployment rates were estimated and the corresponding CAPI subgroup estimates were subtracted. The resulting differences were forced to be constant over subgroups with a margin of 1%, e.g. all subgroups show a downward effect between 0% and 1%. As a last constraint, the sample size was given an upper limit in order not to deplete Statistics Netherlands' sampling frame. Relevant subgroups were formed based on a mix of age, gender, household size, ethnicity, and registered unemployment. A regression tree was conducted to predict unemployment using these five variables, leading to nine strata. For the nine strata, response propensities, costs per sample unit, and mode effects relative to CAPI were estimated for a range of designs with or without CATI and CAPI.

Table 11.1 shows the estimated weighted mode effects relative to CAPI. For example, for stratum 1 there is an estimated 1.5% higher unemployment rate in web relative CAPI and for stratum 5 there is a 4.5% lower unemployment rate. Since CAPI is the benchmark, obviously, mode effects of CAPI are 0%. In general, the web-CAPI sequential strategy is closest to CAPI, which is not surprising as about half of the answers are CAPI.

The adaptive survey design was optimized by formulating it as a mathematical optimization problem and then by searching the global minimum in weighted mode effect towards CAPI through numeric optimization routines. As the optimization problem is nonlinear and nonconvex due to the absolute signs and the constraint on the subgroup modes effects, there is no guarantee of a global optimum. For this reason, several starting values were used in the numeric routines. One way to create starting values was to let them correspond to uniform designs with different modes. Another way to find starting values was to first solve simpler optimization problems in which one of the constraints was omitted. In the study of the budget, the subgroup mode effect and sample size upper limits were varied in order to explore robustness of optimal solutions. Table 11.2 shows the solution to the optimization problem where the budget was set at the original level, the subgroup mode effect at 1%, and the sample size at 10,000. For three out of the nine strata, web-CATI is the strategy that is allocated (the allocation probabilities are equal to one). All other strata get a mix of strategy. CATI, CAPI, web-CATI, and web-CAPI are allocated. The resulting design has an exotic mix of strategies. In subsequent optimizations, the solution space was restricted. For example, by forcing all allocation probabilities to be either 0 or 1.

The strata with a mix of strategies are subject to randomization. For example, for stratum 1 each sample unit has a probability of 40% to be assigned to CAPI and 60% to be

**TABLE 11.1**

Estimated Mode Effects for Nine Strata Based on Age, Ethnicity, Household Size, and Registered Unemployment Relative to CAPI for a Range of Designs

|          | *Stratum* | | | | | | | | |
|----------|------|------|------|------|------|------|------|------|------|
|          | *1*  | *2*  | *3*  | *4*  | *5*  | *6*  | *7*  | *8*  | *9*  |
| Web      | 1,5  | 0,0  | −2,3 | −4,5* | 0,9 | −0,4 | −2,2* | 0,6 | −0,4 |
|          | (1.3)| (0.0)| (1.4)| (1.8)| (1.9)| (1.3)| (0.8)| (1.0)| (0.9)|
| CATI     | −0,1 | −0,1 | −2,3* | −4,9* | −0,6 | −1,0 | −0,8 | −0,2 | −1,2 |
|          | (0.7)| (0.2)| (0.9)| (1.2)| (1.1)| (0.9)| (1.2)| (1.4)| (0.6)|
| CAPI     | 0,0  | 0,0  | 0,0  | 0,0  | 0,0  | 0,0  | 0,0  | 0,0  | 0,0  |
|          | (0.0)| (0.0)| (0.0)| (0.0)| (0.0)| (0.0)| (0.0)| (0.0)| (0.0)|
| web-CATI | 0,9  | −0,1 | −3,7* | −1,7 | 0,5 | −0,7 | −3,0 | 0,6 | −0,4 |
|          | (1.2)| (0.0)| (1.4)| (2.3)| (1.8)| (1.3)| (1.6)| (2.0)| (0.9)|
| web-CAPI | 0,9  | 0,0  | −1,2 | −2,0 | 0,6 | −0,3 | −1,2 | 0,4 | −0,2 |
|          | (1.3)| (0.0)| (1.4)| (1.8)| (1.9)| (1.3)| (1.6)| (2.0)| (0.9)|

Standard errors between brackets. Asterisks denote estimated mode effects significant at the 1% level.

**TABLE 11.2**

Optimal Stratum Allocation Probabilities for the LFS Adaptive Survey Design

| | Stratum | | | | | | | | |
|---|---|---|---|---|---|---|---|---|---|
| | *1* | *2* | *3* | *4* | *5* | *6* | *7* | *8* | *9* |
| Web | | | | | | | | | |
| CATI | | | 0.3 | 0.3 | 0.5 | | | | |
| CAPI | 0.4 | | 0.7 | 0.7 | | | 0.7 | | |
| Web-CATI | | 1.0 | | | 0.5 | 0.4 | | 1.0 | 1.0 |
| Web-CAPI | 0.6 | | | | | 0.6 | 0.3 | | |

assigned to web-CAPI. In order to fix workloads, this sampling can be done with a fixed size, i.e. when stratum 1 consists of 1000 persons, say, then a simple random sample without replacement of 400 is drawn.

In monitoring the LFS, a dashboard has been created that considers response and costs but also survey estimates for various auxiliary variables after web, after web plus CATI and after web plus CATI plus CAPI.

## 11.3.2 The Dutch HS

Since 2018, the HS employs a static adaptive survey design in which CAPI follow-up to online nonresponse is varied. We describe the various steps in the HS design.

The HS is a repeated, cross-sectional survey with monthly samples, but statistics is produced on an annual basis only. Main survey estimates closely resemble the EHIS, but the HS is shorter and less elaborate and there are subtle differences in definitions and wording of questions. The monthly samples follow a simple random sampling without replacement design and vary in size between 1000 and 1500 persons. The HS employs two modes in sequence, web followed by CAPI. All sample units are invited to participate online through a paper invitation letter containing a web link. Two paper reminders are sent to those that do not respond. After four weeks, online participation is no longer possible and nonrespondents can be forwarded to CAPI. CAPI interviews start on the first day of the month after the next month, e.g. if a person does not respond online in January then he/ she is approached by CAPI beginning of March. The target population consists of all registered persons except those in institutions. For persons below 16 years, parental consent is required. For persons below 12 years, parents act as proxy respondents.

The design feature of interest in the adaptive survey design is the CAPI follow-up. A preselected part of the online nonrespondents receives a follow-up by CAPI. We now follow the first five steps of Section 2.3: priorities, risks, indicators, decision rules, and monitoring.

The main priorities in the HS are acceptable and similar response rates among relevant population subgroups, sufficient precision on annual survey estimates, and costs satisfying a specified budget. In 2010, Statistics Netherlands decided to cut budget and a single mode CAPI HS was no longer feasible. In 2011, web was introduced as a sequential mode, preceding CAPI. Over all Statistics Netherlands' surveys, CAPI workload has decreased over the last ten years due to the introduction of web, leading to higher travel costs per sample unit. The increased costs raised the question whether a CAPI interviewer force, covering the whole country, is sustainable on the longer run. The CAPI discussion formed one of the main starting points for the consideration of adaptive survey designs. Since CAPI is relatively expensive, it was deemed imperative to weigh the benefits more carefully against the costs. The HS is the first survey for which adaptation has been introduced.

Three risks have been put forward. The first risk is that of incomparability in time, i.e. time series breaks in survey estimates. These breaks can be caused by a change of mode choice and mode allocation. Since the HS focusses on time change in survey estimates rather than on absolute levels of the estimates, such breaks can be devastating to the utility of the survey. The second risk is an unpredictable CAPI workload due to varying monthly and annual web response rates. This risk was not identified at first but followed after implementation. It is, therefore, a good example of the plan-do-check-act cycle in which it turned out that structural variation in response and costs was too volatile. The third risk concerns incomparability between different population subgroups of interest. The shares of web response and CAPI response vary, for example, for age groups, ethnicity groups, and income groups. When mode impacts measurement error, then differences in survey estimates between subgroups may (partially) be the result of mode effects.

At the start of the HS redesign, a range of quality and cost constraints were imposed. All quality indicators were directed at nonresponse and precision. Mode-specific measurement error was not considered for the first migration to an adaptive survey design but remains a subject of discussion and a risk. As the main quality objective, the coefficient of variation (CV) of response propensities was chosen; see, e.g. Schouten, Cobben, and Bethlehem (2009). The CV is defined as the ratio of the estimated design-weighted standard deviation of response propensities over the design-weighted response rate. The CV has an indirect relation to nonresponse bias (Schouten, Cobben, and Bethlehem, 2009); the smaller the CV, the better. A smaller CV may be achieved by a higher response rate and/or a smaller variation in response propensities. Importantly, the CV depends on the set of auxiliary variables that are used to estimate response propensities. As a precision constraint, a minimum total number of about 9500 respondents was requested. A second constraint was set on costs. An upper limit of 8000 was imposed to the number of non-respondents that are sent to CAPI, as a proxy for a budget constraint. This number was roughly 10,000 in the original design. A third practical upper limit of 18,000 persons was set to the sample size in order not to deplete the Dutch sampling frame. As mentioned, the CV depends on the stratification of the population. The exact stratification was created by running a classification tree on the 2017 historic HS online response using age, ethnicity, urbanization, and income as explanatory variables. Nine strata were formed. They are shown in Figure 11.4 and are taken from Van Berkel, Van der Doef, and Schouten (2020). Since 2017, the classification trees have been recreated for each new calendar year of the HS. The stratification has changed slightly over these years.

The next step is the set of decision rules or optimization approach. A mathematical optimization problem was formulated with the CV as an objective function and subject to the three constraints on response size, CAPI workload, and sample size. The nine population strata of Figure 11.4 were sorted based on their web response propensities. Subsequently, the expected CV was estimated while moving the strata to CAPI one by one starting with the stratum with the lowest propensity. Doing so, a range of CVs was estimated and the best allocation was selected. Since in real data collection the web response rates vary and the CAPI workload is fixed, the allocation of strata to CAPI varies slightly from one month to the other. The resulting adaptive survey design has an average CV of 0.116, whereas the original uniform design had a CV of 0.158. In the adaptive survey design, nonrespondents in some subgroups are never allocated to CAPI and are fully observed through web, while nonrespondents in others are always allocated to CAPI. There are also subgroups that are only partly allocated to CAPI. Web nonrespondents in these subgroups are randomly sub-sampled with equal inclusion probabilities. It must be noted, however, that the optimization does not account for measurement bias differences between web and CAPI.

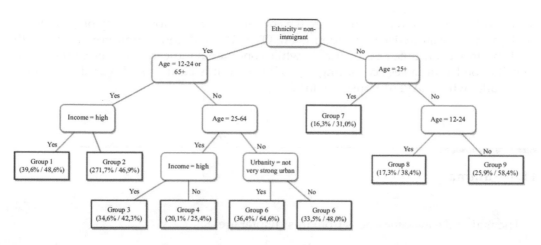

**FIGURE 11.4**
Classification tree for online response in the HS. The nodes also give the web-CAPI response rates when the whole stratum nonresponse is allocated to CAPI.

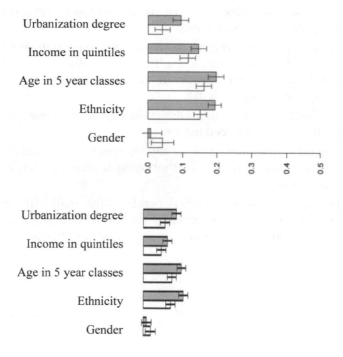

**FIGURE 11.5**
Variable-level partial CV for web and web plus CAPI for the Dutch Health Survey of 2015 for five auxiliary variables. The grey and white bars represent unconditional and conditional values, respectively. 95% confidence intervals based on normal approximation are provided.

Finally, a dashboard has been developed using a Shiny app in R in which the response and costs of monthly data collection can be monitored at any given time per mode and per subgroup. At the end of each month, the web subgroup response rates are computed and form the basis for allocation of the strata to CAPI one month later. Figure 11.5 shows the so-called unconditional and conditional variable-level partial CVs for web and for web plus CAPI. Variable-level CVs measure the contribution to the total CV by one variable in analogy to a variance decomposition in ANOVA. Unconditional indicators measure the total variance contributed by a variable, while conditional indicators represent the unique contribution by variables accounting for collinearity. It is clear that all variable-level CVs are smaller when CAPI response is added.

## 11.4 Summary

The main takeaway messages from this chapter are:

- Adaptive mixed-mode survey designs allocate different mode strategies to different sample subgroups or strata based on explicit quality-cost functions through an optimization strategy.
- The main decision variables in adaptive mixed-mode survey design are mode strategy allocation probabilities.
- Sample strata can be formed based on linked auxiliary data available at the start of the survey data collection, leading to static adaptive survey design, and/or based on paradata coming in during data collection, leading to dynamic adaptive survey design.
- Quality and cost metrics should address both nonresponse and measurement error differences between modes.
- Quality metrics accounting for modes-specific measurement error depend on the survey variable, so that optimization problems tend to be multi-dimensional.
- The choice of optimization strategy depends on the availability of auxiliary data, the availability of historic survey data, the flexibility of data collection staff and systems, and the length of the data collection period.

# 12

## The Future of Mixed-Mode Surveys

### 12.1 Introduction

This final chapter takes a more general look at mixed-mode surveys and how they may be designed and analyzed in the near and more distant future. Mixed-mode surveys have changed drastically in the last decades and are very likely to undergo further changes. Information and communication technology innovations are developing at a rapid speed. Also, sensor technology and smart devices have been introduced at a large scale in daily life. Finally, there are various types of data generated for different purposes that may be donated or linked by survey respondents. What are the current developments and innovations in survey modes, and more generally, surveys originating from such innovations, and will they affect how mixed-mode surveys are designed and/or analyzed? Can survey methodologists still keep track of these changes in the survey climate or should they already anticipate different types of surveys? In the following subsections, these questions are discussed with an official statistics' setting in mind.

### 12.2 What Are the Current Developments?

To discuss relevant developments for mixed-mode surveys, a broader view of the purpose and implementation of surveys is needed. Surveys have two defining characteristics: a population sample and a questionnaire. A survey starts from the need for a coherent set of population parameters that allows for identification of unknown associations and/or evaluation of known associations in a relevant social, political, or research area. The information is, however, not yet available and needs to be collected. The information is retrieved through a series of questions. To reduce costs and response burden, to improve timeliness, and to avoid complex logistics, a subset of the population is invited. Since population units are not aware of the survey, communication channels or modes are chosen to facilitate contact and to enable data entry. To date, many general population surveys are still based on choices made several years ago. While the need for coherent information is still there, today none of the subsequent choices are as obvious as they were in the past. First, there is a lot more information available, as governments and other entities have massively started to move to digital archiving of information. Second, questions are not the only way to collect data, and sensors have become ubiquitous in society both in private devices and in public systems. Third, there have been big shifts in how persons communicate and how they contact each other. Finally, given the availability of more data and the option to

DOI: 10.1201/9780429461156-12

measure/sense, the logic of what modes to use for data entry has changed as well. Because of these developments, the cost, burden, timeliness, and logistics context for using samples have changed. In sum, the options for surveys have widened and diversified and in the slipstream also the motivation for using certain mixes of modes and devices.

Various developments can be linked to the observed wider range of data, devices, and types of measurements.

When restricting measurements to asking questions, the diversification of online devices in terms of screen size and navigation is a challenging questionnaire design, especially for general population official surveys. Such surveys were traditionally administered by interviewers and tend to be relatively long. The still increasing use and population coverage of mobile devices is urging designers to rethink the length and format of questionnaires when they mix the web mode with other modes.

The growing diversity of modes is an incentive to the implementation of adaptive survey designs; they imply that the number of 'levers' is growing. The ever-decreasing response rates, and, consequently, growing costs per respondent give an even stronger push towards designs that efficiently allocate budget. To date, adaptive survey designs still focus mostly on interviewer effort, but the large cost-quality differential of modes inevitably makes the survey mode one of the prime design features for differentiation.

The greater diversity in modes goes hand in hand with more individual communication. Telephones have become individual devices. Persons can go online anytime, anywhere on their own devices. And persons are creating their own private data spaces. This stronger individualism in communication stimulates personalized applications and measurements. It paves the way for more intimate data collection through mobile device apps and sensor measurements.

The greater presence of all kinds of data is questioning the reliance on self-instigated survey measurements alone. An important development is the inclusion of existing data in surveys, either at the estimation stage or at the data collection stage. In the estimation stage, existing data may be used in time series models or small area estimations. In the data collection stage, respondents may be asked to donate existing personal data by consenting to automated linkage or by linking and aggregating data themselves. Such data flows may be viewed as additional modes and also affect the choice of modes as some modes lend themselves better to data donation than others.

In Chapter 10, Mixed-Device Surveys, the term 'smart surveys' was introduced. Smart surveys employ one or more features supported by smart devices. They are a direct response to the wider availability of data, both public and respondent-owned, and the growing presence of sensors, either in devices themselves or in sensor systems nearby. They are a form of hybrid data collection that mixes modes, supplements questions by measurements, and combines existing data and new data. Doing so they are a fusion between primary data collection and secondary data collection. Smart surveys are, however, a recent development that still needs to prove a strong business case at the scale of general population surveys. If they do, they will strongly affect mixed-mode surveys.

Moving to smart devices, communication and data entry are changing. Smart devices can make pictures and record sound. This means answers to survey questions can be recorded by making pictures and applying computer vision, such as text recognition or image extraction. Survey question answering may also be facilitated by speech to text, possibly moderated through a chatbot, smart speaker, or voice assistant. While these developments are predominantly related to the web mode, they mimic interviewer assistance and may be used by interviewers directly as well.

All in all, the current developments point at both greater diversity and a stronger potential to tailor mixed-mode surveys. So what are the most pressing open research areas in mixed-mode surveys to keep up with these developments?

## 12.3 What Are the Open Areas in the Methodology of Mixed-Mode Surveys

There are at least six open areas in the methodology of mixed-mode surveys:

Effects of Devices on Measurement: The screen size and type of navigation of online devices have become more diverse. There is no strong evidence in the literature that the type of device has an impact on measurement error. However, to date there are few empirical studies in the context of general population official surveys. It is this type of survey that is often less flexible in adapting and revising questionnaires because of long statistical time series and an interviewer-assisted tradition. Official statistics lean heavily on comparability in time and comparability between relevant population subgroups. Major revisions of questionnaires to make them fit for smaller screens and different types of navigation are not attractive from a comparability point of view. Official statistics also legitimately aim for a high and balanced response which implies that interviewer assistance may be warranted, making it more natural to stick to longer questionnaires. Research is needed to understand device effects in general population surveys that are less flexible in adapting. This research is complicated by the confounding of device selection and device measurement, which leads to the second open area.

Estimating and Adjusting for Mode-Specific Measurement Effects: To understand and remedy mode-specific measurement effects, it is imperative to disentangle selection from measurement. In the Dutch Health Survey, for example, the difference in the proportion smoking between web and face-to-face is close to 10%. This large gap is likely to be a result of both selection and measurement differences. The actual size of the measurement difference is, however, important in comparing subgroups and in making decisions in adaptive mode follow-up. Over the last ten years, various scholars have ventured into the decomposition of mode effects. To date, in general population surveys these methods and tactics are still rarely being used; even if terms of mean square error, it is advantageous to do so. More research is needed to understand in what circumstances assumptions underlying mode effect estimation methods are valid, for example, when is it reasonable to assume that a re-interview does not strongly affect response and answering behavior. This research cannot be seen separately from the third research area on time series modelling.

Time Series Models to Strengthen Comparability in Time: As mentioned, general population official surveys tend to be repeated and to produce statistics in a historic context. These time series can borrow strength from other time series, especially when they are combined or co-integrated with time series on administrative data or big data. The COVID-19 pandemic in 2020 pointed very clearly at the importance of understanding the contribution of different modes to the overall statistics that are published. Face-to-face interviewing was temporarily stopped and necessitated the adjustment of statistics for the resulting loss in response. The only options were to consider time series in which absence of face-to-face was simulated and to consider other time series auxiliary to the survey. Structural time series models including trends, seasonal components, and interventions are needed to do so, and they may also be used outside extreme events such as COVID-19. More research in this area is needed. The loss of face-to-face assistance also urges survey designers to think of alternatives that to some extent mimic face-to-face. One option that was tried was video conference interviewing and this brings us to the fourth open area.

Reducing Mode Measurement Differences Through Chatbots and Other Interactive Features: Mode features impact answer behavior, and survey designers have attempted to transfer features from one mode to another. Examples were mentioned in Chapter 3 on Mode-Specific Measurement Effects, such as showcards. Artificial intelligence offers new possibilities to mimic interviewer assistance in self-administered surveys. Chatbots may be designed to understand and reply to questions respondents may have before or during the survey. They can be trained to understand a range of questions about terminology and definitions in survey questions. They may also answer more general questions on the purpose and use of the survey. Such interaction may be text-only, but may also be voice-to-text. Apart from the option to interact, the format in which questions are presented on the screen may be copied from contemporary applications as a chat. There are, however, clear risks that such 'gamification'-like features affect answer behavior, which must be investigated. A more interactive design of self-administered surveys also brings into question whether the length of a survey should be adjusted or not, which is the fifth open area.

Split Questionnaire Design: Official surveys tend to be relatively long, up to 45 minutes and longer. The longer format again results from a tradition of a mix of face-to-face interviewing and paper questionnaires in which longer interviews are less problematic in terms of measurement. Longer surveys may not be suited for communication channels that are normally used for relatively short interactions such as telephone and mobile devices. This means that surveys need to be shortened with the obvious consequence of fewer questions and less information being collected. To overcome the loss of information, several decades ago scholars already proposed planned missing designs in which questionnaires are split and different parts are randomly allocated to different respondents. Also a modular or alternating approach in which certain themes or block of questions are included in a rotating design, i.e. included every other year or month, has been proposed as an option to reduce questionnaire length. These modular and split questionnaire designs have never gained much popularity due to complex trade-offs in costs, response rates, and context effects. The rise of mobile devices, however, lifted the attention for questionnaire length. The trade-off in split questionnaire or modular designs is a complicated one that evaluates higher response and completion rates due to shorter surveys, larger required sample sizes, smaller measurement error due to a smaller respondent burden, and larger context and order effects due to varying allocations of question modules. The main gain may need to come from improvement in measurement as higher response rates will likely not outweigh the costs of larger samples. Research is needed to understand the trade-off in the context of multi-device and multi-mode surveys. The increased attention for multiple devices brings us to the last open research area of smart surveys.

Smart Surveys: Smart surveys are the broadest and most far reaching of the research areas mentioned here. They imply a different view at surveys; respondents still play a central role in collecting and submitting information, but data become a mix of survey questions, sensor measurements, and existing data being donated. Such hybrid forms of data collection even imply a renewed look at the concepts of interest and how they are operationalized in the survey. The potential of smart surveys is described in Chapter 11, Mixed-Device Surveys, in greater detail and is not repeated here. The main open research questions are in easy-to-use user interfaces, effective recruitment and motivation strategies, trade-offs in active and passive data collection, and estimation strategies integrating different modes and different types of data that measure variables or indicators of common social phenomena to produce valid statistics.

## 12.4 Open Areas in Logistics and Implementation of Mixed-Mode Surveys

Mixed-mode surveys are much more challenging in logistics and operations. The following open areas exist in implementing mixed-mode surveys:

Mixed-Mode Case Management: As discussed in Chapters 2, Designing Mixed-Mode Surveys, and 5, Mixed-Mode Data Collection Design, mixed-mode surveys mix modes concurrently and/or sequentially. This means survey response may come in at different time points and through different channels. Typically, modes have their own case management systems that need to be integrated into an overarching system. Identification of units across modes, arranging of reminders and call-backs, booking of intermediate and final case codes, and integrating data into an overall database are all tasks that can go wrong when combining modes. One example is the mail survey mode. Paper questionnaires can be sent in at any time, even after a mode switch took place. This means interviewers may contact a sample unit that has submitted a questionnaire. Another example is the telephone mode where telephone numbers may turn out to be no longer in use or false and a switch to another mode is needed. Mixed-mode case management is probably the biggest logistical challenge in mixed-mode surveys.

Interviewer Workloads: A special example of mixed-mode case management is the allocation of workload of face-to-face interviewers. Face-to-face often is sequential to other modes for cost reasons. The implication is that the interviewer workloads become highly unpredictable. Sampling variation affects both sizes of samples remaining after other modes as well as their topographical spread across interviewer areas. Given that face-to-face interviewers work in specific regions of a country, it is attractive to cluster geographically through multi-stage sampling designs. However, such clustering is much harder to control when other modes precede face-to-face. It is an open research area how to efficiently draw samples under this uncertainty.

Mixed-Mode Monitoring Dashboards: Closely related to case management systems are monitoring dashboards. As was explained in Chapter 2, Designing Mixed-Mode Surveys, dashboards play two roles in the survey plan-do-check-act cycle. They play a role during data collection in which unexpected events or incidents may require change of plans and follow-up actions. They also play a role in redesigning a survey in time across waves. Since mixed-mode implies that responses and nonresponses may come in through different channels and at different points in time, careful designing of dashboards is needed. Furthermore, different modes have different kinds of paradata; in web one may monitor break-off rates, while in telephone and face-to-face one may monitor timing and numbers of calls and visits. Dashboards need to be flexible to handle different orders and sequences of modes and different types of paradata. The most challenging survey error in monitoring is measurement error. In single mode surveys, measurement error is rarely monitored explicitly. In mixed-mode surveys, lack of complete comparability in time and between population subgroups can be a real threat due to mode-specific measurement biases. Monitoring differences in answering behavior across modes is complicated, however, by the variability in paradata. It still is an open area how to monitor answer behavior across modes.

Push-to-Web and Push-to-App Implementation: Survey designers that prefer response through the web survey mode and/or through personal mobile devices in mixed-mode settings may adopt push-to-web or push-to-app strategies. Such strategies have to face the wide range and constant and rapid development of different operating systems, browsers, screen sizes, and other technical specifications of devices. Telephone, face-to-face, and paper modes do not have the variety and changing environment of the online mode. The implication is that constant maintenance and updating is needed and both front-office and back-office may quickly become outdated without constant care. The heavy maintenance poses questions as to how to make trade-offs in what is supported or stimulated and what is discouraged or even blocked. Survey institutes have yet to find the right balance and it is likely that they will be challenged for the coming years in ways that are hard to predict.

Smart Survey Data Processing: The last open area is linked to smart surveys. The mix of different types and sources of data may be viewed as an extreme form of mixing modes. However, other than survey modes, sensor data and forms of existing data donated to a survey have completely different formats and definitions. Handling of such data, i.e. monitoring, editing, and integrating must be tailored. This means case management systems and monitoring dashboards need to be even more diverse and flexible. Also in smart surveys, survey designers may choose to actively involve respondents, meaning that data is shown to respondents for validation. Such respondent feedback is unprecedented in surveys and poses new challenges and trade-offs.

## 12.5 Will There Be Mixed-Mode Surveys in the Future?

Without doubt, surveys will remain very important tools in policy making in official statistics. Surveys provide coherence and specificity in collected information and control over definitions and implementation. Given the increasing diversity in communication channels, there is also no doubt that surveys will keep using different modes to contact target populations and to administer data collection. But what will these surveys look like?

A clear trend that affects surveys in general, and mixed-mode surveys in particular, is the steady, gradual decline in response rates. It is not possible to attribute this trend to willingness to participate in surveys alone as declining response rates have led to increasing costs per respondent which in turn have led to cost reductions through different mixes of modes. In other words, there is an interaction between the type of survey design and the response rates. Nonetheless, the prospect for future survey response rates is relatively grim. This prospect has already been forcing survey designers to tailor and adapt design features, and the survey mode is the most powerful of features. It is likely that a range of communication channels will remain to be needed in order to attract and recruit a diverse set of respondents.

While face-to-face interviewing is perhaps the one mode that has been relatively resistant to changes in response rates, it is questionable whether it can be a dominant mode in the longer run. This is not because it is outdated as a communication channel, but because it becomes less and less attractive from the cost perspective. To date, face-to-face often is a mode that is added in sequence to cheaper modes. This means that the relatively hard cases are sent to interviewers, leading to lower interviewer response rates and thus higher costs per respondent. Furthermore, due to the earlier mentioned unpredictable topographical variation in interviewer workloads, clustering of addresses within interviewers is complicated and face-to-face travel times may increase. These two consequences of mixed-mode surveys make face-to-face even more expensive and may be the start of a downward spiraling of face-to-face: When face-to-face becomes more expensive, then it may only be affordable for certain surveys or certain underrepresented population subgroups. This then leads to a smaller number of interviewers as workloads get smaller. Smaller numbers of interviewers lead to larger interviewer travel distances and travel costs, which in turn make face-to-face again more expensive. At Statistics Netherlands, this downward trend in face-to-face is already visible, despite being the most powerful mode still to recruit respondents.

The online mode is likely to undergo further changes as new and more powerful devices may emerge. Currently, the small screen sizes of mobile devices are still a barrier and a challenge in designing surveys and user-friendly user interfaces. It is foreseeable, however, that the drawback of small screens may be overcome by clever design, new navigation options or speech-to-text, and other forms of additional communication. If so, these developments will force questionnaire designers to rethink how they construct question-answer processes. Although survey general populations have become more and more online populations, there is likely to remain some digital divide that forces the same designers to align the potential of devices with more traditional modes.

Finally, it seems a natural avenue that surveys will more and more make use of existing data and of other means of collecting data besides asking questions. Existing data may be added in the estimation stage to strengthen time series. Existing data may also be added by respondents themselves as a form of data donation. Other forms of measurements, in particular data coming from mobile device sensors and other sensor systems, offer powerful options to further enrich survey data. Doing so, surveys move towards big data and big data move to surveys, creating hybrid forms of data collection in which the respondent still is a key person.

# References

AAPOR (2010). *New considerations for survey researchers when planning and conducting RDD telephone surveys in the U.S. with respondents reached via cell phone numbers.* AAPOR Cell Phone Task Force. Retrieved from 2010aaporcellphonetfreport.pdf

AAPOR (2016). *Address-based sampling.* AAPOR Task Force. Retrieved from https://www.aapor.org/Education-Resources/Reports/Address-based-Sampling.aspx

AAPOR (2017). *The future of U.S. general population telephone survey research.* AAPOR Task Force. Retrieved from https://www.aapor.org/Education-Resources/Reports/The-Future-Of-U-S-General-Population-Telephone-Sur.aspx

Aichholzer, J. (2013). Intra-individual variation of extreme response style in mixed-mode panel studies. *Social Science Research,* 42, 957–970.

Alwin, D.F. (2007). *Margins of error: A study of reliability in survey measurement.* New York: Wiley.

Alwin, D.F., and Krosnick, J.A. (1991). The reliability of survey attitude measurement. The influence of question and respondent attributes. *Sociological Methods and Research,* 20, 139–181.

Andrews, F.M., and Herzog, A.R. (1986). The quality of survey data as related to age of respondent. *Journal of the American Statistical Association,* 81(394), 403–410.

Antoun, C., Couper, M.P., and Conrad, F.G. (2017a). Effects of mobile versus PC Web on survey response quality: A crossover experiment in a probability web panel. *Public Opinion Quarterly,* 81, 280–306.

Antoun, C., Katz, J., Argueta, J., and Wang, L. (2017b). Design heuristics for effective smartphone questionnaires. *Social Science Computer Review,* 36(5), 1–18.

Bachman, J.G., and O'Malley, P.M. (1984a). Black–white differences in self-esteem: Are they affected by response styles? *American Journal of Sociology,* 90, 624–639.

Bachman, J.G., and O'Malley, P.M. (1984b). Yea-saying, nay-saying, and going to: Black–White differences in response styles. *Public Opinion Quarterly,* 48, 491–509.

Bais, F., Schouten, B., Lugtig, P., Toepoel, V., Arends-Tóth, J., Douhou, S., Kieruj, N., Morren, M., and Vis, C. (2017). Can survey item characteristics relevant to measurement error be coded reliably? A case study on eleven Dutch general population surveys. *Sociological Methods and Research,* 48, 1–33.

Bais, F., Schouten, B., and Toepoel, V. (2019). *Is undesirable answer behaviour consistent across surveys?* Statistics Netherlands Discussion paper.

Bais, F., Schouten, B., and Toepoel, V. (2020). Investigating response patterns across surveys. Do respondents show consistency in answering behaviour over multiple surveys. *Bulletin of Sociological Methodology,* 147–148, 150–168.

Bakker, B.F.M. (2012). Estimating the validity of administrative variables. *Statistica Neerlandica,* 66, 8–17.

Bandilla, W., Couper, M.P., and Kaczmirek, L. (2014). The effectiveness of mailed invitations for web surveys and the representativeness of mixed-mode versus Internet only samples. *Survey Practice,* 7(4). Retrieved July 2018 at http://www.surveypractice.org/article/2863

Battese, G.E., Harter, R.M., and Fuller, W.A. (1988). An error components model for prediction of county crop areas using survey and satellite data. *Journal of the American Statistical Association,* 83, 28–36.

Baumgartner, H., and Steenkamp, J.E.M. (2001). Response styles in marketing research: A cross-national investigation. *Journal of Marketing Research,* 28, 143–156.

Beatty, P., Cosenza, C., and Fowler, F. (2019). Experiments on the design and evaluation of complex survey questions. In P.J. Lavrakas, M.W. Traugott, C. Kennedy, A.L. Holbrook, E.D. De Leeuw, and B.T. West (eds.), *Experimental methods in survey research. Techniques that combine random sampling with random assignment.* New York: Wiley, pp. 113–129.

Beatty, P., and Herrmann, D. (2002). To answer or not to answer: Decision processes related to survey item nonresponse. In R.M. Groves, D.A. Dillman, J.L. Eltinge, and R.J.A. Little (eds.), *Survey nonresponse*. 1st edition. New York: Wiley, pp. 71–86.

Beebe, T.J., Locke, G.R., Barnes, S.A., Davern, M.E., and Anderson, K.J. (2007). Mixing web and mail methods in a survey of physicians. *Health Services Research*, 42, 1219–1234. doi:10.1111/j.1475-6773.2006.00652.x

Beebe, T.J., McAlpine, D.D., Ziegenfuss, J.Y., Jenkins, S., Haas, L., and Davern, M.E. (2012). Deployment of a mixed-mode data collection strategy does not reduce nonresponse bias in a general population health survey. *Health Services Research*, 47(4), 1739–1754. doi:10.1111/j.1475-773.2011.01369.x

Benavent, R., and Morales, D. (2016). Multivariate Fay-Herriot models for small area estimation. *Computational Statistics and Data Analysis*, 94, 372–390.

Berkel, K. van, van der Doef, S., and Schouten, B. (2020). Implementing adaptive survey design with an application to the Dutch health survey *Journal of Official Statistics*, 36, 609–629.

Berlin, M., Mohadjer, L., Waksberg, J., Kolstad, A., Kirsch, I., Rock, D., and Yamamoto, K. (1992). An experiment in monetary incentives. In *JSM Proceedings*, 393–98. Alexandria, VA: American Statistical Association.

Bernardi, R.A. (2006). Associations between Hofstedes cultural constructs and social desirability response bias. *Journal of Business Ethics*, 65(1), 43–53.

Bethlehem, J. (2009). *Applied survey methods: A statistical perspective*. New York: Wiley and Sons.

Bethlehem, J. (2015). Web surveys in official statistics. In Engel, U., Jann, B., Lynn, P., Scherpenzeel, A., and Sturgis, P. (eds.), *Improving survey methods. Lessons from recent research*. New York: Routledge, pp. 156–169.

Bethlehem, J., and Biffignandi, S. (2012). *Handbook of web surveys*. Hoboken, NJ: Wiley.

Bethlehem, J., Cobben, F., and Schouten, B. (2011). *Handbook of nonresponse in household surveys* (Vol. 568). Hoboken, NJ: John Wiley & Sons.

Beukenhorst, D., Buelens, B., Engelen, F., Van der Laan, J., Meertens, V., and Schouten, B. (2014). *The impact of survey item characteristics on mode-specific measurement bias in the crime victimisation survey*. CBS Discussion paper 2014-16. Statistics Netherlands, The Hague.

Bianchi, A., Biffignandi, S., and Lynn, P. (2017). Web-face-to-face mixed-mode design in a longitudinal survey: Effects on participation rates, sample composition, and costs. *Journal of Official Statistics*, 33, 385–408.

Biemer, P. (2001). Nonresponse bias and measurement bias in a comparison of face-to-face and telephone interviewing. *Journal of Official Statistics*, 17, 295–320.

Biemer, P.P. (2010). Overview of design issues: Total survey error. In P.V. Marsden and J.D. Wright (eds.), *Handbook of survey research*. Bingley: Emerald, pp. 27–57.

Biemer, P.P., and Lyberg, L.E. (2003). *Introduction to survey quality*. Hoboken, NJ: John Wiley and Sons.

Biemer, P.P., Murphy, J., Zimmer, S., Berry, C., Deng, G., and Lewis, K. (2018). Using bonus monetary incentives to encourage web response in mixed-mode household surveys. *Journal of Survey Statistics and Methodology*, 6(2), 240–261. doi:10.1093/jssam/smx015

Biemer, P.P., and Stokes, L. (1991). Approaches to the modeling of measurement errors. In P.P. Biemer, R.M. Groves, L.E. Lyberg, N.A. Mathiowetz, and S. Sudman (eds.), *Measurement errors in surveys*. Hoboken, NJ: Wiley and Sons, pp. 487–517.

Bietz, M., Patrick, K., and Bloss, C. (2019), Data donation as a model for citizen science health research. *Citizen Science. Theory and Practice*, 4(1), 6. Retrieved from https://theoryandpractice.citizenscienceassociation.org/articles/10.5334/cstp.178/

Billiet, J.B., and McClendon, J.M. (2000). Modeling acquiescence in measurement models for two balanced sets of items. *Structural Equation Modeling*, 7, 608–628.

Blanke, K., and Luiten, A. (2014). Query on data collection for social surveys. ESSnet project. *Data collection for social surveys using multiple modes*. http://ec.europa.eu/eurostat/cros/system/files/Query_report_DCSS.pdf_en

Blom, A., and West, B. (2017). Explaining interviewer effects: A research synthesis. *Journal of Survey Statistics and Methodology*, 6, 175–211.

Bloom, J. (2008). The speech IVR as a survey interviewing methodology. In F.G. Conrad, and M.F. Schober (eds.), *Envisioning the survey interview of the future*. New York: Wiley, pp. 119–136.

Blumberg, S.J., and Luke, J.V. (2007). Coverage bias in traditional telephone surveys of low-income and young adults. *Public Opinion Quarterly*, 71, 734–749.

Blumberg, S.J., and Luke, J. (2016). Wireless substitution: Early release from the National Health Interview Survey, January–June 2016. Washington, DC: Centers for Disease Control and Prevention, National Center for Health Statistics. Available at https://www.cdc.gov/nchs/data/nhis/earlyrelease/wireless201612.pdf

Bollineni-Balabay, O. Van den Brakel, J.A., and Palm, F. (2016). Multivariate state-space approach to variance reduction in series with level and variance breaks due to sampling redesigns. *Journal of the Royal Statistical Society, A series*, 179, 377–402.

Boonstra, H.J. (2012). *hbsae: Hierarchical Bayesian small area estimation*. R package version 1.0. CRAN - Package hbsae (r-project.org)

Bosnjak, M., and Tuten, T.L. (2003). Prepaid and promised incentives in web surveys - An experiment. *Social Science Computer Review*, 21, 208–217.

Brancato, G., Macchia, S., Murgia, M., Signore, M., Simeoni, G., Blanke, K., Körner, T., Nimmergut, A., Lima, P., Paulino, R., and Hoffmeyer-Zlotnik, J. (2006). *Handbook of recommended practices for questionnaire development and testing in the European statistical system*. European Commission Grant. https://ec.europa.eu/eurostat/documents/64157/4374310/13-Handbook-recommend ed-practices-questionnaire-development-and-testing-methods-2005.pdf/52bd85c2-2dc5-44ad-8f5d-0c6ccb2c55a0

Brick, J.M., Montaquila, J., Hagedorn, M.C., Roth, S.B., and Chapman, C. (2005). Implications for RDD design from an incentive experiment. *Journal of Official Statistics*, 21, 571–589.

Buelens, B., Burger, J., and van den Brakel, J.A. (2015). *Van mixed-mode naar uni-mode web surveys: Twee case studies*. Internal report, Statistics Netherlands, Heerlen.

Buelens, B., and van den Brakel, J.A. (2015). Measurement error calibration in mixed-mode sample surveys. *Sociological Methods & Research*, 44(3), 391–426.

Buelens, B., and van den Brakel, J.A. (2017). Comparing two inferential approaches to handling measurement error in mixed-mode surveys. *Journal of Official Statistics*, 33(2), 513–531.

Buisman, M., and Houtkoop, W. (2014). Laaggeletterdheid in kaart. (Low literacy mapped). ECBO en Stichting Lezen en Schrijven. Retrieved from https://www.lezenenschrijven.nl/uploads/editor/Laaggeletterdheid_in_Kaart_(2014).pdf

Busse, B., and Fuchs, M. (2012). The components of landline telephone survey coverage bias. The relative importance of no-phone and mobile-only populations. *Quality and Quantity*, 46, 1209–1225.

Butler, R.J., and McDonald, J.B. (1987). Interdistributional income inequality. *Journal of Business and Economic Statistics*, 5, 13–18.

Buuren, S. van. (2012). *Flexible imputation of missing data*. New York: Chapman and Hall/CRC Press.

Calinescu, M., and Schouten, B. (2015a). *Selecting adaptive survey design strata with partial R-indicators*. Statistics Netherlands discussion paper.

Calinescu, M., and Schouten, B. (2015b). Adaptive survey designs to minimize mode effects. A case study on the Dutch Labour Force Survey, *Survey Methodology*, 41(2), 403–425.

Calinescu, M., and Schouten, B. (2016). Adaptive survey designs for nonresponse and measurement error in multi-purpose surveys. *Survey Research Methods*, 10, 35–47. doi:10.18148/srm/2016.v10i1.6157

Campanelli, P., Blake, M., Mackie, M., and Hope, S. (2015). Mixed modes and measurement error: Using cognitive interviewing to explore the results of a mixed modes experiment. ISER Working Paper Series, No. 2015–18, University of Essex, Institute for Social and Economic Research (ISER), Colchester. https://www.iser.essex.ac.uk/research/publications/working-p apers/iser/2015-18

Campanelli, P., Nicolaas, G., Jäckle, A., Lynn, P., Hope, S., Blake, M., and Gray, M. (2011). *A classification of question characteristics relevant to measurement (error) and consequently important for mixed mode questionnaire design*. Paper presented at the Royal Statistical Society, October 11, London, UK.

Cantor, D., O'Hare, B.C., and O'Connor, K.S. (2008). The use of monetary incentives to reduce nonresponse in random digit dial telephone surveys. In J.M. Lepkowski, C. Tucker, J.M. Brick, E. de Leeuw, L. Japec, P. Lavrakas, M.W. Link, and R.L. Sangster (eds.), *Advances in telephone survey methodology.* New York: Wiley, pp. 471–498.

Cernat, A. (2015). *Evaluating mode differences in longitudinal data: Moving to a mixed mode paradigm of survey methodology.* PhD Thesis, University of Essex. Available at http://repository.essex.ac.uk/15739/

Cernat, A., Couper, M.P., Ofstedal, M.B. (2016). Estimation of mode effects in the health and retirement study using measurement models. *Journal of Survey Statistics and Methodology*, 4(4), 501–524. https://doi.org/10.1093/jssam/smw021

Cernat, A., and Lynn, P. (2014). The role of email addresses and email contact in encouraging web response in a mixed mode design. In Understanding Society at the Institute for Social and Economic Research, editor. Working Paper. Essex 2014.

Cernat, A., and Sakshaug, J. (2020). The impact of mixed modes on multiple types of measurement error. *Survey Research Methods*, 14, 79–91.

Chang, L., and Krosnick, J.A. (2009). National surveys via RDD telephone interviewing versus the Internet: Comparing sample representativeness and response quality. *Public Opinion Quarterly*, 4, 641–678.

Chen, C., Lee, S., and Stevenson, H.W. (1995). Response style and cross-cultural comparisons of rating scales Among East Asian and North American students. *Psychological Science*, 6, 170–175.

Cheung, G.W., and Rensvold, R.B. (2000). Assessing extreme and acquiescence response sets in cross-cultural research using structural equations modeling. *Journal of CrossCultural Psychology*, 31, 187–212.

Chun, K.T., Campbell, J.B., and Yoo, J.H. (1974). Extreme response style in cross-cultural research: A reminder. *Journal of Cross-Cultural Psychology*, 5, 465–480.

Church, A.H. (1993). Estimating the effect of incentives on mail survey response rates: A meta-analysis. *Public Opinion Quarterly*, 57, 26–79.

Cialdini, R.B. (2007). *Influence. The psychology of persuasion.* New York: Harper Collins World.

Cobben, F. (2009). *Nonresponse in sample surveys. Methods for analysis and adjustment.* PhD Dissertation, University of Amsterdam. Available at https://www.cbs.nl/NR/rdonlyres/2C300D9D-C65D-4B44-B7F3-377BB6CEA066/0/2009x11cobben.pdf

Cobben, F., and Bethlehem, J. (2005). Adjusting undercoverage and nonresponse bias in telephone surveys. Statistics Netherlands discussion paper 05006. Retrieved from https://www.research gate.net/publication/242207187_Adjusting_Undercoverage_and_Non-Response_Bias_in_Telephone_Surveys

Cochran, W. (1977). *Sampling theory* (2nd ed.). New York: Wiley and Sons.

Cole, J.S., Sarraf, S.A, and Wang, X. (2015). *Does use of survey incentives degrade data quality?* Paper presented at the Association for Institutional Research Annual Forum, Denver, CO.

Collins, D. (ed.). (2015). *Cognitive interviewing practice.* London: SAGE Publications Ltd.

Commandeur, J.J.F., and Koopman, S.J. (2007). *An introduction to state space time series analysis.* Oxford: Oxford University Press.

Conrad, F.G., Schober, M.F., Jans, M., Orlowski, R.A., Nielsen, D., and Levenstein, R. (2015). Comprehension and engagement in survey interviews with virtual agents. *Frontiers in Psychology*, 6, Oct 2015.

Conrad, F.G., Tourangeau, R., Couper, M.P., Zhang, C. (2017). Reducing speeding in web surveys by providing immediate feedback. *Survey Research Methods*, 11(1), 45–61.

Cornesse, C., and Bosnjak, M. (2018). Is there an association between survey characteristics and representativeness? A meta-analsyis. *Survey Research Methods*, 12(1), 1–13. https://ojs.ub.uni-k onstanz.de/srm/article/view/7205

Couper, M. (2008). *Designing effective web surveys.* Cambridge: Cambridge University Press.

Couper, M., Antoun, C., and Mavletova, A. (2017). Mobile web surveys: A total survey error perspective. In P. Biemer, E. de Leeuw, S. Eckman, B. Edwards, F. Kreuter, L. Lyberg, and B. West (eds.), *Total survey error in practice.* Hoboken, NJ: John Wiley & Sons, pp. 133–154.

Couper, M.P. (2000). Web surveys: A review of Issues and approaches. *Public Opinion Quarterly*, 64, 464–494.

Couper, M.P. (2011). The future of modes of data collection. *Public Opinion Quarterly*, 75, 889–908.

Couper, M.P., Antoun, C., and Mavletova, A. (2017). Mobile web surveys: A total survey error perspective. In P. Biemer, E. de Leeuw, S. Eckman, B. Edwards, F. Kreuter, L. Lyberg, and B. West (eds.), *Total survey error in practice*. Hoboken, NJ: John Wiley and Sons, pp. 133–154.

Couper, M.P., Tourangeau, R., Conrad, F., and Zhang, C. (2013). The design of grids in web surveys. *Social Science Computer Review*, 31(3), 322–345.

Cremers, M. (2016). *Survey practices: Omni mode design at Statistics Netherlands*. Presentation at the Second International Conference on Questionnaire Design, Development, Evaluation, and Testing (QDET2), Florida, Miami, 9–13 November 2016. https://ww2.amstat.org/meetings/qdet2/OnlineProgram/ViewPresentation.cfm?file=303388.pptx

d'Ardenne, J., and Collins, D. (2020). Combining multiple question evaluation methods: What does it mean when the data appear in conflict? In Beatty, P., Collins, D., Kaye, L., Padilla, J.L., Willis, G.B., Wilmot, A. (eds.), *Advances in questionnaire design, development, evaluation and testing. Proceedings of the 2nd Conference on Questionnaire Design, Development, Evaluation, and Testing (QDET2)*, 9–13 November 2016, Miami, FL. Hoboken, NJ: Wiley, 91–116.

de Bruijne, M., and Wijnant, A. (2013). Comparing survey results obtained via mobile devices and computers: An experiment with a mobile web survey on a heterogeneous group of mobile devices versus a computer-assisted web survey. *Social Science Computer Review*, 31, 482–504.

de Leeuw, E.D. (1992). *Data quality in mail, telephone and face-to-face surveys*. PhD dissertation, Free University of Amsterdam.

de Leeuw, E.D. (2005). To mix or not to mix data collection modes in surveys. *Journal of Official Statistics*, 21(2), 233–255.

de Leeuw, E.D. (2008). Choosing the method of data collection. In E.D. de Leeuw, J.J. Hox, and D.A. Dillman (eds.), *International handbook of survey methodology*. New York: Routledge, Taylor & Francis, European Association of Methodology (EAM) Methodology Series, pp. 113–135.

de Leeuw, E.D. (2018). Mixed-mode: Past, present, and future. *Survey Research Methods*, 12(2), 75–89. doi:10.18148/srm/2018.v12i2.7402

de Leeuw, E.D., and Berzelak, N. (2016). Survey mode or survey modes? In C. Wolf, D. Joye, T. Smith, and Y.-C. Fu (eds.), *The Sage handbook of survey methodology*. London: Sage, pp. 142–156.

de Leeuw, E.D., Callegaro, M., Hox, J., Korendijk, E., and Lensvelt-Mulders, G. (2007). The influence of advance letters on response in telephone surveys: A meta-analysis. *Public Opinion Quarterly*, 71, 413–443.

de Leeuw, E.D., Hox, J.J., and Boeve, A. (2016). Handling do-not-know answers: Exploring new approaches in online and mixed-mode surveys. *Social Science Computer Review*, 34, 116–132.

de Leeuw, E.D, Hox, J., and Luiten, A. (2018). International nonresponse trends across countries and years. *Survey Methods: Insights from the Field*. https://surveyinsights.org/?p=10452

de Regt, S. (2020). *Resultaten experiment incentives ODIN 2020* (Results of the incentive experiments Transporation Survey, in Dutch). Internal CBS Report.

Dijkstra, W., and Ongena, Y. (2006). Question-answer sequences in survey-interviews. *Quality and Quantity*, 40, 983–1011.

Dillman, D.A. (1996). Token financial incentives and the reduction of nonresponse error in mail surveys. In *Proceedings of the Government Statistics Section of the American Statistical Association*.

Dillman, D. (2017). The promise and challenges of pushing respondents to the web in mixed-mode surveys. *Survey Methodology*, 43, 1. Retrieved from https://www150.statcan.gc.ca/n1/pub/12-001-x/2017001/article/14836-eng.htm

Dillman, D.A. (2017). The promise and challenge of pushing respondents to the web in mixed-mode surveys. *Survey Methodology*, 43, 3–30.

Dillman, D.A., and Christian, L.M. (2005). Survey mode as a source of instability in responses across surveys. *Field Methods*, 17(1), 30–52.

Dillman, D.A., and Edwards, M.L. (2016). Designing a mixed-mode survey. In C. Wild D. Joye, T.W. Smith, and Y.-C. Fu (eds.), *The SAGE handbook of survey methodology*. London: Sage, pp. 255–269.

Dillman, D.A., Phelps, G., Tortora, R., Swift, K., Kohrell, J., Berck, J., and Messer, B.L. (2009). Response rate and measurement differences in mixed-mode surveys using mail, telephone, interactive voice response (IVR) and the internet. *Social Science Research*, 38(1), 1–18.

Dillman, D.A., Smyth, J.D., and Christian, L.M. (2014). *Internet, phone, mail, and mixed-mode surveys: The tailor design method* (4th ed.). New York: Wiley and Sons.

Dippo, C.S., Kostanich, D.L., and Polivka, A.E. (1994). Effects of methodological change in the current population survey. *Proceedings of the Section on Survey Research Methods*, American Statistical Association, 260–262.

Dolnicar, S., and Grün, B. (2007). Cross-cultural differences in survey response patterns. *International Marketing Review*, 31(2), 127–143.

Doornik, J.A. (2009). *An Object-oriented Matrix Programming Language Ox 6*. London: Timberlake Consultants Press.

Durbin, J., and Koopman, S.J. (2012). *Time series analysis by state space methods*. Oxford: Oxford University Press. Second edition.

Dutwin, D., and Lavrakas, P. (2016). Trends in telephone outcomes, 2008–2015. *Survey Practice*, 9. Available at http://www.surveypractice.org/

Dykema, J., Stevenson, J., Klein, L., Kim, Y., and Day, B. (2013). Effects of e-mailed versus mailed invitations and incentives on response rates, data quality, and costs in a web survey of university faculty. *Social Science Computer Review*, 31(3), 359–370.

Eckman, S. (2010). *Errors in housing unit listing and their effect on survey estimates*. PhD dissertation, University of Maryland.

Eckman, S., and Kreuter, F. (2017). The undercoverage-nonresponse tradeoff. In Biemer, P.P., De Leeuw, E., Eckman, S., Edwards, B., Kreuter, F., Lyberg, L.E., Tucker, C.N, and West, B.T. (eds.), *Total survey error in practice*. Hoboken, NJ: John Wiley and Sons.

Edwards, P., Roberts, I., Clarke, M., DiGuiseppi, P.S., Wentz, R., and Kwan, I. (2002). Increasing response rates to postal questionnaires: Systematic review. *BMJ*, 324, 1183–1192.

ELINET - European Literacy Policy Network (2015). Facts and figures. Retrieved from http://www.eli-net.eu/fileadmin/ELINET/Redaktion/Factsheet-Literacy_in_Europe-A3.pdf. Retrieved 20180521.

European Statistical System Committee. (2018). *European statistics code of practice*, revised edition 2017. https://ec.europa.eu/eurostat/web/quality/european-statistics-code-of-practice

Eurostat (2017a). http://ec.europa.eu/eurostat/statistics-explained/index.php/Digital_economy_and_society_statistics_-_households_and_individuals. Retrieved 20180427.

Eurostat (2017b). http://ec.europa.eu/eurostat/statistics-explained/index.php?title=Quality_of_life_indicators_-_education#Computer_and_language_skills. Retrieved 20180427.

Eurostat. (2018). http://appsso.eurostat.ec.europa.eu/nui/show.do?dataset=isoc_ci_ifp_iu&lang=en. Retrieved 20180706.

Fay, R.E., and Herriot, R.A. (1979). Estimation of income for small places: An application of James-Stein procedures to census data. *Journal of the American Statistical Association*, 74, 268–277.

Fessler, P., Kasy, M., and Lindner, P. (2018). Survey mode effects on measured income inequality. *The Journal of Economic Inequality*, 16(4), 1–19.

Fienberg, S.E., and Tanur, J.M. (1987). Experimental and sampling structures: Parallels diverging and meeting. *International Statistical Review*, 55, 75–96.

Fienberg, S.E., and Tanur, J.M. (1988). From the inside out and the outside in: Combining experimental and sampling structures. *Canadian Journal of Statistics*, 16, 135–151.

Fienberg, S.E., and Tanur, J.M. (1989). Combining cognitive and statistical approaches to survey design. *Science*, 243, 1017–1022.

Fowler, F. (1995). *Improving survey questions: Design and evaluation*. Applied Social Research Methods Series, vol. 38, Thousand Oaks, CA: Sage Publications.

Fowler, F., and Cosenza, C. (2008). Writing effective survey questions. In De Leeuw, E., Hox, J., and Dillman D. (eds.), *International handbook of survey methodology*. New York: Lawrence Erlbaum, pp. 136–160.

Fox, R.J., Crask, M.R., and Kim, J. (1988). Mail survey response rates: A meta-analysis of selected techniques for inducing response. *Public Opinion Quarterly*, 52, 467–491.

Fricker, S., Galesic, M., Tourangeau, R., and Yan, T. (2005). An experimental comparison of web and telephone surveys. *Public Opinion Quarterly*, 69, 370–392.

Friese, C.R., Lee, C.S., O'Brien, S., and Crawford, S.D. (2010). Multi-mode and method experiment in a study of nurses. *Survey Practice*, 3. http://surveypractice.org/2010/10/27/multi-mode-nurses-survey

Fuchs, M. (2009). Gender-of-interviewer effects in a video-enhanced web survey. Results from a randomized field experiment. *Social Psychology*, 40, 37–42.

Fuchs, M., Couper, M., and Hansen, S.E. (2000). Technology efects: Do CAPI or PAPI interviews take longer? *Journal of Official Statistics*, 16, 273–286.

Fuchs, M., and Funke, F. (2007). Video web survey - results of an experiment with a text-based web survey. In M. Trotman et al (eds.), *The challenge of a changing world*. Proceedings of the 2007 Association for Survey Computing.

Furse, D.H., and Stewart, D.W. (1982). Monetary incentives versus promised contribution to charity: New evidence on mail survey response. *Journal of Marketing Research*, 19, 375–380.

Gajic, A., Cameron, D., and Hurley, J. (2010). The cost-effectiveness of cash versus lottery incentives for a web based, stated preference community survey. *European Journal of Health Economics*, 13(6), 789–799. doi:10.1007/s10198-011-0332-0

Galesic, M., and Bosnjak, M. (2009). Effects of questionnaire length on participation and indicators of response quality in a web survey. *Public Opinion Quarterly*, 73(2), 349–360.

Gallhofer, I.N., Scherpenzeel, A., and Saris, W.E. (2007). The code-book for the SQP program, available at (http://www.europeansocialsurvey.org/methodology/sqpcoding.html).

Geisen, E., and Murphy, J. (2020). A compendium of web and mobile survey pretesting methods. In Beatty, P., Collins, D., Kaye, L., Padilla, J.L., Willis, G.B., and Wilmot, A. (eds.), *Advances in questionnaire design, development, evaluation and testing*. Proceedings of the 2nd Conference on Questionnaire Design, Development, Evaluation, and Testing (QDET2); 9–13 November 2016, Miami, FL. Hoboken, NJ: Wiley, pp. 287–314.

Geisen, E., and Romano Bergstrom, J. (2017). *Usability testing for survey research*. Boston, MA: Morgan Kaufmann.

Gerber, E.R., and Wellens, T.R. (1995). Literacy and the self-administered form in special populations: A primer. In *Proceedings of the survey research methods section, American Statistical Association*. Alexandria, VA: American Statistical Association, pp. 1087–1092.

Giesen, D., Meertens, V., Vis-Visschers, R., and Beukenhorst, D. (2012). *Questionnaire development*. Heerlen: Statistics Netherlands. Questionnaire development (cbs.nl)

Goetz, E.G., Tyler, T.R., and Cook, F.L. (1984). Promised incentives in media research: A look at data quality, sample representativeness, and response rate. *Journal of Marketing Research*, 21, 148–154.

Göritz, A.S. (2006). Incentives in web studies: Methodological issues and a review. *International Journal of Internet Science*, 1, 58–70.

Göritz, A.S. (2015). Incentive effects. In Engel, U., Jann, B., Lynn, P., Scherpenzeel, A., and Sturgis, P. (eds.), *Improving survey methods. Lessons from recent research*. London: Routledge, pp. 339–350.

Göritz, A.S., and Luthe, S.C. (2013). Effects of lotteries on response behavior in online panels. *Field Methods*, 25, 219–237.

Gravem, D., Meertens, V., Luiten, A., Giesen, D., Berg, N., Bakker, J., and Schouten, B. (2019a). *Final methodological report presenting results of usability tests on selected ESS surveys and census. Smartphone fitness of ESS surveys – Case studies on the ICT survey and the LFS*. Deliverable 5.4 of Mixed Mode Designs for Social Surveys – MIMOD. https://www.istat.it/en/research-activity/international-research-activity/essnet-and-grants

Gravem, D., Meertens, V., Luiten, A., Giesen, D., Berg, N., Bakker, J., and Schouten, B. (2019b). Smartphone fitness of ESS surveys – Case studies on the ICT survey and the LFS, ESSnet MIMOD. Deliverable 4 of Work package 5, available at https://www.istat.it/en/research-activity/international-research-activity/essnet-and-grants

Gravem, D.F., Berg, N., Berglund, F., Lund, K., and Roßbach, K. (2018a). *Test report from Statistics Norway's cognitive and usability testing of questions and questionnaires*. Appendix B of WP4 Deliverable 3 Deliverable 5.4 of Mixed Mode Designs for Social Surveys – MIMOD. https://www.istat.it/en/research-activity/international-research-activity/essnet-and-grants

Gravem, D.F., Falnes-Dalheim, E., Demofonti, S., and Signore, M. (2018b). Survey communication in mixed-mode ESS surveys. Essnet MIMOD deliverable. Available at Istat.it - Projects funded by Eurostat: Essnet and Grants.

Gravem, D.F, Holseter, C., Falnes-Dalheim, E., Macchia, S., Barcherini, S., Luiten, A., and Meertens, V. (2018c). Mixed-mode experiences of European NSIs. ESSnet MIMOD deliverable. Available at Istat.it - Projects funded by Eurostat: Essnet and Grants.

Gray, M. (2015). Survey mode and its implication for cognitive interviewing. In Collins, D. (ed.), *Cognitive interviewing practice*. London: SAGE Publications Ltd., pp. 197–219.

Greene, J., Speizer, H., and Wiitala, W. (2008). Telephone and web: Mixed-mode challenge. *Health Survey Research*, 43, 230–248.

Greenleaf, E.A. (1992). Measuring extreme response style. *Public Opinion Quarterly*, 56(3), 328–351.

Groves, R.M. (1989). *Survey errors and survey costs*. Hoboken, NJ: John Wiley & Sons.

Groves, R.M., and Couper, M.P. (1998). *Nonresponse in household interview surveys*. Hoboken, NJ: Wiley and Sons.

Groves, R.M., Dillman, D.A., Eltinge, J.L., and Little, R.J.A. (2002). *Survey nonresponse*. New York: Wiley.

Groves, R.M., Fowler, F.J., Couper, M.P., Lepkowski, J.M., Singer, E., and Tourangeau, R. (2004). *Survey methodology*. New York: Wiley.

Groves, R.M., Fowler, F.J., Couper, M., Lepkowski, J.M., Singer, E., and Tourangeau, R. (2009). *Survey methodology* (2nd ed.). Hoboken, NJ: Wiley and Sons.

Groves, R.M., and Heeringa, S.G. (2006). Responsive design for household surveys: Tools for actively controlling survey errors and costs. *Journal of the Royal Statistical Society: Series A*, 169, 439–457.

Groves, R.M., and Lyberg, L. (2010). Total survey error. Past, present and future. *Public Opinion Quarterly*, 74(5), 849–879.

Guo, Y., and Little, R.J.A. (2013). Bayesian multiple imputation for assay data subject to measurement error. *Journal of Statistical Theory and Practice*, 7, 219–232.

Hájek, J. (1971). Comment on a paper by D. Basu. In V.P. Godambe and D.A. Sprott (eds.), *Foundations of statistical inference*. Toronto, ON: Holt, Rinehart and Winston, p. 236.

Harris, L., Weinberger, M., and Tierney, W. (1997). Assessing inner-city patients' hospital experiences: A controlled trial of telephone interviews versus mailed surveys. *Medical Care*, 35, 70–76. http://www.jstor.org/stable/3766839

Hartley, H.O., and Rao, J.N.K. (1978). Estimation of nonsampling variance components in sample surveys. In N.K. Namboodiri (ed.), *Survey sampling and measurement*. New York: Academic Press, pp. 35–43.

Harvey, A.C. (1989). *Forecasting, structural time series models and the Kalman filter*. Cambridge: Cambridge University Press.

Harvey, A.C., and Durbin, J. (1986). The effects of seat belt legislation on British road casualties: A case study in structural time series modelling. *Journal of the Royal Statistical Society, Series A*, 149, 187–227.

Harzing, A.-W. (2006). Response styles in cross-national survey research: A 26-country study. *International Journal of Cross Cultural Management*, 6(2), 243–266.

He, J., and Van de Vijver, F.J.R. (2013). A general response style factor: Evidence from a multi-ethnic study in the Netherlands. *Personality and Individual Differences*, 55, 794–800.

He, J., Van de Vijver, F.J.R., Espinosa, A.D., and Mui, P.H. (2014). Toward a unification of acquiescent, extreme, and midpoint response styles: A multilevel study. *International Journal of Cross Cultural Management*, 14, 306–322.

Heerwegh, D., and Loosveldt, G. (2011). Assessing mode effects in a national crime victimization survey using structural equation models: Social desirability bias and acquiescence. *Journal of Official Statistics*, 27, 49–63.

Helske, J. (2021). Documentation R package 'KFAS'. https://cran.r-project.org/web/packages/KFAS/KFAS.pdf

Hinkelmann, K., and Kempthorne, O. (1994). *Design and analysis of experiments, volume 1: Introduction to experimental design*. New York: John Wiley.

Hinkelmann, K., and Kempthorne, O. (2005). *Design and analysis of experiments, volume 2: Advanced experimental design*. New York: John Wiley.

Holbrook, A.L., Green, M.C., and Krosnick, J.A. (2003). Telephone versus face-to-face interviewing of national probability samples with long questionnaires: Comparisons of respondent satisficing and social desirability response bias. *Public Opinion Quarterly*, 67, 79–125.

Holbrook, A.L., Krosnick, J.A., Moore, D., and Tourangeau R. (2007). Response order effects in dichotomous categorical questions presented orally: The impact of question and respondent attributes. *Public Opinion Quarterly*, 71(3), 325–348.

Holmberg, A., Lorenc, B., and Werner, P. (2010). Contact strategies to improve participation via the web in a mixed-mode mail and web survey. *Journal of Official Statistics*, 26, 465–480.

Hope, S., Campanelli, P. Nicolaas, G., Lynn, P., and Jäckle, A. (2014). The role of the interviewer in producing mode effects: Results from a mixed modes experiment comparing face-to-face, telephone and web administration. ISER Working Paper Series. Available at https://www.iser.essex.ac.uk/research/publications/working-papers/iser/2014-20.pdf

Hopkins, K.D., and Gullickson, A.R. (1992). Response rates in survey research: A meta-analysis of the effects of monetary gratuities. *Journal of Experimental Education*, 61, 52–62.

Hox, J.J., De Leeuw, E., and Kreft, I.G. (1991). The effect of interviewer and respondent characteristics on the quality of survey data: A multilevel model. In P.P. Biemer, R.M. Groves, L.E. Lyberg, N.A. Mathiowetz, and S. Sudman (eds.), *Measurement errors in surveys*. New York: Wiley, pp. 439–461.

Hox, J., De Leeuw, E.D., and Zijlmans, E.A.O. (2015). Measurement equivalence in mixed mode surveys. *Frontiers in Psychology*, 6, 1–11.

Huang, E.T., and Fuller, W.A. (1978). Nonnegative regression estimation for survey data. *Proceedings of the Social Statistics Session*, American Statistical Association, 300–305, Aug. 14–17 1978, San Diego, California.

Hui, C.H., and Triandis, H.C. (1989). Effects of culture and response format on extreme style. *Journal of Cross-Cultural Psychology*, 20, 296–309.

Ilic, G., Lugtig, P., Schouten, B., Mulder, J., Streefkerk, M., Kumar, P., and Höcük, S. (2020). Pictures instead of questions. An experimental investigation of the feasibility of using pictures in a housing survey, Discussion paper, Statistics Netherlands, The Netherlands, www.cbs.nl.

Israel, G.D. (2009). Obtaining responses by mail or web: Response rates and data consequences. *Survey Practice*, 2, 1–2. doi:10.29115/SP-2009-0021.

Israel, G.D. (2012). Combining mail and email contacts to facilitate participation in mixed-mode surveys. *Social Science Computer Review*, 31(9), 346–358. doi:10.1177/0894439312464942

Jäckle, A., Lynn, P., and Burton, J. (2015). Going online with a face-to-face household panel: Effects of a mixed mode design on item and unit non-response. *Survey Research Methods*, 9, 57–70. doi:10.18148/srm/2015.v9i1.5475

Jäckle, A., Roberts, C., and Lynn, P. (2010). Assessing the effect of data collection mode on measurement. *International Statistical Review*, 78, 3–20.

James, J.M., and Bolstein, R. (1990). The effect of monetary incentives and follow-up mailings on the response rate and response quality in mail surveys. *Public Opinion Quarterly*, 54, 346–361.

Jans, M., Sirkis, R., and Morgan, D. (2013). Managing data quality indicators with paradata-based statistical quality control tools. In Kreuter, F. (ed.), *Improving surveys with paradata: Making use of process information*. New York: John Wiley & Sons, pp. 191–229.

Janssen, B. (2006). Web data collection in a mixed mode approach: An experiment. Proceedings of the Q2006. Available at https://ec.europa.eu/eurostat/documents/64157/4374310/24-WEB-DATA-COLLECTION-IN-A-MIXED-MODE-APPROACH-EXPERIMENT_2006.pdf/61101 2cd-9c37-4432-afad-942ea7df36fc

Jobber, D., Saunders, J., and Mitchell, V. (2004). Prepaid monetary incentive effects on mail survey response. *Journal of Business Research*, 57, 21–25.

Johnson, T., and Van de Vijver, F.J.R. (2002). Social desirability in crosscultural research. In J. Harness, F.J.R. van de Vijver, and P. Mohler (eds.), *Cross-cultural survey methods*. New York: Wiley, pp. 193–202.

Joye, D., Pollien, A., Sapin, M., and Stähli, M.E. (2012). Who can be contacted by phone? Lessons from Switzerland. In Häder, S., Häder, M., and Kühne, M. (eds.), *Telephone survey in Europe. Research and practice.* Berlin: Springer, pp. 85–102.

Kaplowitz, M.D., Lupi, F., Couper, M.P., and Thorp, L. (2012). The effect of invitation design on web survey response rates. *Social Science Computer Review*, 30, 339–349.

Kappelhof, J.W. (2015). Face-to-face or sequential mixed-mode surveys among non-western minorities in the Netherlands: The effect of different survey designs on the possibility of nonresponse bias. *Journal of Official Statistics*, 31, 1–30. doi:10.1515/jos-2015-0001

Kellogg, R.T. (2007). *Fundamentals of cognitive psychology.* Los Angeles, CA: SAGE Publications.

Keusch, F., Struminskaya, B., Antoun, C., Couper, M.P., and Kreuter, F. (2019). Willingness to participate in passive mobile data collection. *Public Opinion Quarterly*, 82, 210–235.

Keusch, F., and Zhang, C. (2017). A review of issues in gamified surveys. *Social Science Computer Review*, 35, 147–166.

Kieruj, N.D., and Moors, G. (2013). Response style behavior: Question format dependent or personal style? *Quality and Quantity*, 47, 193–211.

Kindermann, C., and Lynch, J. (1997). *Effects of the redesign on vistimization estimates.* Technical report, US Department of Justice, Bureau of Justice Statistics. http://www.ojp.usdoj.gov/bjs/abstract/erve.htm

Klausch, T. (2014). *Informed design of mixed-mode surveys: Evaluating mode effects on measurement and selection error.* PhD thesis, University of Utrecht.

Klausch, L.T., Hox, J.J., and Schouten, B. (2013). Measurement effects of survey mode on the equivalence of attitudinal rating scale questions. *Sociological Methods & Research*, 42, 227–263.

Klausch, T., Hox, J., and Schouten, B. (2015). Selection error in single- and mixed mode surveys of the Dutch general population. *Journal of the Royal Statistical Society: Series A (Statistics in Society)*, 178, 945–961. doi:10.1111/rssa.

Klausch, T., Schouten, B. Buelens, B., and van den Brakel, J.A. (2017). Adjusting measurement bias in sequential mixed-mode surveys using re-interview data. *Journal of Survey Statistics and Methodology*, 5, 409–432.

Kolenikov, S., and Kennedy, C. (2014). Evaluating three approaches to statistically adjust for mode effects. *Journal of Survey Statistics and Methodology*, 2(2), 126–158.

Koopman, S.J., Harvey, A.C., Doornik, J.A., and Shephard, N. (2007). *STAMP 8: Structural time series analyser, modeller and predictor.* London: Timberlake Consultants Press.

Koopman, S.J., Shephard, N., and Doornik, J.A. (1999). Statistical algorithms for models in state space form using Ssfpack 2.2. *Econometrics Journal*, 2, 113–166.

Koopman, S.J., Shephard, N., and Doornik, J.A. (2008). *SsfPack 3.0: Statistical algorithms for models in state space form.* London: Timberlake Consultants Press.

Kreuter, F. (2013). *Improving surveys with paradata. Analytic uses of process information.* Hoboken, NJ: Wiley.

Kreuter, F., Haas, G.C., Keusch, F., Bähr, S., and Trappmann, M. (2018). Collecting survey and smartphone sensor data with an app. Opportunities and challenges around privacy and informed consent. *Social science Computer Review*, 36(1), 1–17.

Kreuter, F., Müller, G., and Trappmann, M. (2010). Nonresponse and measurement error in employment research: Making use of administrative data. *Public Opinion Quarterly*, 74, 880–906.

Kreuter, F., Presser, S., and Tourangeau, R. (2008). Social desirability bias in CATI IVR and Web surveys: The effects of mode and question sensitivity. *Public Opinion Quarterly*, 72(5), 847–865.

Krosnick, J.A. (1991). Response strategies for coping with the cognitive demands of attitude measures in surveys. *Applied Cognitive Psychology*, 5, 213–36.

Krosnick, J.A., and Alwin, D.F. (1987). An evaluation of a cognitive theory of response order effects in survey measurement. *Public Opinion Quarterly*, 51, 201–219.

Krosnick, J.A., Holbrook, A.L., Berent, M.K., Carson, R.T., Hanemann, W.M., Kopp, R.J., Mitchell, R.C., Presser, S., Ruud, P.A., Smith, V.K., Moody, W.R., Green, M.C., and Conaway, M. (2002). The impact of 'no opinion' response options on data quality: Non-attitude reduction or an invitation to satisfice? *Public Opinion Quarterly*, 66, 371–403.

Krosnick, J.A., and Pressner, S. (2010). Question and questionnaire design. In P.V. Marsden, and J.D. Wright (eds.), *Handbook of survey research*. Bingley, UK: Emerald, pp. 263–313.

Krumpal, I. (2013). Determinants of social desirability bias in sensitive surveys: A literature review. *Q&Q*, 47, 2025–2047.

Kunz, T., and Fuchs, M. (2019). Using experiments to assess interactive feedback that improves response quality in web surveys. In P. Lavrakas, M. Traugott, C. Kennedy, A. Holbrook, E. de Leeuw, and B. West (eds.), *Experimental methods in survey research: Techniques that combine random sampling with random assignment*. Hoboken: John Wiley and Sons, pp. 247–273.

Lagerstrøm, B. (2008). *Cost efficiency in a mixed-mode survey - Evidence from the Norwegian Rent Market Survey*. Paper presented at the 19th Workshop on Household Nonresponse, Ljubljana, 15 September 2008.

Laguilles, J.S., Williams, E.A., and Saunders, D. (2011). Can lottery incentives boost web survey response rates? Findings from four experiments. *Research in Higher Education*, 52, 537–553. doi:10.1007/s11162-010-9203-2

Laurie, H., and Lynn, P. (2008). The use of respondent incentives on longitudinal surveys. Institute for Social and Economic Research, University of Essex. Working paper no. 2008-42. https://www.iser.essex.ac.uk/research/publications/working-papers/iser/2008-42.pdf

Lavrakas, P.J., and the AAPOR Cell Phone Task Force. (2010). New considerations for survey researchers when planning and conducting RDD telephone surveys in the U.S. with respondents reached via cell phone numbers. Available at http://www.aapor.org

Lavrakas, P.J., Tompson, T.N., and Benford, R. (2010). *Investigating data quality in cell phone surveying*. Paper presented at the Annual American Association for Public Opinion Research Conference, Chicago, IL.

Lemaître, G., and Dufour, J. (1987). An integrated method for weighting persons and families. *Survey Methodology*, 13, 199–207.

Lemcke, J., Schmich, P., and Albrecht, S. (2018). The impact of incentives on data quality in a representative national health survey. Presentation for the *General Online Research (GOR) Conference*, Köln, Germany, February 28–March 2.

Lesser, V., Newton, L., Yang, D., and Sifneos, J. (2016). Mixed-mode surveys compared with single mode surveys: Trends in responses and methods to improve completion. *Journal of Rural Social Sciences*, 31(3), 7–34. https://egrove.olemiss.edu/jrss/vol31/iss3/2

Link, M.W., Murphy, J., Schober, M.F., Buskirk, T.D., Childs, J.H., and Tesfaye, C.L. (2014). Mobile technologies for conducting, augmenting and potentially replacing surveys: Report of the AAPOR task force on emerging technologies in public opinion research. REVISED_Mobile_Technology_Report_Final_revised10June14.pdf.aspx (aapor.org)

Lipps, O., Pekari, N., Roberts, C. (2015). Undercoverage and nonresponse in a list-sampled telephone election survey. *Survey Research Methods*, 9, 71–82. doi:10.18148/srm/2015.v9i2.6139

Little, R.J.A., and Rubin, D.B. (2002). *Statistical analysis with missing data* (2nd ed.). Hoboken, NJ: Wiley & Sons.

Little, R., and Vartivarian, S. (2005). Does weighting for nonresponse increase the variance of survey means? *Survey Methodology*, 31, 161–168.

Lo Conte, M., Murgia, M., Coppola, L., Frattarola, D., Fratoni, A., Luiten, A., and Schouten, B. (2019). Mixed-mode strategies for social surveys: How to best combine data collection modes. Deliverable 2. Work package 1 of the ESSnet on Cooperation on multi-mode data collection – Mixed mode designs for social surveys – MIMOD. Available at Istat.it – Projects funded by Eurostat: Essnet and Grants.

Lo Conte, M., Murgia, M., Luiten, A., and Schouten, B. (2019). Deciding the mixed mode strategy. Available at Istat.it - Projects funded by Eurostat: Essnet and Grants.

Lohr, S. (2008). Coverage and sampling. In De Leeuw, E.D., Hox, J., and Dillman, D.A. (eds.), *International handbook of survey methodology*. New York: Laurence Erlbaum, pp. 97–112.

Lozar Manfreda, K., Bosnjak, M., Berzelak, J., Haas, I., and Vehovar, V. (2008). Web surveys versus other survey modes: A meta-analysis comparing response rates. *International Journal of Market Research*, 50, 79–104.

Lugtig, P. (2017). The relative size of measurement error and attrition error in a panel survey. Comparing them with new Multi-Trait Multi-Method model. *Survey Research Methods*, 11(4), 369–382.

Luiten, A., Båshus, T., Buelens, B., Gravem, D., Körner, T., Lagerstrøm, B., Merad, S., Pohjanpää, K., and Schouten, B. (2015). *WP III summary*, Deliverable for the ESSnet Project "Data Collection for Social Surveys using Multiple Modes" (DCSS). Retrieved from https://ec.europa.eu/eurostat/cros/system/files/D3_7_WPIII_Final_report_rev_20150129.pdf

Luiten, A., and Groffen, D. (2018). The effect of a lottery incentive on response, representativeness and data quality in the LFS. Statistics Netherlands discussion paper. Available at https://www.cbs.nl/en-gb/our-services/methods/papers

Luiten, A., Hox, J., and de Leeuw, E. (2020). Survey nonresponse trends and fieldwork effort in the 21st century: Results of an international study across countries and surveys. *Journal of Official Statistics*, 36, 469–487. doi:10.2478/jos-2020-0025

Luiten, A., and Schouten, B. (2013). Tailored fieldwork design to increase representative response. An experiment in the survey of consumer satisfaction. *Journal of the Royal Statistical Society, Series A*, 176, 169–191.

Lumley, T. (2010). *Complex surveys: A guide to analysis using R*. John Wiley and Sons.

Lundquist, P., and Särndal, C.E. (2013). Aspects of responsive design with applications to the Swedish living conditions survey. *Journal of Official Statistics*, 29, 557–582.

Lynn, P. (ed.). (2009). *Methodology of longitudinal surveys*. Chichester, UK: Wiley.

Lynn, P. (2012). Mode-switch protocols: How a seemingly small design difference can affect attrition rates and attrition bias. ISER Working Paper Series, No. 2012–28, University of Essex, Institute for Social and Economic Research (ISER), Colchester. Retrieved from https://www.econstor.eu/bitstream/10419/91697/1/731811518.pdf

Lynn, P. (2013). Alternative sequential mixed-mode designs: Effects on attrition rates, attrition bias, and costs. *Journal of Survey Statistics and Methodology*, 1, 183–205. doi:10.1093/jssam/smt015

Lynn, P. (2020). Evaluating push-to-web methodology for mixed-mode surveys using address-based samples. *Survey Research Methods*, 14, 19–30. doi:10.18148/srm/2020.v.14i1.7591

Lynn, P., and Kaminska, O. (2012). The impact of mobile phones on survey measurement error. *Public Opinion Quarterly*, 77, 586–605.

Mack, S., Huggins, V., Keathley, D., and Sundukchi, M. (1998). Do monetary incentives improve response rates in the survey of income and program participation? In *JSM Proceedings*, 529–34. Alexandria, VA: American Statistical Association.

Mahalanobis, P.C. (1946). Recent experiments in statistical sampling in the Indian statistical institute. *Journal of the Royal Statistical Society*, 109, 325–370.

Maitland, A., and Presser, S. (2020). A comparison of five question evaluation methods in predicting the validity of respondent answers to factual items. In Beatty, P., Collins, D., Kaye, L., Padilla, J.L., Willis, G.B., and Wilmot, A. (eds.), *Advances in questionnaire design, development, evaluation and testing. Proceedings of the 2nd Conference on Questionnaire Design, Development, Evaluation, and Testing (QDET2)*, 9–13 November 2019, Miami, FL. Hoboken, NJ: Wiley, pp. 75–90.

Mamedova, S., and Pawlowski, E. (2018). A description of U.S. adults who are not digitally literate. National Centre for Education Statistics. Retrieved from https://nces.ed.gov/pubs2018/2018161.pdf

Manfreda, K.L., Bosnjak, M., Berzelak, J., Haas, I., and Vehovar, V. (2008). Web surveys versus other survey methods: A meta-analysis comparing response rates. *International Journal of Market Research*, 50, 79–104.

Marín, G., Gamba, R.J., and Marín, B.V. (1992). Extreme response style and acquiescence among hispanics. *Journal of Cross-Cultural Psychology*, 23, 498–509.

Marshall, R., and Lee, C. (1998). A cross-cultural, between-gender study of extreme response Style. In B.G. Englis and A. Olofsson (eds.), *European advances in consumer research*. Volume 3. Provo, UT: Association for Consumer Research, pp. 90–95. http://acrwebsite.org/volumes/11158/volumes/e03/E-03

Martin, E., Childs, J., DeMaio, T., Reiser, C., Gerber, E., Styles, K., and Dillman, D. (2007). *Guidelines for designing questionnaires for administration in different modes*. Washington, DC: US Census Bureau.

Mauz, E., Hoffman, R., Houben, R., Krause, L., Kamtsiuris, P., and Gößwald, A. (2018a). Mode equivalence of health indicators between data collection modes and mixed-mode survey designs in population-based health interview surveys for children and adolescents: Methodological study. *Journal of Medical Internet Research*, 20, e64. Retrieved from http://www.jmir.org/2018/3/e64/

Mauz, E., von der Lippe, E., Allen, J., Schilling, R., Müters, S., Hoebel, J., Schmich, P., Wetztein, M., Kamtsiuris, P., and Lange, C. (2018b). Mixing modes in a population-based interview survey: Comparison of a sequential and a concurrent mixed-mode design for public health research. *Archives of Public Health*, 76(8), 1–17. doi:10.1186/s13690-017-0237-1

Mavletova, A. (2015). Web surveys among children and adolescents: Is there a gamification effect? *Social Science Computer Review*. 33, 372–398.

McClendon, M.J. (1986). Response-order effects for dichotomous questions. *Social Science Quarterly*, 67, 205–211.

McClendon, M.J. (1991). Acquiescence and recency response-order effects in interview surveys. *Sociological Methods and Research*, 20, 60–103.

McCool, D., Lugtig, P., Schouten, B., and Mussmann, O. (2021). Longitudinal smartphone data for general population mobility studies, to appear in Journal of Official Statistics.

Medway, R. (2012). *Beyond response rates: The effect of prepaid incentives on measurement error*. PhD Dissertation, University of Maryland.

Medway, R., and Tourangeau, R. (2015). Response quality in telephone surveys. Do pre-paid cash incentives make a difference? *Public Opinion Quarterly*, 79, 524–543.

Medway, R.L., and Fulton, J. (2012). When more gets you less: A meta-analysis of the effect of concurrent web options on mail survey response rates. *Public Opinion Quarterly*, 76(4), 733–746. doi:10.1093/poq/nfs047

Meisenberg, G., and Williams, A. (2008). Are acquiescent and extreme response styles related to low intelligence and education? *Personality and Individual Differences*, 44, 1539–1550.

Messer, B.L., and Dillman, D.A. (2011). Surveying the general public over the internet using address-based sampling and mail contact procedures. *Public Opinion Quarterly*, 75, 429–457. doi:10.1093/poq/nfr021

Millar, M.M., and Dillman, D.A. (2011). Improving response to mixed-mode surveys. *Public Opinion Quarterly*, 75(2), 249–269.

Miller, R.G. (1986). *Beyond ANOVA, basics of applied statistics*. New York: John Wiley.

Millsap, R.E. (2011). *Statistical approaches to measurement invariance*. New York: Taylor and Francis Group.

Mohorko, A., de Leeuw, E., and Hox, J. (2013a). Internet coverage and coverage bias in Europe: Developments across countries and over time. *Journal of Official Statistics*, 29, 609–622.

Mohorko, A., de Leeuw, E., and Hox, J. (2013b). Coverage bias in European telephone surveys: Developments of landline and mobile phone coverage across countries and over time. *Survey Methods: Insights from the Field*. Retrieved from https://surveyinsights.org/?p=828, doi:10.13094/SMIF-2013-00002

Montgomery, D.C. (2001). *Design and analysis of experiments*. New York: Wiley & Sons.

Murgia, M., LoConte, M., Coppola, L, Frattarola, D., Fratoni, A., Luiten, A., and Schouten, B. (2019). Mixed-mode strategies for social surveys: How to best combine data collection modes. ESSnet MIMOD Deliverable. Available at Istat.it - Projects funded by Eurostat: Essnet and Grants.

Murgia, M., Lo Conte, N., and Gravem, D. (2018). Report on MIMOD survey on the state of the art of mixed mode for EU social surveys. ESSnet MIMOD, WP1 - Deliverable 1, ISTAT. Available at https://www.istat.it/en/research-activity/international-research-activity/essnet-and-grants

Narayan, S., and Krosnick, J.A. (1996). Education moderates some response effects in attitude measurement. *Public Opinion Quarterly*, 60, 58–88.

National Centre for Educational Statistics (NCES) (2015). Summary of PIAAC results. Retrieved from https://nces.ed.gov/surveys/piaac/results/summary.aspx

Nichols, E. (2012). The 2011 American Community Survey Internet Test: Attitutes and behavior study follow up. U.S. Census Bureau: Decennial Statistical Studies Division. American Community Survey Research Report Memorandum Series. #ASC12-RER-16.

Nichols, E., Olmsted-Hawala, E., Holland, T., and Riemer, A.A. (2020). Usability testing online questionnaires: Experiences at the U.S. Census Bureau. In Beatty, P., Collins, D., Kaye, L., Padilla, J.L., Willis, G.B., Wilmot, A. (eds.), *Advances in questionnaire design, development, evaluation and testing. Proceedings of the 2nd Conference on Questionnaire Design, Development, Evaluation, and Testing (QDET2)*, 9–13 November 2019, Miami, FL. Hoboken, NJ: Wiley, pp. 315–348.

Nocedal, J., and Wright, S.J. (2006). *Numerical optimization.* New York: Springer.

OECD (2013). OECD Skills outlook 2013. First results from the Survey of Adult Skills. Retrieved from Skills volume 1 (eng)--full v12--eBook (04 11 2013).pdf (oecd.org)

Olson, K. (2007). *An investigation of the nonresponse – Measurement error nexus.* PhD dissertation, University of Michigan, USA.

Olson, K., and Parkhurst, B. (2013). Collecting paradata for measurement error evaluations. In F. Kreuter (ed.), *Improving surveys with paradata: Analytic uses of process information.* Hoboken, NJ: John Wiley and Sons, pp. 43–72.

Olson, K., and Smyth, J.D. (2014). Accuracy of within-household selection in web and mail surveys of the general population. *Field Methods*, 26(1), 56–69. doi:10.1177/1525822X13507865

Olson, K., Smyth, J.D., Dykema, J. (2020). *Interviewer effects from a total survey error perspective.* New York: Chapman and Hall/CRC.

Olson, K., Smyth, J.D., and Ganshert, A. (2019). The effects of respondent and question characteristics on respondent answering behaviors in telephone interviews. *Journal of Survey Statistics and Methodology*, 7, 275–308

Olson, K., Smyth, J.D., and Wood, H.M. (2012). Does giving people their preferred survey mode actually increase survey participation rates? An experimental examination. *Public Opinion Quarterly*, 76(4), 611–635. doi:10.1093/poq/nfs024

Olson, K., Wagner, J., and Anderson, R. (2020). Survey costs: Where are we and what is the way forward? *Journal of Survey Statistics and Methodology*, smaa014, https://doi.org/10.1093/jssam/smaa014

O'Muircheartaigh, C., Krosnick, J.A., and Helic, A. (2000). Middle alternatives, acquiescence, and the quality of questionnaire data. Retrieved, October 1, 2009, from http://harrisschool.uchicago.edu/About/publications.

Park, S., Kim, J.K., and Park, S. (2016). An imputation approach for handling mixed-mode surveys. *The Annals of Applied Statistics*, 10(2), 1063–1085.

Patrick, M.E., Couper, M.A., Laetz, V.B., Schulenberg, J.E., O'Malley, P.M., Johnston, L.D., and Miech, R.A. (2018). A sequential mixed-mode experiment in the U.S. national monitoring the future study. *Journal of Survey Statistics and Methodology*, 6(1), 72–97. doi:10.1093/jssam/smx011

Peterson, G., Griffin, J., LaFrance, J., and Li, J. (2017). Smartphone participation in websurveys: Choosing between the potential for coverage, nonresponse, and measurement error. In P. Biemer, E. de Leeuw, S. Eckman, B. Edwards, F. Kreuter, L. Lyberg, N.C. Tucker, and B. West (eds.), *Total survey error in practice.* Hoboken, NJ: Wiley & Sons, pp. 203–234.

Petris, G. (2010). An R package for dynamic linear models. *Journal of Statistical Software*, 36, 1–16.

Petrolia, D.R., and Bhattacharjee, S. (2009). Revisiting incentive effects: Evidence from a random sample mail survey on consumer preferences for fuel ethanol. *Public Opinion Quarterly*, 73, 537–550.

Peytchev, A., and Neely, B. (2013). RDD telephone surveys: Toward a single frame cell-phone design. *Public Opinion Quarterly*, 77, 283–304.

Pfeffermann, D.A. (2002). Small area estimation – New developments and directions. *International Statistical Review*, 70, 125–143.

Pfeffermann, D.A. (2013). New important developments in small area estimation. *Statistical Science*, 28, 40–68.

Pfeffermann, D.A. (2017). Bayes-based non-Bayesian inference on finite populations from non-representative samples: A unified approach. *Calcutta Statistical Association Bulletin*, 69(1), 35–63.

Pforr, K., Blohm, M., Blom, A., Erdel, B., Felderer, B., Frässdorf, M., Hajek, K.M. Helmschrott, S., Kleinert, C., Koch, A., Krieger, U., Kroh, M., Martin, S., Sassenroth, D., Smiedeberg, C., Trüdfiner, E.M., and Rammstedt, B. (2015). Are incentive effects on response rates and nonresponse bias in large-scale, face-to-face surveys generalizable to Germany? Evidence from ten experiments. *Public Opinion Quarterly*, 79, 740–768.

Pickery, J., and Loosveldt, G. (1998). The impact of respondent and interviewer characteristics on the number of 'no opinion' answers. A multilevel model for count data. *Quality and Quantity*, 32, 31–45.

Pierzchala, M. (2006). Disparate modes and their effect on instrument design. In *Proceedings of the 10th international Blaise Conference*. Arnhem, the Netherlands, 9-12 May.

R Core Team. (2017). *R: A language and environment for statistical computing*. Vienna, Austria: R Foundation for Statistical Computing.

Rao, J.N.K., and Molina, I. (2015). *Small area estimation*. New York: Wiley-Interscience.

Redline, C., and Dillman, D. (2002). The influence of alternative visual designs on respondents' performance with branching instructions in self-administered questionnaires. In R. Groves, D. Dillman, J. Eltinge, and R. Little (eds.), *Survey nonresponse*. New York: Wiley, 179–195.

Reep, C. (2013). *Responsgedrag bij de Veiligheidsmonitor. Onderzoek naar responsgedrag en selectiviteit bij inzet van internet en papier (Response behaviour in the Safety Monitor. Research onto response behaviour and selectivity while using the internet and paper. In Dutch)*. Statistics Netherlands Discussion Paper.

Ricciato, F., Wirthmann, A., Giannakouris, K., Reis, F., and Skaliotis, M. (2019). Trusted smart statistics: Motivations and principles. *Statistical Journal of the IAOS*, 35(4), 589–603.

Roberts, A., and Bakker, J. (2018). *Mobile device login and break-off in individual surveys of Statistics Netherlands*. Statistics Netherlands Discussion Paper.

Roberts, A., Groffen, D., and Paulissen, R. (2017). Analyse prioriteitsklassen telefoonnummers (respons en selectiviteit). (in Dutch: *Analysis of priority classes telephone numbers: Response and selectivity)*. Heerlen, Internal Statistics Netherlands' report.

Roberts, C. (2016). Response styles in surveys: Understanding their causes and mitigating their impact on data quality. In Christof Wolf, Dominique Joye, Tom W. Smith, and Yang-Chih Fu (eds.), *The Sage handbook of survey methodology*. London: Sage Publications, pp. 579–596.

Roberts, C., Gilbert, E., Allum, N., and Eisner, L. (2019). Research synthesis: Satisficing in surveys: A systematic review of the literature. *Public Opinion Quarterly*, 83(3), 598–626. doi:10.1093/poq/nfz035

Robinson, G.K. (2000). *Practical strategies for experimenting*. New York: Wiley & Sons.

Rosenbaum, P.R., and Rubin, D.B. (1983). The central role of the propensity score in observational studies for causal effects. *Biometrika*, 70, 41–55. doi:10.1093/biomet/70.1.41

Rosenbaum, P.R., and Rubin, D.B. (1985). The bias due to incomplete matching. *Biometrics*, 41(1), 103–116. doi:10.2307/2530647

Rubin, D.B. (1987). *Multiple imputation for nonresponse in surveys*. New York: Wiley and Sons.

Rubin, D.B. (2005). Causal inference using potential outcomes: Design, modeling, decisions. *Journal of the American Statistical Association*, 100(469), 322–331.

Sakshaug, J.W., Cernat, A., and Raghhunatan, T.E. (2019). Do sequential mixed-mode surveys decrease nonresponse bias, measurement error bias, and total bias? An experimental study. *Journal of Survey Statistics and Methodology*, 7, 545–571. doi:10.1093/jssam/smy024

Sala, E., and Lillini, R. (2015). The impact of unlisted and no-landline respondents on non-coverage bias. The Italian case. *International Journal of Public Opinion Research*, 29, 133–156. doi:10.1093/ijpor/edv033

Saris, W.E. (2009). *The MTMM approach to coping with measurement errors in survey research*. RECSM working paper, University Pompeu-Fabra, Research Expertise Centre for Survey Methodology, Barcelona, Spain.

Saris, W.E., and Gallhofer, I.N. (2007). Estimation of the effects of measurement characteristics on the quality of survey questions. *Survey Research Methods*, 1(1), 29–43.

Saris, W.E., Revilla, M., Krosnick, J.A., and Shaeffer, E. (2010). Comparing questions with agree/disagree response options to questions with item-specific response options. *Survey Research Methods*, 4(1), 61–79.

Särndal, C.E., and Lundström, S. (2005). *Estimation in surveys with nonresponse*. Chichester: Wiley and Sons.

Särndal, C.-E., Swensson, B., and Wretman, J. (1992). *Model assisted survey sampling*. New York: Springer-Verlag.

Sauermann, H., and Roach, M. (2013). Increasing web survey response rates in innovative research: An experimental study of static and dynamic contact design features. *Research Policy*, 42, 273–286.

Schaefer, D., and Dillman, D. (1998). Development of a standard e-mail methodology: Results of an experiment. *Public Opinion Quarterly*, 62, 378–397.

Schaeffer, N.C., and Dykema, J. (2020). Advances in the science of asking questions. *Annual Review of Sociology*, 46, 37–60.

Schonlau, M., and Toepoel, V. (2015). Straightlining in web survey panels over time. *Survey Research Methods*, 9, 125–137.

Schouten, B., Blanke, K., Gravem, D., Luiten, A., Meertens, V., and Paulus, O. (2018). *Assessment of fitness of ESS surveys for smartphones WP5: Challenges for phone and tablet respondents within CAWI*. Deliverable 5.1 of Mixed Mode Designs for Social Surveys – MIMOD. https://www.istat.it/en/research-activity/international-research-activity/essnet-and-grants

Schouten, B., Brakel, J.A. van den, Buelens, B., and Klausch, T. (2013). Disentangling mode-specific selection bias and measurement bias in social surveys. *Social Science Research*, 42, 1555–1570.

Schouten, B., and Calinescu, M. (2013). Paradata as input to monitoring representativeness and measurement profiles: A case study of the Dutch labour force survey. In F. Kreuter (ed.), *Improving surveys with paradata: Analytic uses of process information*. Hoboken, NJ: Wiley, pp. 231–258.

Schouten, B., Calinescu, M., and Luiten, A. (2013). Optimizing quality of response through adaptive survey designs. *Survey Methodology*, 39, 29–58.

Schouten, B., Peytchev, A., and Wagner, J. (2017). *Adaptive survey design*. New York: Chapman and Hall.

Schouten, B., and Shlomo, N. (2017). Selecting adaptive survey design strata with partial R-indicators. *International Statistical Review*, 85, 143–146.

Schouten, J.G., Cobben, F., and Bethlehem, J. (2009). Indicators for the representativeness of survey response. *Survey Methodology*, 35, 101–113.

Schräpler, J.P. (2004). Response behavior in panel studies: A case study for income nonresponse by means of the German socio-economic panel (SOEP). *Sociological Methods and Research*, 33, 118–156.

Schuman, H., and Presser, S. (1981). *Questions and answers in attitude surveys: Experiments on question form, wording and context*. New York: Academic Press.

Schwarz, N., Strack, F., Hippler, H.J., and Bishop, G. (1991). The Impact of Administrative Mode on Response Effects in Survey Measurement. *Applied Cognitive Psychology*, 5, 193–212.

Sheehan, K.B. (2001). Email survey response rates: A review. *Journal of Computer-Mediated Communication*, 6 (2), JCMC621. doi:10.1111/j.1083-6101.2001.tb00117.x

Shih, T.H., and Fan, X. (2008). Comparing response rates from web and mail surveys: A meta-analysis. *Field Methods*, 20, 249–271. doi:10.1177/1525822X08317085.

Si, S.X., and Cullen, J.B. (1998). Response categories and potential cultural bias: Effects of an explicit middle point in cross-cultural surveys. *International Journal of Organizational Analysis*, 6, 218–230.

Signore. (2019). Final Report Final methodological report summarizing the results of WP 1–5. Available at MIMOD-project-Final-report-WP1-WP5.pdf (istat.it)

Singer, E., and Kulka, R.A. (2002). Paying respondents for survey participation. In Ploeg, M., Moffitt, R.A., and Citro, C.F. (eds.), *Studies of welfare populations: Data collection and research issues*, Washington, DC: National Academy Press, pp. 105–128.

Singer, E., Van Hoewyk, J., Gebler, N., Raghunathan, T., and McGonagle, K. (1999). The effect of incentives on response rates in interviewer-mediated surveys. *Journal of Official Statistics*, 15, 217–230.

Singer, E., Van Hoewyk, J., and Maher, M.P. (2000). Experiments with incentives in telephone surveys. *Public Opinion Quarterly*, 64, 171–188.

Singer, E., and Ye, C. (2013). The use and effects of incentives in surveys. *The Annals of the American Academy of Political and Social Science*, 645, 112–141.

Slavec, A., and Toninelli, D. (2015). An overview of mobile CATI issues in Europe. In Toninelli, D., Pinter, R., and de Pedraza, P. (eds.), *Mobile research methods: Opportunities and challenges of mobile research methodologies*. London: Ubiquity Press, pp. 41–62. doi:10.5334/bar.d.License: CC-BY 4.0

Smeets, L., Lugtig, P., and Schouten, B. (2019). Automatic travel mode prediction in a national travel survey, Discussion paper, December, Statistics Netherlands, The Netherlands, www.cbs.nl

Smith, P.B. (2004). Acquiescent response bias as an aspect of cross-cultural communication style. *Journal of Cross-Cultural Psychology*, 35(1), 50–61.

Smyth, J.D. (2016). Designing questions and questionnaires. In C. Wolf, D. Joye, T.W. Smith, and Y.-C. Fu (eds.), *The Sage handbook of survey methodology*. London: Sage, pp. 218–235.

Smyth, J.D., Dillman, D.A., Christian, L.M., and O'Neill, A.C. (2010). Using the Internet to survey small towns and communities: Limitations and possibilities in the early 21st century. *American Behavioral Scientist*, 53, 1423–1448. doi:10.1177/0002764210361695

Stening, B.W., and Everett, J.E. (1984). Response styles in a cross-cultural managerial study. *Journal of Social Psychology*, 122, 151–156.

Stevenson, J., Dykema, J., Cyffka, K., Klein, L., and Goldrick-Rab, S. (2012). What are the odds? Lotteries versus cash incentives. Response rates, costs and data quality for a web survey of low-income former and current college students. Paper presented at the *AAPOR* Conference, Orlando, Florida, May 18, 2012.

Storms, V., and Loosveldt, G. (2004). Who responds to incentives? *Field Methods*, 16, 414–421. doi:10.1177/1525822X04266358

Struminskaya, B., Toepoel, V., Lugtig, P., Haan, M., Luiten, A., and Schouten, B. (2020). Mechanisms of willingness to collect smartphone sensor data and longitudinal consent: Evidence from the general population in the Netherlands, to appear in Public Opinion Quarterly.

Suzer-Gurtekin, Z.T. (2013). *Investigating the bias properties of alternative statistical inference methods in mixed-mode surveys*. Ph.D. thesis, University of Michigan, Michigan.

Suzer-Gurtekin, Z.T., Elkasabi, M., Lepkowski, J., Liu, M., and Curtin, R. (2019). Randomized experiments for web-mail surveys conducted using address-based samples of the general population. In Lavrakas, P., Traugott, M., Kennedy, C., Holbrook, A., De Leeuw, E., and West, B (eds.), *Experimental methods in survey research: Techniques that combine random sampling with random assignment*. New York: Wiley, pp. 275–289. doi:10.1002/9781119083771.ch14

Suzer-Gurtekin, Z.T., Heeringa, S., and Vaillant, R. (2012). Investigating the bias of alternative statistical inference methods in sequential mixed-mode surveys. *Proceedings of the JSM, Section on Survey Research Methods*, 4711–2.

Tancreto, J.G., Zelenak, M.F., Davis, M., Ruiter, M., and Matthews B. (2011). American community survey internet tests: Results from first test in April 2011. *American community survey research and evaluation report memorandum series*. Washington, DC: U.S: Census Bureau.

Toepoel, V., de Leeuw, E., and Hox, J. (2020). Single- and mixed-mode survey data collection. In P. Atkinson, S. Delamont, A. Cernat, J.W. Sakshaug, and R.A. Williams (Eds.), *SAGE Research Methods Foundations*. London and Los Angeles and Thousand Oakes: SAGE. https://www.doi.org/10.4135/9781526421036876933

Toepoel, V., Das, J.W.M., and van Soest, A.H.O. (2009). Design of web questionnaires: The effect of layout in rating scales. *Journal of Official Statistics*, 25(4), 509–528.

Toepoel, V., and Dillman, D.A. (2011). Words, numbers, and visual heuristics in web surveys: Is there a hierarchy of importance? *Social Science Computer Review*, 29(2), 193. doi:10.1177/0894439310370070

Tourangeau, R. (2017). Mixing modes. Tradeoffs among coverage, nonresponse and measurement error. In Biemer, P.P., de Leeuw, E.D., Eckman, S., Edwards, B., Kreuter, F., Lyberg, L., Tucker, N.C., and West, B.T. (eds.), *Total survey error in practice*. New York: Wiley, pp. 115–132.

Tourangeau, R., Conrad, F.G., and Couper, M.P. (2013). *The science of web surveys*. New York: Oxford University Press.

Tourangeau, R., Couper, M.P., and Conrad, F.G. (2004). Spacing, position, and order. Interpretive heuristics for visual features of survey questions. *Public Opinion Quarterly*, 68, 368–393.

Tourangeau, R., Couper, M.P., and Conrad, F.G. (2007). Colors, labels, and interpretive heuristics for response scales. *Public Opinion Quarterly*, 71, 91–112.

Tourangeau, R., Couper, M.P., and Conrad, F.G. (2013). "Up means good": The effect of screen position on evaluative ratings in web surveys. *Public Opinion Quarterly*, 77, 69–88.

Tourangeau, R., Groves, R.M., Kennedy, C., and Yan, T. (2009). The presentation of a web survey, nonresponse and measurement error among members of web panel. *Journal of Official Statistics*, 25(3), 299–321.

Tourangeau, R., Maitland, A., Rivero, G., Sun, H., Williams, D., and Yan, T. (2017). Web surveys by smartphone and tablets: Effects on survey responses. *Public Opinion Quarterly*, 81, 896–929. doi:10.1093/poq/nfx035

Tourangeau, R., Maitland, A., Steiger, D., and Yan, T. (2020). A framework for making decisions about question evaluation methods. In Beatty, P., Collins, D., Kaye, L., Padilla, J.L., Willis, G.B., and Wilmot, A. (eds.), *Advances in questionnaire design, development, evaluation and testing. Proceedings of the 2nd Conference on Questionnaire Design, Development, Evaluation, and Testing (QDET2)*, 9–13 November 2019, Miami, FL. Hoboken, NJ: Wiley, pp. 27–73.

Tourangeau, R., and Rasinski, K.A. (1988). Cognitive processes underlying context effects in attitude measurement. *Psychological Bulletin*, 103, 299–314.

Tourangeau, R., Rips, L., and Rasinski, K. (2000). *The psychology of survey response*. Cambridge, UK: Cambridge University Press.

Tourangeau, R., and Yan, T. (2007). Sensitive questions in surveys. *Psychological Bulletin*, 133(5), 859–883.

Tuten, T.L., Galesic, M., and Bosnjak, M. (2004). Effects of immediate versus delayed notification of prize draw results on response behavior in web surveys: An experiment. *Social Science Computer Review*, 22, 377–384. doi:10.1177/0894439304265640

Twickler, T.B., Hoogstraaten, E., Reuwer, A.Q., Singels, L., Stronks, K., and Essink-Bot, M.L. (2009). Laaggeletterdheid en beperkte gezondheidsvaardigheden vragen om een antwoord in de zorg (in Dutch: *Low literacy and limited health literacy require health care measures*). *Nederlands tijdschrift voor de Geneeskunde*, 153: A250.

U.S Office of Management and Budget. (2006). Standards and guidelines for statistical surveys. https://unstats.un.org/unsd/dnss/docs-nqaf/USA_standards_stat_surveys.pdf

U.S. Office of Management and Budget. (2016a). OMB statistical policy working paper 47: Evaluating survey questions: An inventory of methods. Statistical and Science Policy Office, Office of Information and Regulatory Affairs, Office of Management and Budget. Retrieved from https://s3.amazonaws.com/sitesusa/wp-content/uploads/ sites/242/2014/04/spwp47.pdf

U.S. Office of Management and Budget. (2016b). OMB statistical policy directive no. 2 addendum: Standards and guidelines for cognitive interviews. Retrieved from https://obamawhitehouse.archives.gov/sites/default/files/omb/inforeg/directive2/final_addendum_to_stat_policy_dir_2.pdf

Vaida, F., and Blanchard, S. (2005). Conditional Akaike information formixed-e_ects models. *Biometrika*, 92, 351–370.

Van De Schoot, R., Schmidt, P., De Beuckelaer, A., Kimberley, L., and Zondervan-Zwijnenburg, M. (2015). Editorial: Measurement Invariance. *Frontiers in Psychology*, 6, 1064.

Van den Brakel, J.A. (2008). Design-based analysis of experiments with applications in the Dutch Labour Force Survey. *Journal of the Royal Statistical Society, Series A*, 171, 581–613.

Van den Brakel, J.A. (2013). Design based analysis of factorial designs embedded in probability samples. *Survey Methodology*, 39, 323–349.

Van den Brakel, J.A., Buelens, B., and Boonstra, H.J. (2016). Small area estimation to quantify discontinuities in sample surveys. *Journal of the Royal Statistical Society A Series*, 179, 229–250.

Van den Brakel, J.A., and Renssen, R.H. (1998). Design and analysis of experiments embedded in sample surveys. *Journal of Official Statistics*, 14, 277–295.

Van den Brakel, J.A., and Renssen, R.H. (2005). Analysis of experiments embedded in complex sampling designs. *Survey Methodology*, 31, 23–40.

Van den Brakel, J.A., and Roels, J. (2010). Intervention analysis with state-space models to estimate discontinuities due to a survey redesign. *Annals of Applied Statistics*, 4, 1105–1138.

Van den Brakel, J.A., Smith, P.A., and Compton, S. (2008). Quality procedures for survey transitions - Experiments, time series and discontinuities. *Survey Research Methods*, 2, 123–141.

Van Herk, H., Poortinga, Y.H., and Verhallen, T.M.M. (2004). Response styles in rating scales: Evidence of method bias in data from six EU countries. *Journal of Cross-Cultural Psychology*, 35, 346–360.

Vandenberg, R.J., and Lance, C.E. (2000). A review and synthesis of the measurement invariance literature: Suggestions, practices, and recommendations for organizational research. *Organizational Research Methods*, 3, 4–70.

Vannieuwenhuyze, J. (2013). Some basic concepts of mixed-mode surveys. Chapter 2 in *Mixed-mode data collections: Basic concepts and analysis of mode effects*, PhD dissertation, Catholic University of Leuven, Belgium.

Vannieuwenhuyze, J. (2014). On the relative advantage of mixed-mode versus single-mode surveys. *Survey Research Methods*, 8, 31–42.

Vannieuwenhuyze, J. (2015). Mode effects on variances, covariances, standard deviations, and correlations. *Journal of Survey Statistics and Methodology*, 3, 296–316.

Vannieuwenhuyze, J., and Loosveldt, G. (2013). Evaluating relative mode effects in mixed-mode surveys: Three methods to disentangle selection and measurement effects. *Sociological Methods and Research*, 42, 82–104.

Vannieuwenhuyze, J., Loosveldt, G., and Molenberghs, G. (2010). A method for evaluating mode effects in mixed-mode surveys. *Public Opinion Quarterly*, 74, 1027–1045.

Vannieuwenhuyze, J., Loosveldt, G., and Molenberghs, G. (2012). A method to evaluate mode effects on the mean and variance of a continuous variable in mixed-mode surveys. *International Statistical Review*, 80, 306–322.

Vannieuwenhuyze, J.T., Loosveldt, G., and Molenberghs, G. (2014). Evaluating mode effects in mixed-mode survey data using covariate adjustment models. *Journal of official statistics*, 30(1), 1–21.

Van Rosmalen, J., Van Herk, H., and Groenen, P.J. (2010). Identifying unknown response styles: A latent-class bilinear multinomial logit model. *Journal of Marketing Research*, 47(1), 157–172.

Vicente, P., and Reis, E. (2012). Coverage error in internet surveys. Can fixed phones fix it? *International Journal of Market Research*, 54, 323–345.

Villar, A., and Fitzgerald, R. (2017). Using mixed modes in survey data research: Results from six experiments. In M. Breen (ed.), *Values and identities in Europe: Evidence from the European social survey*. Routledge, pp. 273–310. Accessed from https://openaccess.city.ac.uk/id/eprint/16770/1/Villar_and_Fitzgerald_chapter16.pdf

Vis-Visschers, R., and Meertens, V. (2016). How question format and mode can affect the number of bicycles in a household: An evaluation and pretest satudy. Presentation at *Second International Conference on Questionnaire Design, Development, Evaluation, and Testing (QDET2)*, Florida, Miami 9–13 November 2016. https://ww2.amstat.org/meetings/qdet2/OnlineProgram/ViewPresentation.cfm?file=303371.pptx

Vogl, S., Parsons, J.A., Owens, L.K., and Lavrakas, P.J. (2020). Experiments on the effects of advance letters in surveys. In Lavrakas, P.J., M.W. Traugott, C. Kennedy, A.L. Holbrook, E.D. De Leeuw, and B.T. West (eds.), *Experimental methods in survey research. Techniques that combine random sampling with random assignment*. New York: Wiley, pp. 89–110.

Voogt, R.J., and Saris, W.E. (2005). Mixed mode designs: Finding the balance between nonresponse bias and mode effects. *Journal of Official Statistics*, 21(3), 367.

Wagner, J. (2012). A comparison of alternative indicators for the risk of nonresponse bias. *Public Opinion Quarterly*, 76, 555–575.

Wagner, J., Arrieta, A., Guyer, H., and Ofstedal, M.B. (2014). Does sequence matter in multi-mode surveys: Results from an experiment. *Field Methods*, 26(2), 141–155. doi:10.1177/1525822X13491863

Wagner, J., Schroeder, H.M., Piskorowski, A., Ursano, R.J., Murray, B.S., Heeringa, S.G., and Colpe, L.J. (2017). Timing the mode switch in a sequential mixed-mode survey: An experimental evaluation of the impact on final response rates, key estimates, and costs. *Social Science Computer Review*, 35(2), 262–276. doi:10.1177/0894439316654611

Watkins, D., and Cheung, S. (1995). Culture, gender, and response bias: An analysis of responses to the self-description questionnaire. *Journal of Cross-Cultural Psychology*, 26, 490–504.

Weigold, A., Weigold, I.K., and Natera, S.N. (2018). Response rates for surveys completed with paper-and-pencil and computers: Using meta-analysis to assess equivalence. *Social Science Computer Review*, 37(5), 649–668. doi:10.1177/0894439318783435

Weijters, B., Schillewaert, N. and Geuens, M. (2008). Assessing response styles across modes of data collection. *Journal of the Academy of Marketing Science*, 36, 409–422.

West, B., and Wagner, J. (2017). *Checklist for responsive survey design*. Working paper, University of Michigan, USA.

Wetzels, W., Schmeets, H., van den Brakel, J., and Feskens, R. (2008). Impact of prepaid incentives in face-toface surveys: A large-scale experiment with postal stamps. *International Journal of Public Opinion Research*, 20, 507–516. doi:10.1093/ijpor/edn050

Willis, G. (2005). *Cognitive interviewing: A tool for improving questionnaire design*. London: Sage Publications.

Worldbank (2018). Table: *Individuals using the internet* https://data.worldbank.org/indicator/it.net.user.zs?end=2016&start=1960. Retrieved 20180521.

Yan, T., and Tourangeau, R. (2008). Fast times and easy questions: The effects of age, experience and question complexity on web survey response times. *Applied Cognitive Psychology*, 22, 51–68.

Ybarra, L.M.R. and Lohr, S.L. (2008). Small area estimation when auxiliary information is measured with error. *Biometrika*, 95, 919–931.

Ye, C., Fulton, J., and Tourangeau, R. (2011). More positive or more extreme? A meta-analysis of mode differences in response choice. *Public Opinion Quarterly*, 75(2), 349–365.

Zang, C., Lonn, S., and Teasley, S. (2016). Understanding the impact of lottery incentives on web survey participation and response quality: A leverage-salience theory perspective. *Field Methods*, 29, 1–19. doi:10.1177/1525822X16647932

Zax, M., and Takahashi, S. (1967). Cultural influences on response style: Comparisons of Japanese and American college students. *Journal of Social Psychology*, 71, 3–10.

Zhang, C. (2013). *Satisficing in web surveys: Implications for data quality and strategies for reduction* PhD Dissertation, Ann Arbor, MI: University of Michigan.

Zhang, C., and Conrad, F.G. (2014). Speeding in web surveys: The tendency to answer very fast and its association with straightlining. *Survey Research Methods*, 8, 127–135.

Zhang, L.-C. (2011). Topics of statistical theory for register-based statistics and data integration. *Statistica Neerlandica*, 66(1), 41–63.

# Index

Taylor & Francis Group
an **informa** business

# Taylor & Francis eBooks

www.taylorfrancis.com

A single destination for eBooks from Taylor & Francis
with increased functionality and an improved user
experience to meet the needs of our customers.

90,000+ eBooks of award-winning academic content in
Humanities, Social Science, Science, Technology, Engineering,
and Medical written by a global network of editors and authors.

## TAYLOR & FRANCIS EBOOKS OFFERS:

A streamlined
experience for
our library
customers

A single point
of discovery
for all of our
eBook content

Improved
search and
discovery of
content at both
book and
chapter level

## REQUEST A FREE TRIAL
support@taylorfrancis.com

 Routledge
Taylor & Francis Group

CRC CRC Press
Taylor & Francis Group